PROBLEMS FOR INORGANIC CHEMISTRY

Bodie E. Douglas, *University of Pittsburgh*

Darl H. McDaniel, *University of Cincinnati*

John J. Alexander, *University of Cincinnati*

John Wiley & Sons, Inc.
New York Chichester Brisbane Toronto Singapore

ISBN 0-471-89505-9
Printed in the United States of America

10 9 8 7 6 5 4 3 2 1

Preface

This problems book is designed as a self-study aid. Detailed solutions are presented for the problems in our text, CONCEPTS AND MODELS OF INORGANIC CHEMISTRY, Second Edition, Wiley, 1983, with a few additional problems added to each set to provide more complete coverage of an area or to give a greater challenge. The topics covered are usually included in upper level undergraduate or graduate inorganic courses. References to the literature and to our text (cited as DMA) are given for elaboration on topics covered by the problems. Complete solutions are presented here. In cases where similar examples appear in DMA, reference to DMA is given for background and related information. In order to illustrate how chemists think about a problem, some solutions are more detailed than the reader might be expected to provide.

The necessary data and figures are included so that the problems book can be used independently or with another text. Where unique approaches have been used in DMA, the essentials are presented briefly, with DMA references or easily accessible literature references given. Examples of this are the notation for describing structures of simple inorganic solids (Chapter 6), the calculation of d-orbital energies for various symmetries (Chapter 7), and Pourbaix (E^{o}-pH) diagrams (Chapter 11).

Bodie E. Douglas
Darl H. McDaniel
John J. Alexander

Contents

Atomic Structure and the Periodic Table

1.1 Problem: Bohr postulated that lines in the emission spectrum from hydrogen in highly excited states, such as n = 20, would not be observed under ordinary laboratory conditions, because of the large size of such atoms and the much greater probability of atom collision deactivation as compared with radiation deactivation. Using the Bohr model, calculate the ratio of the cross-sectional area of a hydrogen atom in the n = 20 state to that of one in the n = 1 state.

1.1 Solution: According to the Bohr model, the radius of a hydrogen-like atom, r, is proportional to the square of the quantum number, n. The cross-sectional area is given by πr^2, thus the cross-sectional area is proportional to n^4. Hence a hydrogen atom in the n = 20 state would have a cross-sectional area 20^4, or 160,000, times that of the ground state.

States of an atom or molecule which have a highly excited electron are called Rydberg states, and are spectroscopically similar to highly excited atomic hydrogen. (For a fascinating discussion of recent research on these "floppy, fragile, and huge" atoms see D. Kleppner, M.G. Littman, and M.S. Zimmerman, Sci. Am. 1981, May, 130.)

1.2 Problem: According to the Bohr model of the atom, what would be the size of an Ne^{9+} ion? What would the ionization energy be for this ion? What would the excitation energy be for the first excited state?

1.2 Solution: The Ne^{9+} ion is a hydrogen-like atom; it has one electron and a nuclear charge, Z, of +10. According to the Bohr model (and without introducing the correction for the reduced mass) the radius of such atoms is proportional to n^2/Z and the energy to Z^2/n^2. Thus the ground state of Ne^{9+} would be predicted to have a radius of $(1^2/10)\ a_0$ or 0.1 a_0 and an energy of $-[(10^2)/1^2]\ E_0$ where a_0 and E_0 are atomic units equal to the ground state radius and energy of the hydrogen atom. The ionization energy is the energy required to remove the electron from its ground state, and is thus 1.36×10^3 eV. The energy required to excite the Ne^{9+} to the first excited state would be $1.36 \times 10^3\ (1/1^2 - 1/2^2)$ eV or 1020 eV.

During the 1920's R.A. Millikan did extensive research on the spectra of "stripped atoms." Such studies not only confirmed Bohr's predictions for hydrogen-like atoms, but demonstrated that isoelectronic atomic species had similar spectra which were displaced to higher energies with an increase in Z in accordance with Moseley's law for characteristic x rays. For an account conveying the excitement of these discoveries see Millikan's "Electrons (+ and -), Protons, Photons, Neutrons, and Cosmic Rays," University of Chicago Press, 1935.

1.3 Problem: Calculate the energy (eV) released in the transition of a hydrogen atom from the state n = 3 to n = 1. The wavelength of the radiation emitted in this transition may be found from the relation λ (in A) = 12398/E(eV) (often remembered as the approximate λ = 12345/E(eV). Calculate the wavelengths of all spectral lines that could be observed from a collection of hydrogen atoms excited by a potential of 12 V.

1.3 Solution: $\Delta E = \left(\frac{1}{n_i^2} - \frac{1}{n_f^2}\right) E_0 = \left(\frac{1}{3^2} - \frac{1}{1^2}\right)(-13.60 \text{ eV})$

ΔE = 12.08 eV released in going from the n = 3 state to the ground state; 12.08 V is the excitation potential required to populate the n = 3 state. If exactly 12 V were used to excite the collection of H atoms only the n = 2 state would be populated.

$$\Delta E = 3/4(13.60) = 10.20 \text{ eV}$$

Wavelength of emitted radiation = $\frac{12398}{10.20}$ = 1215 A.

1.4 Problem: Assuming a screening of 1/2 for an s electron, calculate the wavelength expected for the K_α x-ray line of Tc.

1.4 Solution: For Tc, Z = 43; for $\sigma = 1/2$, $Z_{eff} = 42.5$. The K_α line arises from a transition of a 2s electron to the ls level

$$\Delta E \simeq (13.6)(42.5)^2\left(\frac{1}{1^2} - \frac{1}{2^2}\right) = 18424 \text{ eV}$$

$$\lambda \simeq \frac{12398}{18424} \text{ A} = 0.673 \text{ A}$$

This calculation ignores the change in σ for a 2s compared to a ls electron, which would give ~ 0.65 A.

1.5 Problem: The Balmer series in the hydrogen spectrum originates from transitions between n = 2 states and higher states. Compare the wavelengths for the first three lines in the Balmer series with those expected for similar transitions in Li^{2+}.

1.5 Solution: If the reduced mass for H and Li is ignored, the increase in Z will cause the corresponding transitions in Li^{2+} to occur at an energy 9 times greater, or a wavelength of one ninth, the transitions for H.

A more refined value could be obtained using the Rydberg constant 109677.581 cm^{-1} for ^1H and the value 109737.31 cm^{-1} (assuming infinite nuclear mass) for Li. Still better, a reduced mass for an ^7Li nucleus and an electron gives a Rydberg constant of 109729 cm^{-1}. Thus, the following wavelengths may be calculated for the indicated transitions

$$\omega = RZ^2\left(\frac{1}{n_f^2} - \frac{1}{n_i^2}\right)cm^{-1} \qquad \lambda = \frac{1}{\omega}$$

$$\lambda(A) = \frac{10^8}{\omega cm^{-1}}$$

Transition ($n_f \leftarrow n_i$)	$2 \leftarrow 3$	$2 \leftarrow 4$	$2 \leftarrow 5$
H (Z = 1) $R = R_H$	6562 A	4863	4342
Li (Z = 3) $R = R_{Li}$	729	540	482

1.6 Problem: Obtain expressions for $(\partial^2E/\partial Z^2)_n$ using the relationship $E = -2\pi^2mZ^2e^4/n^2h^2$ obtained from the Bohr model. Using the ionization energies for isoelectronic sequences of atoms and ions given below, and finite differences to obtain $(\Delta E/\Delta Z)_n$ and $(\Delta^2E/\Delta Z^2)_n$, show that the n = 1 level is filled when the electron occupancy is 2. How would you estimate unknown ionization energies using a finite-difference approximation? Estimate the electron affinities of F and O by this procedure.

Ionization energies of the elements (in eV; 1 eV/atom = 96.4869 kJ/mole)

Z	Element	I	II	III	IV	V	VI	VII	VIII
1	H	13.598							
2	He	24.587	54.416						
3	Li	5.392	75.638	122.451					
4	Be	9.322	18.211	153.893	217.713				
5	B	8.298	25.154	37.930	259.368	340.217			
6	C	11.260	24.383	47.887	64.492	392.077	489.981		
7	N	14.534	29.601	47.448	77.472	97.888	552.057	667.029	
8	O	13.618	35.116	54.934	77.412	113.896	138.116	739.315	871.387
9	F	17.422	34.970	62.707	87.138	114.240	157.161	185.182	953.886
10	Ne	21.564	40.962	63.45	97.11	126.21	157.93	207.27	239.09
11	Na	5.139	47.286	71.64	98.91	138.39	172.15	208.47	264.18
12	Mg	7.646	15.035	80.143	109.24	141.26	186.50	224.94	265.90
13	Al	5.986	18.828	28.447	119.99	153.71	190.47	241.43	284.59
14	Si	8.151	16.345	33.492	45.141	166.77	205.05	246.52	303.17
15	P	10.486	19.725	30.18	51.37	65.023	220.43	263.22	309.41

1.6 Solution: The constants in the Bohr equation may be grouped together to give:

$$E = \frac{-13.6Z^2}{n^2}$$

for hydrogen-like atoms. For ionization energies from the nth level we have

$$\mathrm{IE} = \frac{13.6Z^2}{n^2}\ \mathrm{eV}$$

$$\left(\frac{\partial(\mathrm{IE})}{\partial Z}\right)_n = \frac{2(13.6)Z}{n^2} \qquad \left(\frac{\partial^2(\mathrm{IE})}{\partial Z^2}\right)_n = \frac{27.2}{n^2}$$

n = quantum level occupied; in an isoelectronic sequence, the screening may be assumed constant, so $\Delta Z_{eff} = \Delta Z$.

Using ionization energy data for 1-electron atoms (hydrogen-like atoms) we have the following

	H		He^+		Li^{2+}		Be^{3+}		B^{4+}		C^{5+}
	13.598		54.416		122.451		217.713		340.217		489.981
$\frac{\Delta IE}{\Delta Z}$		40.818		68.035		95.262		122.504		149.764	
$\left(\frac{\Delta^2 IE}{\Delta Z^2}\right)_n = \frac{27.2}{n^2}$			27.22		27.23		27.24		27.26		
n =			0.9996		0.9995		0.9996		0.9989		

For 2-electron atoms we have

	He		Li^+		Be^{2+}		B^{3+}		C^{4+}		N^{5+}
IE	24.587		75.638		153.893		259.368		392.077		552.057
$\frac{\Delta IE}{\Delta Z}$		51.051		78.255		105.475		132.709		159.980	
$\left(\frac{\Delta^2 IE}{\Delta Z^2}\right)_n = \frac{27.2}{n^2}$			27.204		27.220		27.234		27.271		
n =			0.99993		0.99963		0.99938		0.99870		

For 3-electron atoms we have

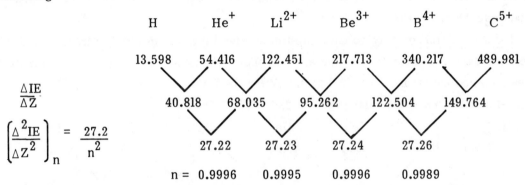

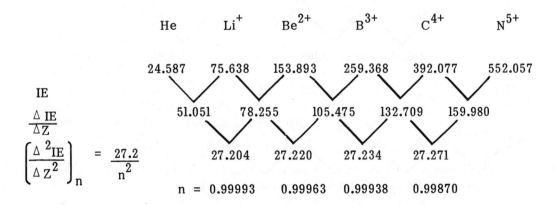

	Li		Be^+		B^{2+}		C^{3+}		N^{4+}		O^{5+}
IE	5.392		18.211		37.930		64.492		97.888		138.116
$\frac{\Delta IE}{\Delta Z}$		12.819		19.719		26.562		33.396		40.228	
$\left(\frac{\Delta^2 IE}{\Delta Z^2}\right)_n = \frac{27.2}{n^2}$			6.900		6.843		6.834		6.832		
		n =	1.985		1.993		1.995		1.995		

From the above it is clear that the electron undergoing ionization has an n quantum number of 1 for 1-electron and 2-electron atoms, but an n quantum number of 2 for 3-electron atoms. For further discussion of trends in ionization numbers see G.P. Haight, J. Chem. Educ. 1967, 44, 468.

Electron affinities may be estimated in similar fashion. In our illustration, a third difference has been used to compensate for slight shifts in shielding. The estimated values (in parentheses) are off by 0.2 to 0.3 eV for the EA's.

-1		0		1st		2nd		3rd				
O^-		F		Ne		Na		Mg		Al		Si
(-7.7 eV)		(3.15)		21.564		47.286		80.143		119.99		166.77
	(11.56)		(18.41)		25.722		32.857		39.85		46.78	
		(7.56)		(7.31)		7.12		6.99		6.93		
			(-0.25)		(-0.19)		-0.13		-0.06			

0		1st		2nd		3rd		4th		5th
O		F		Ne		Na		Mg		Al
(1.28)		17.422		40.962		71.64		109.24		153.71
	(16.14)		23.54		30.678		37.60		44.47	
		(7.40)		7.14		6.92		6.87		

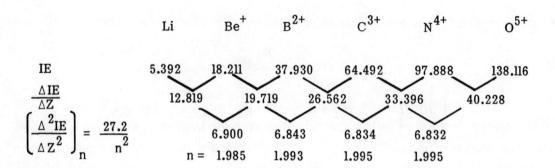

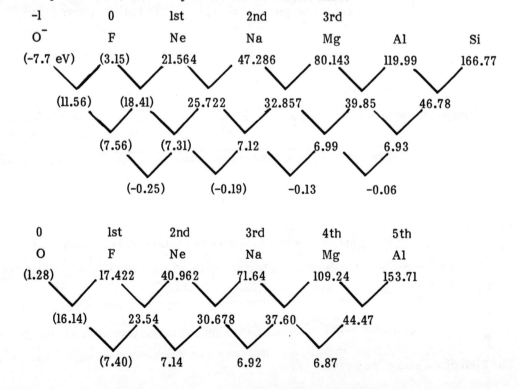

1.7 Problem: Explain briefly the observation that the energy difference between the $1s^2 2s^1\ ^2S_{1/2}$ state and the $1s^2 2p^1\ ^2P_{1/2}$ state for Li is 14904 cm^{-1}, whereas for Li^{2+} the $2s^1\ ^2S_{1/2}$ and the $2p^1\ ^2P_{1/2}$ states differ by only 2.4 cm^{-1}.

1.7 Solution: The Li^{2+} ion is a 1-electron atom; Z_{eff} is equal to Z for all quantum states and the energy depends to a first approximation on n alone. The Li atom has a filled $1s^2$ core. The energy difference tells us that Z_{eff} is higher for the $2s^2 2s^1$ configuration than for the $2s^2 2p^1$ configuration. This may be rationalized on the basis of the greater degree of penetration of the $1s^2$ core by the $2s^1$ electron than by the $2p^1$ electron. The Bohr-Sommerfeld model readily predicts greater penetration of the $1s^2$ core by the electron in an elliptical 2s orbit than by an electron in a circular 2p orbit. Wave mechanics yields a similar result (see page 14 DMA).

1.8 Problem: Using the p orbitals for an example, distinguish between the angular part of the probability function, the radial part of the probability function, and a probability contour. Draw simple sketches to illustrate. How would each of these be affected by a change in the principal quantum number, n?

1.8 Solution: The angular part of the probability function, $\Theta^2\Phi^2$, is a scaling factor which depends only on the angular orientation of a point in space with respect to a framework located on the nucleus. It is represented by a surface; the magnitude of $\Theta^2\Phi^2$ is the magnitude of a vector extending from the origin to the surface. It is independent of n, but does depend on ℓ and m_ℓ. For angular parts of the wave function that contain imaginary numbers, the probability function is obtained by multiplying the wave function by its complex conjugate — that is, by a function in which the sign of each term containing an imaginary number is multiplied by minus one.

The radial probability function, R^2, depends on the quantum numbers n and ℓ. It varies with the distance from the nucleus, r, but not with angular orientation. It has zero values, or nodes at $r = 0$ and $r = \infty$ and $(n - \ell - 1)$ points in between.

The total probability function is the product of the angular probability function and the radial probability function, evaluated at every point in space. Contour lines are drawn by connecting points having the same preselected value of $R^2\Theta^2\Phi^2$.

Boundary surface diagrams are often depicted for orbitals. These enclose volumes within which a given probability (usually 90%) of finding the electron exists.

Sketches of each of these functions are shown below for a $3p_z$ orbital.

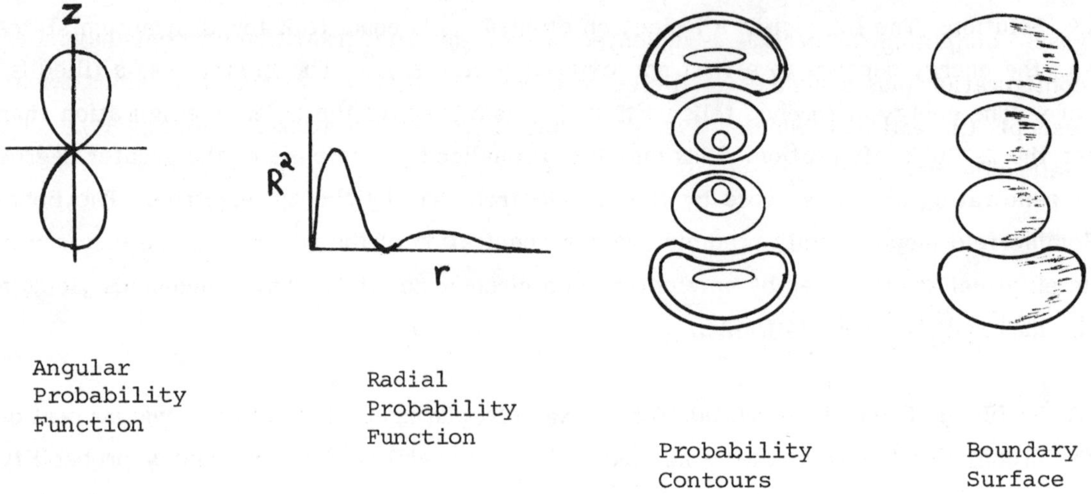

Sketches of Functions for $3p_z$ orbital

1.9 Problem: Give the characteristic valence shell configuration for the following periodic groups (e.g., ns^1 for the alkali metals).

(a) Noble gases
(b) Halogens
(c) Coinage metals (Group IIB)
(d) Ti family
(e) N family

1.9 Solution: The answers are shown below. See also Solution 1.10.

(a) ns^2np^6 except for He, $1s^2$
(b) ns^2np^5
(c) $(n - 1)d^{10}ns^1$
(d) $(n - 1)d^2ns^2$
(e) ns^2np^3

1.10 Problem: Write the electron configuration beyond a noble gas core for (for example, F, $(He)2s^22p^5$) Rb, La, Cr, Fe, Cu, Tl, Po, Gd, and Lu.

1.10 Solution: The periodic table provides at a glance a guide to electron configurations. The period structure is indicated in the figure below. With minor variations, the outermost electron configuration is determined by the group — main group elements having a number of outermost s and p electrons equal to their group number -- the noble gases completing the periods as indicated below. The first transition elements have a $4s^2$ configuration plus a number of d-electrons equal to their atomic number minus twenty (except Cr and Cu which have d^5s^1 and $d^{10}s^1$ configurations, respectively). The lanthanides begin filling the 4f after La which has a $6s^25d^1$ configuration.

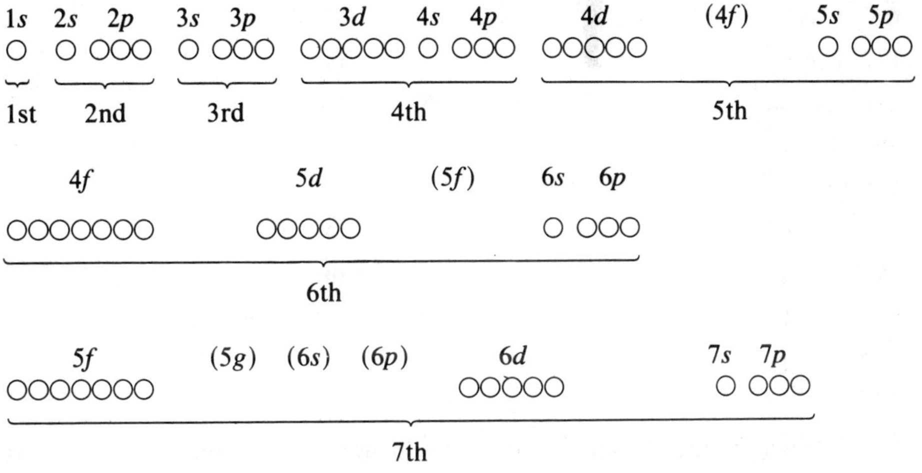

Makeup of the conventional long form of the periodic table.

Rb, (Kr) $5s^1$

La, (Xe) $5d^16s^2$

Cr, (Ar) $3d^54s^1$

Fe, (Ar) $3d^64s^2$

Tl, (Xe) $4f^{14}5d^{10}6s^26p^1$

Po, (Xe) $4f^{14}5d^{10}6s^26p^4$

Gd, (Xe) $4f^75d^16s^2$

Lu, (Xe) $4f^{14}5d^16s^2$

1.11 Problem: Write the electron configuration beyond a noble gas core and give the number of unpaired electrons for (for example, F^-, (He)$2s^22p^6$) K^+, Ti^{3+}, Cr^{3+}, Fe^{2+}, Cu^{2+}, Sb^{3+}, Se^{2-}, Sn^{4+}, Ce^{4+}, Eu^{2+}, and Lu^{3+}.

1.11 Solution: In forming ions, atoms will lose electrons from levels of highest principal quantum number, n; with levels of equal n, the electron will be lost from the orbital of highest ℓ. Electrons in degenerate orbitals will occupy orbitals singly to the maximum permissible extent (Hund's rule of maximum multiplicity). These considerations lead to the following answers.

Configuration	No. of Unpaired Electrons
$K^{+}(Ar)$	(0)
$Ti^{3+}(Ar)3d^{1}$	(1)
$Cr^{3+}(Ar)3d^{3}$	(3)
$Fe^{2+}(Ar)3d^{6}$	(4)
$Cu^{2+}(Ar)3d^{9}$	(1)
$Sb^{3+}(Kr)4d^{10}5s^{2}$	(0)
$Se^{2-}(Ar)3d^{10}4s^{2}4p^{6}$	(0)
$Sn^{4+}(Kr)4d^{10}$	(0)
$Ce^{4+}(Xe)$	(0)
$Eu^{2+}(Xe)4f^{7}$	(7)
$Lu^{3+}(Xe)4f^{14}$	(0)

1.12 Problem: If an atom is in an energy state in which L = 3 and S = 2, how would the state be described in spectroscopic notation?

1.12 Solution: The term notation and the corresponding L values are

	S	P	D	F	G	H	I	K	L	...
L =	0	1	2	3	4	5	6	7	8	...

For L = 3 and S = 2 the term would be ^{5}F where the spin multiplicity 5 is obtained from the value of 2S + 1.

1.13 Problem: What is the number of microstates for an f^{3} configuration? Which of these is unique to the ground state term arising from this configuration.

1.13 Solution: The number of microstates = $\frac{n!}{e!\,h!}$ where n is the number of spin orbitals (that is, twice the number of orbitals), e is the number of electrons, and h is the number of holes (equal to n - e). For an f^3 configuration

$$\text{no. of microstates} = \frac{14!}{3!\,11!} = \frac{14 \times 13 \times 12}{3 \times 2} = 364$$

The microstates uniquely belonging to the ground state are represented below

m_ℓ

	3	2	1	0	-1	-2	-3	
(	↑	↑	↑					
	↓	↓	↓					
					↑	↑	↑	
					↓	↓	↓	)

These are just four of the (2L+1)(2S+1) or 28 microstates of the array for the ground state ^4I term. The first microstate is written such that the $\sum m_\ell$ = L and $\sum m_s$ = S for the ground state term.

1.14 Problem: Derive the spectral terms for the d^7 configuration. Identify the ground state term. Give all J values for the ground state term and indicate which is lowest in energy.

1.14 Solution: Using the spin factoring technique (DMA, Page 30, or D.H. McDaniel, _J. Chem. Educ._ 1977, _54_, 147) we find for d^7 the possible spin sets of $\alpha^5\beta^2$ (or $\alpha^2\beta^5$) and $\alpha^4\beta^3$ (or $\alpha^3\beta^4$). The partial terms (see references) for d-orbital occupancy by electrons of a single spin set are

d^0	d^1	d^2	d^3	d^4	d^5
S	D	P,F	P,F	D	S

For bookkeeping purposes we may note that d^7 gives 120 microstates (see problem 1.13); d^5_α would give $\frac{5!}{5!\,0!}$ = 1, d^2_β would give $\frac{5!}{2!\,3!}$ = 10, d^4_α would give $\frac{5!}{4!\,1!}$ = 5, d^3_β would give $\frac{5!}{3!\,2!}$ = 10— thus the total would be

$$\alpha^5\beta^2 = 1 \times 10 = 10$$
$$\alpha^2\beta^5 = 10 \times 1 = 10$$
$$\alpha^4\beta^3 = 5 \times 10 = 50$$
$$\alpha^3\beta^4 = 10 \times 5 = 50$$
$$\text{Total microstates} = 120$$

Terms

From $\alpha^2\beta^5$ (or $\alpha^2\beta^5$)

(P+F) x (S) → P + F $\quad M_S = \pm 3/2$, ∴ ^4P and ^4F

Note that since the degeneracy of the ^4P and ^4F terms requires 40 microstates (see solution to Problem 1.13) and the spin sets considered yield only 20 microstates, 20 additional microstates corresponding to the $M_S = \pm 1/2$ components of the P and F terms will occur in the remaining terms. For this reason we drop the "^2P" and "^2F" appearing below.

From $\alpha^3\beta^4$ (and $\beta^3\alpha^4$) we obtain

(P+F)(D) = D x P + D x F

D x P = F + D + P

D x F = H + G + F + D + P (see Problem 1.16 for products)

$M_S = \pm 1/2$ yields doublets ^2P, $^2D(2)$, ^2F, ^2G, ^2H

The degeneracy of these states totals 80, which together with the 20 microstates needed to complete the ^4F and ^4P, gives the 100 microstates from the $\alpha^3\beta^4$ and $\alpha^4\beta^3$ configurations.

The ground state term is the one with greatest spin multiplicity and within this restriction, the largest L value, that is, the ^4F term. This has J values running from L + S through L - S— J = 9/2, 7/2, 5/2, 3/2. Since the d-orbitals are more than half filled, the maximum J state is the ground state $^4F_{9/2}$.

1.15 Problem: Derive the spectral terms for the f^2 configuration. Identify the ground state term. Give all J values for the ground state term and indicate which is lowest in energy.

1.15 Solution: Using the spin-factoring technique (see Problem 1.14) and the partial terms for f-orbitals by a single spin set

f^0	f^1	f^2	f^3	f^4	f^5	f^6	f^7
S	F	PFH	SDFGI	SDFGI	PFH	F	S

we obtain the following:

f^2 Term Symbols - spin-factoring

$(21f_\alpha^2)(f_\beta^0)$ and reverse 21 + 21 = 42

$(7f_\alpha^1)(7f_\beta^1)$ 49

91 microstates

$(21f_\alpha^2)(f_\beta^0)$

$(P + F + H) \times S = {}^3P + {}^3F + {}^3H$

Degeneracy 9 + 21 + 33 = 63, but 42 microstates for f_α^2, drops ^1P, ^1F, and ^1H below

$(7f_\alpha^1)(7f_\beta^1)$ 49 microstates

(F) x (F) = I + ~~H~~ + G + ~~F~~ + D + ~~P~~ + S

Leaving ^1I, ^1G, ^1D, ^1S

^3P, ^3F, ^3H

^3H is the ground state term J = 4, 5, 6. Since the orbitals are less than 1/2 filled, $^3H_4$ is lowest in energy.

1.16 Problem: What spectroscopic terms arise from the configuration $1s^2 2s^2 2p^6 3s^2$-$3p^5 3d^1$?

1.16 Solution: This problem asks for the spectroscopic terms for a configuration containing nonequivalent electrons--electrons in orbitals differing from each other in their n and/or ℓ quantum numbers. In such cases the equivalent electrons may be treated individually to give "spectroscopic terms" which may then be multiplied together under the Clebsch-Gordon product rules already used to obtain J values from L and S values. A filled shell yields a ^1S state, one-electron occupancy (or one hole) gives a doublet state with L = ℓ, and other occupancies may be worked out by techniques such as those in Problems 1.14 and 1.15.

For our configuration the terms from equivalent electrons would be:

$1s^2$	^1S	S = 0, L = 0
$2s^2$	^1S	S = 0, L = 0
$2p^6$	^1S	S = 0, L = 0
$3s^2$	^1S	S = 0, L = 0
$3p^5$	^2P	S = 1/2, L = 1
$3d^1$	^2D	S = 1/2, L = 2

The product of $L_1 \times L_2$ gives $(L_1 + L_2)$, $(L_1 + L_2 - 1)$... $|L_1 - L_2|$. The product of S_1 and S_2 gives $(S_1 + S_2)$, $(S_1 + S_2 - 1)$, ...$|S_1 - S_2|$. Both run in integer steps from the sum through to the difference in quantum numbers.

The P x D product would thus give L values of (2 + 1), (2 + 1 - 1), and (2 - 1) or 3, 2, and 1 or F, D, and P terms.

The S values would be (1/2 + 1/2) and (1/2 - 1/2) or 1 and 0.

The result of $^2P \times {}^2D$ is thus ^3F, ^3D, ^3P and ^1F, ^1D, and ^1P, where the superprefix indicates the spin multiplicity, 2S + 1. The product of a ^1S with any other term leaves the other term unchanged.

1.17 Problem: What nodal plane(s) might be expected for the $f_{x^3 - \frac{3}{5}xr^2}$ orbital?

1.17 Solution: Nodal surfaces arise when the function

$$x^3 - \frac{3}{5}xr^2 = 0$$

This occurs when x = 0, giving the yz plane as a nodal plane, and when $x = \sqrt{\frac{3}{5}}r$, a conical surface in which the generatrix makes an angle of 39.2° with the x axis.

1.18 Problem: The angular function of an orbital is described by $\frac{1}{r^6}[x^4(y^2 - z^2) + y^4(z^2 - x^2) + z^4(x^2 - y^2)]$. Is the orbital of g or u symmetry? What is the basis of your assignment?

1.18 Solution: Functions involving even ℓ, hence even exponents of r and of $x^a y^b z^c$ where $a + b + c = \ell$, are gerade. Hence the given orbital (an i orbital) is gerade.

1.19 Problem: The trend in electron affinities for Group VB (N family) is opposite that for VIB (O family). Explain the trend for each family.

1.19 Solution: At the outset it should be stated that there is no universally accepted answer to the question posed, but approaches to the question should cause us to look more closely at our model of the atom.

First we must find the trends alleged to exist in the question. Inspection of a table of electron affinities shows the following orders of electron affinities (values quoted in parentheses are in eV units).

O(1.462) < S(2.0772) > Se(2.0206) > Te(1.9708) > Po(1.9) Series a

N(< 0) < P(0.743) < As(0.80) < Sb(1.05) > Bi(0.947) Series b

The difference in neighboring pairs shows a trend in (EA Group VIB - EA Group VB) as follows

EA for O - N > S - P > Se - As > Te - Sb > Po - Bi Series c

If the latter can be explained, then, perhaps, the initial question may be answered. This trend is similar to that observed for the differences in ionization energies of the pairs isoelectronic to O^- and N^-, that is, F and O, Cl and S, etc.

IE for F - O >> Cl - S > Br - Se > I - Te > At - Po Series d

Here we recognize the "normal" increase in ionization energy on crossing a period, the increase being greatest at the top of the table and decreasing as we descend a group. We attribute such trends to inefficient shielding of the nuclear charge within a set of electrons having the same n and ℓ quantum numbers. We find a "normal" decrease in ionization energy on descending, within a group of the table.

But we also recognize that O, S, and Se have lower ionization energies than that of either of their neighbors in the same period. This we may attribute to the particular stability of the half filled p-orbitals. The interaction involved here is sometimes referred to as spin-exchange interactions and decreases in importance with average separation of the electrons, that is, with the size of the atom.

Both of the above factors work in the same direction to explain Series c. To explain Series b we have to postulate that the spin exchange term which stabilizes the neutral N, P, etc., atom decreases faster than the "normal" decrease in the "zeroth" ionization energy.

Finally, to explain the much smaller electron affinities of the later second period elements than that of the third period elements of similar groups we postulate that the small size of the atoms of the second group elements makes it much more difficult to crowd electrons around than for the larger atoms.

1.20 Problem: The following equations give the angular functions for the set of d-orbitals of well defined m_ℓ values. Show the linear combinations of these functions that lead to the "real" set of d-orbitals.

$$d_{-2} = \sqrt{\frac{3}{8}}\,\frac{(x - iy)^2}{r^2}$$

$$d_{-1} = \sqrt{\frac{3}{2}}\,\frac{z(x - iy)}{r^2}$$

$$d_0 = \sqrt{\frac{1}{4}}\,\frac{3z^2 - r^2}{r^2}$$

$$d_1 = \sqrt{\frac{3}{2}}\,\frac{z(x + iy)}{r^2}$$

$$d_2 = \sqrt{\frac{3}{8}}\,\frac{(x + iy)^2}{r^2}$$

1.20 Solution: Linear combinations of solutions to the wave equation will also be solutions of the wave equation. Imaginary coefficients are permissible. When the angular function for the d-orbitals of well defined m_ℓ values are expanded, the following combinations will readily yield the "real" set of d-orbitals.

$$d_{x^2-y^2} = \frac{1}{\sqrt{2}}(d_2 + d_{-2}) = \frac{\sqrt{3}}{2}\frac{x^2 - y^2}{r^2}$$

$$d_{xy} = -\frac{1}{\sqrt{2}}\,i(d_2 - d_{-2}) = \sqrt{3}\frac{xy}{r^2}$$

$$d_{z^2} = d_0 = \frac{1}{2}\frac{3z^2 - r^2}{r^2}$$

$$d_{xz} = \frac{1}{\sqrt{2}}(d_{-1} - d_1) = \sqrt{3}\frac{xz}{r^2}$$

$$d_{yz} = -\frac{1}{\sqrt{2}}\,i(d_1 + d_{-1}) = \sqrt{3}\frac{yz}{r^2}$$

1.21 Problem: The ionization energies of several Group IV elements are as follows:

	I	II	III	IV	V
Ti	6.82	13.58	27.49	43.27	99.22 eV
Zr	6.84	13.13	22.99	34.34	81.5
Pb	7.42	15.03	31.94	42.32	68.8

What stable oxidation states are expected for compounds of these elements?

1.21 Solution: Stable oxidation states are expected when the difference in successive ionization energies exceeds ~12 eV for valence electrons (see L. H. Ahrens, J. Inorg. Nucl. Chem. 1956, 2, 290, or DMA p. 37). For Ti, Zr, and Pb, these differences are:

	Δ 1,2	Δ 2,3	Δ 3,4	Δ 4,5
Ti	6.8	13.9	15.8	56.0
Zr	6.3	9.9	11.4	47.2
Pb	7.6	16.9	10.4	26.5

From the above we expect to find compounds of Ti(II), Ti(III), and Ti(IV); Zr(IV) only; and

Pb(II) and Pb(IV).

1.22 Problem: The ionization energies of first row atoms increase approximately linearly from B to N. A drop occurs between N and O. However, the energy necessary to remove an electron from the ground state ^3P oxygen to produce O^+ in the excited ^2P state lies on the extrapolated line of B, C, N ionization energies. Explain this observation.

1.22 Solution: The ionization energies of B, C, and N all represent processes which remove a p electron from a singly occupied 2p orbital giving B^+, C^+, and N^+ in their ground states. For O which has the electron configuration $2s^2 2p^4$, the normal ionization energy represents removal of an electron from the doubly occupied p orbital giving a ^4S O which is stabilized by exchange energy. Removal of a 2p electron from a singly occupied 2p orbital giving a ^2P state O is the process comparable to that undergone by B, C, and N. Hence its energy might be expected to fall on the same straight line when plotted vs. atomic number.

1.23 Problem: The emission spectrum of atomic Ca shows a transition from a ^3D state to a ^3P state. If the selection rule permits only $\Delta J = \pm 1$ or 0 (but not J=0 to J=0), how many lines will be observed for this transition?

1.23 Solution: For a ^3D state the S value must be 1 since 2S + 1 = 3, and the L value is 2 for a D term. The J values run from $|L + S|$ through $|L - S|$ in integer steps, and would thus be 3, 2, and 1. For the ^3P state the J values would be 2, 1, and 0. From the chart below it may be seen that six transitions have $\Delta J = \pm 1$ or 0 and would be allowed.

J_{final} \ $J_{initial}$	3	2	1
2	1	0	-1
1	2	1	0
0	3	2	1

II Localized Bonding

2.1 Problem: Indicate the electron configuration expected for the possible covalent compounds MX_n, (where X = Cl or Br) of Sn, At, and Ra. Assuming only σ bonding, predict the geometry of each molecule.

2.1 Solution: To predict possible covalent states, start with the ground state electron configuration of the atom (See Problem 1.10) and obtain the simplest possible covalent state by utilizing only the initially unpaired electrons in covalent bond formation. More covalent bonds may be formed by promoting an initially paired electron into a low energy vacant orbital. Geometry may be predicted by counting both lone pairs and bonding pairs as participating in the sigma bond hybridization.

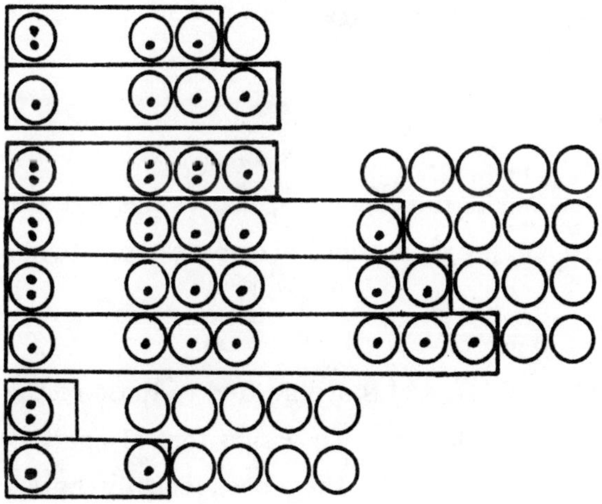

	Valence State Configuration	Hybridization	Geometry
Sn		sp^2	bent
		sp^3	tetrahedral
At		sp^3	lincar
		sp^3d	T-shaped
		sp^3d^2	square pyramidal
		sp^3d^3	pentagonal bipyramid
Ra			
		sd	bent

Note that the shape of the molecule is described by the location of atoms, and predict the atom location in molecules containing lone pairs using the VSEPR model (See DMA, Section 2.3.3). We also note the effect of having the drop in 6d energy below that of 7p for Ra.

2.2 Problem: What ionic states are expected for the elements given in Problem 2.1? What ionic states are expected for Ti, Pr, and Se?

2.2 Solution: The noble-gas configuration ($1s^2$, or s^2p^6) is the configuration for stable, simple negative ions, and is also common for Groups IA, IIA, IIIB (except for most lanthanides and actinides) and Al. The pseudo-noble-gas configuration (underlying d^{10}) and inert s pair configuration ($d^{10}s^2$) are common for the post-transiton metals. Transition metals generally have states that run from +2 to a maximum of their group number. Thus, predicted ionic states are:

Sn^{2+} $d^{10}s^2$ Sn^{4+} d^{10} At^- s^2p^6 Ra^{2+} s^2p^6

Ti^{2+} d^2 and Ti^{3+} d^1 (irregular configurations) Ti^{4+} s^2p^6

Pr^{3+} f^2 Pr^{4+} f^1 (irregular configurations)

Se^{2-} s^2p^6

2.3 Problem: Assign formal charges and oxidation numbers, and evaluate the relative importance of three Lewis structures of OCN^-, cyanate ion. Compare these to the Lewis structures of the fulminate ion, CNO^-.

2.3 Solution: Since the valence orbitals of C, O, and N are limited to those of the 2s and 2p orbitals, we expect the octet rule to be followed. We assign formal charges in the valence bond structures by dividing the electrons in the bonds equally between the participating atoms and calculating the charge on the pseudo-ion that is formed. Formal charges are useful in qualitatively evaluating the contribution of structures to a resonance hybrid. Oxidation numbers are found by assigning both of the electrons in a bond to the more electronegative atoms. Oxidation numbers will be independent of our valence bond structure provided the same atoms are attached to each other. The structures in which the negative charge resides on the more electronegative atom, charges are minimized, and not highly separated, will contribute the most to the resonance hybrid.

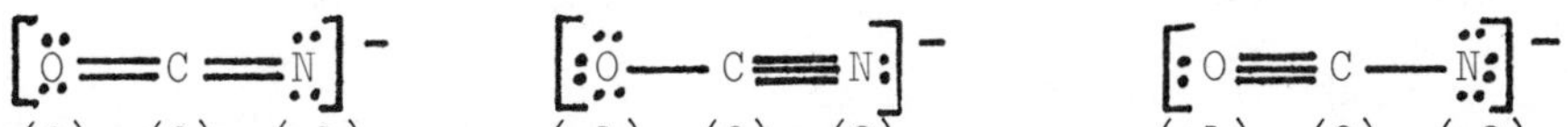

(a)	(b)	(c)	
(0) (0) (-1)	(-1) (0) (0)	(+1) (0) (-2)	formal charges
-II +IV -III	-II +IV -III	-II +IV -III	oxidation numbers

Best charge distribution - (b). Structure (b) should be most important, with structure (a) next. Structure (c) should be unimportant because of poor charge distribution. For fulminate (b) with $[C\text{-}N{\equiv}O]^-$ is the worst structure (formal charges -3, +1, +1) and (c) is best, $[C{\equiv}N\text{-}O]^-$ (formal charges -1, +1, -1). Note that the cyanate ion is stable, fulminate is explosive, and CON^- does not exist — a pattern consistent with increasingly poor charge distribution in these ions.

2.4 Problem: Write a reasonable electron dot structure and assign formal charges and oxidation numbers for each of the following: ClF, ClF_3, ICl_4^-, $HClO_3(HOClO_2)$.

2.4 Solution:

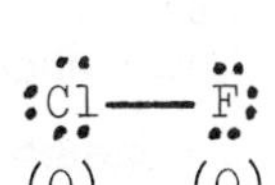

(0) (0)

+I -I

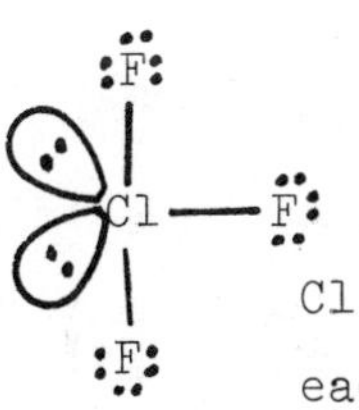

Cl (0) +III

each F (0) -I

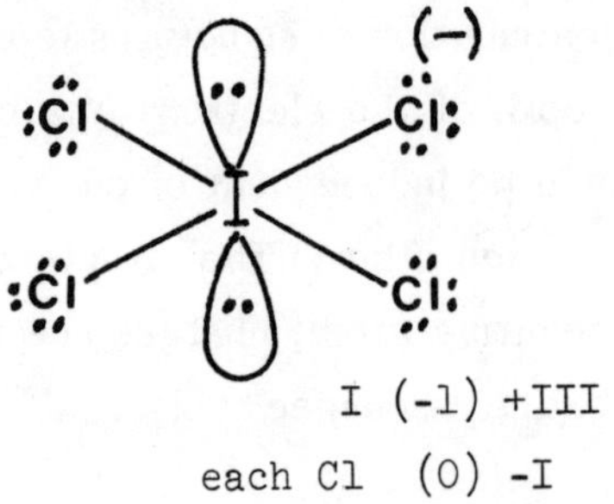

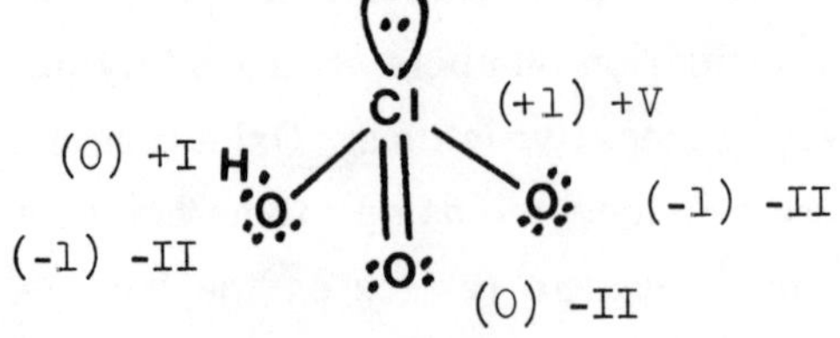

2.5 Problem: Give the oxidation number, formal charge, and hybridization of the central atom in each of the following: NO_3^-, BF_4^-, $S_2O_3^{2-}$, ICl_2^+, ClO_3^-. What are the molecular shapes?

2.5 Solution: We proceed here as in Problem 2.1, except initially we assign extra electrons to the more electronegative atoms and, in the case of positively charged ions, remove electrons from the least electronegative atom. We recognize that only one bond between two atoms may be a σ bond and that further bonding must involve π (or more rarely δ) bonding.

	Oxidation Number	Number of Bonds	Hybrid	Formal Charge	Shape	Comments on Bonding
NO_3^-	+V	3	sp^2	+1	Planar	1 π bond to complete octet of N
BF_4^-	+III	4	sp^3	-1	Tetrahedral	No π bonding possible
$S_2O_3^{2-}$	+V[a]	4	sp^3	+1	Tetrahedral	Partial π bonding to lower charges
ICl_2^+	+III	2	sp^3	+1	Angular	π bonding is not important
ClO_3^-	+V	3	sp^3	+1	Pyramidal	Partial π bonding to lower charges

[a]The oxidation number of the central S of $S_2O_3^{2-}$ is sometimes given as +VI by analogy with SO_4^{2-}. However, for an S-S bond the shared electrons are divided equally, giving +V.

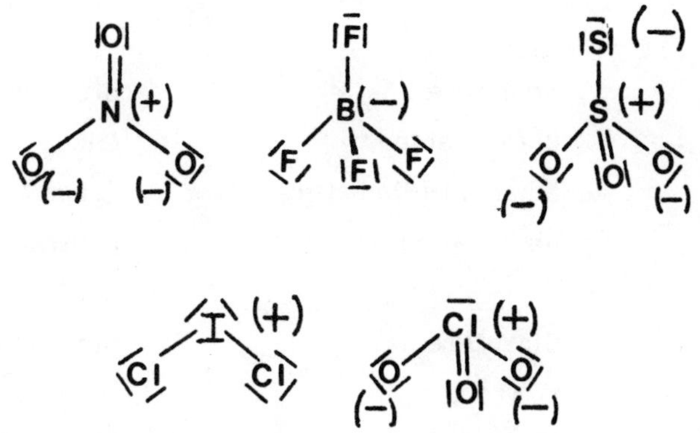

Only one resonance structure is shown in each case. The number of double bonds for $S_2O_3^{2-}$ and ClO_3^- corresponds to favorable charge distribution.

2.6 Problem: Select the reasonable electron-dot structures for each of the following compounds. Indicate what is wrong with each incorrect or unlikely structure.

a. 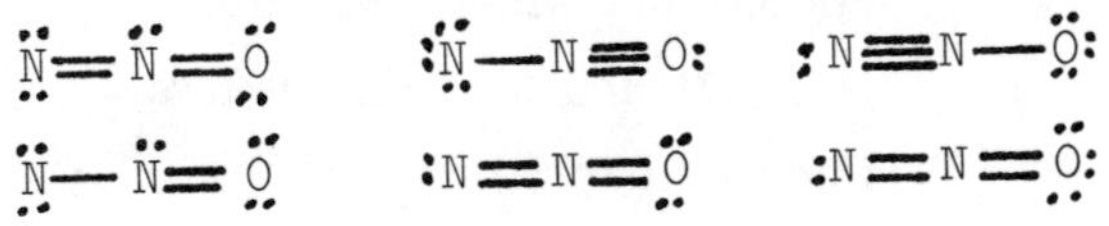

b. 

c. 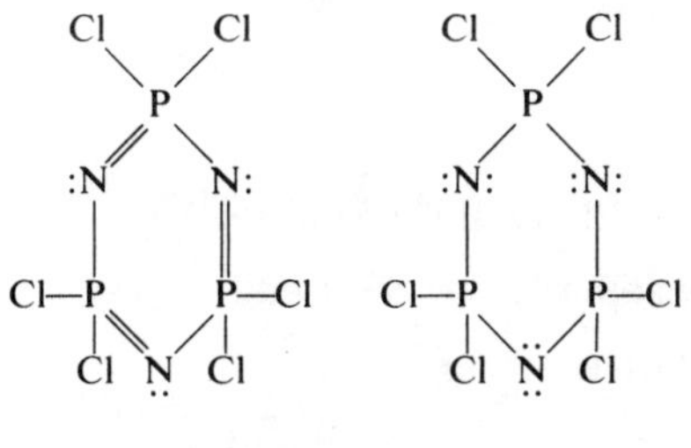

d. 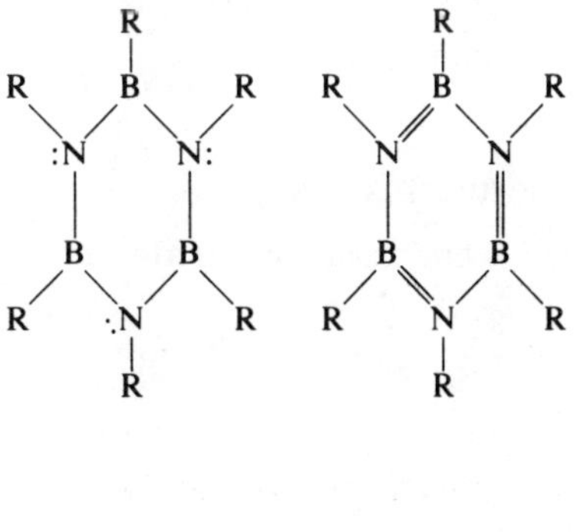

2.6 Solution:

(a)

Structure 1 is wrong, 10 e on N, 18 e total

Structure 2 is unfavorable, adjacent (+) charges, high (-2) charge on N

Structure 3 is OK, O has (-) charge, central N is (+)

Structure 4 is unfavorable, only 3 bonds

Structure 5 is wrong, 14 e total

Structure 6 is wrong, 10 e on O

(b)

Structure 1 is unfavorable (-1) on S, not the more electronegative N

Structure 2 is very unfavorable, (+) charge and only 6e on N and (-2) charge on S

Structure 3 is best for charge distribution with (-1) on N

Structure 4 is very unfavorable, charges are widely separated, N is (-2), and π bonding is less important to S than to N

(c)

Structure 1 has favorable charge distribution (all 0), but π bonding is less important for third period elements

Structure 2 has charge separation, but with (-1) on the more electronegative N

(d)

Structure 1 is good, both B and N are neutral

Structure 2 has unfavorable charge separation, B(-1) and N(+1), but there are more strong bonds and the π bonding is delocalized around the ring

2.7 Problem: Give the expected hybridization of P, O, and Sb in $Cl_3P\text{-}O\text{-}SbCl_5$. The P-O-Sb bond angle is 165°.

2.7 Solution: The P has four σ bonds and is sp^3, Sb has six σ bonds and is d^2sp^3 (octahedral). The hybridization for O could be sp (180°), sp^2 (120°); or even sp^3 (109.5^{o}). The 165° bond angle at O suggests sp to sp^2 hybridization with some degree of π bonding involving donation from the filled O p orbitals into the empty d orbitals of P.

2.8 Problem: Predict both the gross geometry (from the σ orbital hybridization) and the fine geometry (from bond-electron pair repulsion, etc.) of the following species: F_2SeO, $SnCl_2$, I_3^-, and $IO_2F_2^-$.

2.8 Solution: F_2SeO has three σ bonds and a lone pair on Se. The sp^3 hybridization with one lone pair leads to a pyramidal molecule with bond angles less than 109° because of the lone-pair repulsion. The $Se\langle^{F}_{O}$ bond angle is expected to be larger than that of $Se\langle^{F}_{F}$ since π bonding to O is more favorable than to F because of the neutralization of charge for Se=O.

$SnCl_2$ has two σ bonds and one lone pair on Sn, leading to an angular molecule with the bond angle less than 120° because of lone-pair repulsion.

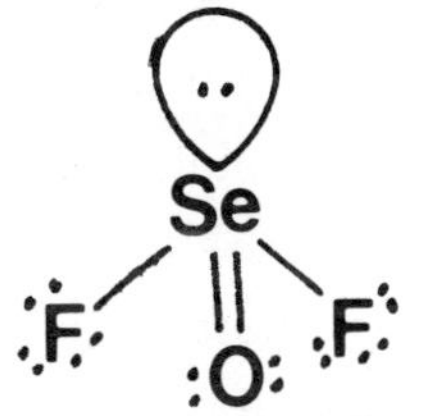

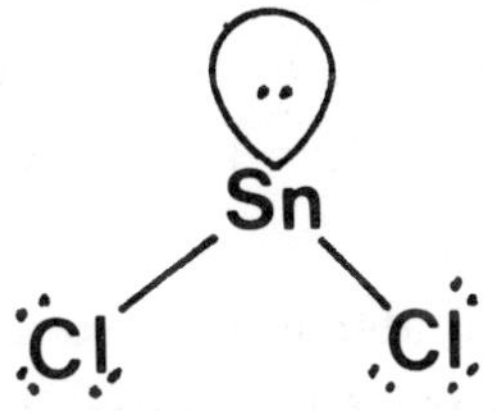

I_3^- has two σ bonds and three lone pairs on I. The three lone pairs are expected to occupy the equatorial positions of a trigonal bipyramid, leading to a linear ion.

$IO_2F_2^-$ has four σ bonds and a lone pair on I. We expect the lone pair in an equatorial position of the trigonal bipyramid to give a "sawhorse." The F-I-F bonds are expected to be essentially linear because of the opposing effects of repulsion involving the lone pair and the double bonds to O. The $I\langle^{O}_{O}$ angle is 100°, suggesting that the lone pair has a greater repulsive effect than the π bonds do.

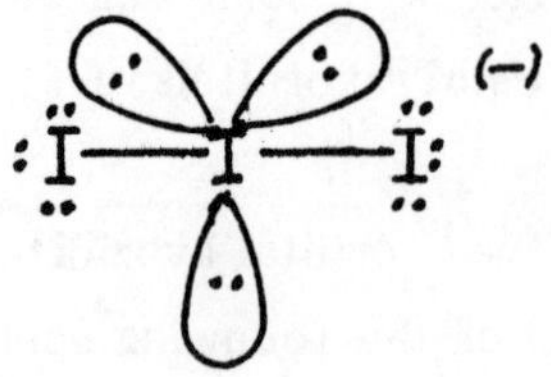

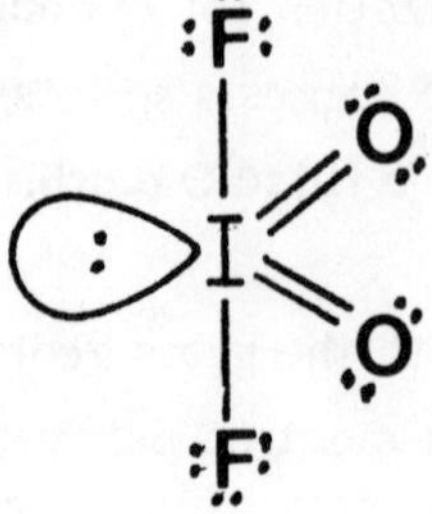

2.9 Problem: Which of the following in each pair will have the larger bond angle? Why? CH_4, NH_3; OF_2, OCl_2; NH_3, NF_3; PH_3, NH_3.

2.9 Solution: The bond angle is smaller in NH_3 than in CH_4 because of the repulsion between the lone pair on N and the bonding pairs. The bond angle is greater in OCl_2 than in OF_2 because there is some π interaction in OCl_2 involving donation from the filled p orbitals on O into the empty d orbitals on Cl. No π bonding is possible for OF_2 because all orbitals are filled on both atoms. The bond angle in NH_3 is larger than that in NF_3 because the N-F bonds are longer than N-H bonds and the electron density is displaced toward the more electronegative F, both effects diminish the bonding pair-bonding pair repulsion.

The bond angle is greater in NH_3 than in PH_3 because the P-H bonds are longer and the lower electronegativity of P permits electron density to be displaced toward H to a greater extent than in the case of NH_3. Both of these effects diminish the bonding pair-bonding pair repulsion.

2.10 Problem: Trimethylphosphine has been reported by Holmes to react with $SbCl_3$ and with $SbCl_5$ to form, respectively, $(Me_3P)(SbCl_3)$ or $(Me_3P)_2(SbCl_3)$ and $(Me_3P)(SbCl_5)$ and $(Me_3P)_2(SbCl_5)$. Suggest valence bond structures for each of these and indicate approximate bond angles around the Sb atom.

2.10 Solution: $Me_3P \rightarrow \ddot{S}bCl_3$. There are five electron pairs on Sb, so the structure is based on a trigonal bipyramid with the lone pair and Me_3P in the equatorial plane. Lone pairs and groups of low electronegativity prefer equatorial positions. The $Sb\genfrac{}{}{0pt}{}{\diagup Cl}{\diagdown P}$ angle should be less than 120° because of the lone-pair repulsion.

$(Me_3P)_2\ddot{S}bCl_3$. There are six electron pairs on Sb, so the structure is based on an octahedron with the PMe_3 trans to one another and cis to the lone pair. The PMe_3 groups are expected to bend away from the lone pair.

$Me_3P \rightarrow SbCl_5$. There are six electron pairs on Sb giving an octahedron. The four equatorial Cl atoms might be expected to bend away from the PMe_3 slightly.

$(Me_3P)_2SbCl_5$. There are seven electron pairs on Sb giving a pentagonal bipyramid with both Me_3P groups in the less crowded axial positions. No distortion is expected for the symmetrical structure.

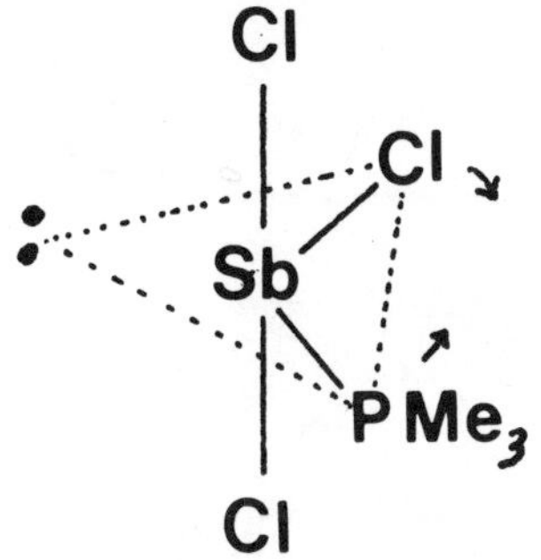

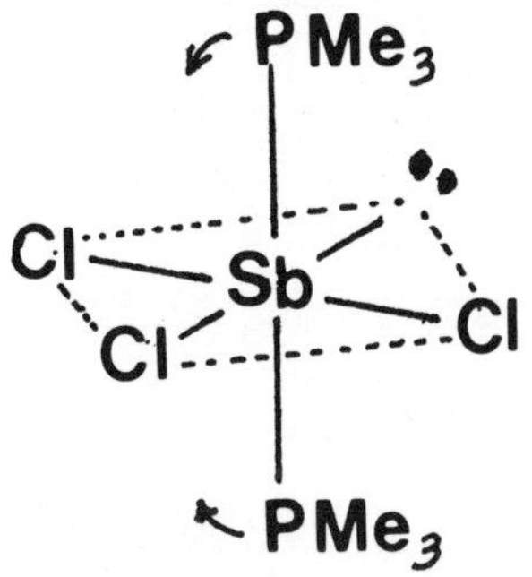

2.11 Problem: Indicate by a sketch the following hybrid orbitals (indicate the sign of the amplitude of the wave function on your sketch): (a) an sd hybrid (b) a pd hybrid.

2.11 Solution: (a) $s + d_z2$ — Combination of d_z2 with s increases the (+) lobes of d_z2 and diminishes the negative donut. Angular hybrids result from combination of s with other d orbitals.

$s + d_{z^2}$

$s - d_{z^2}$

$s + d_{x^2-y^2}$

$s - d_{x^2-y^2}$

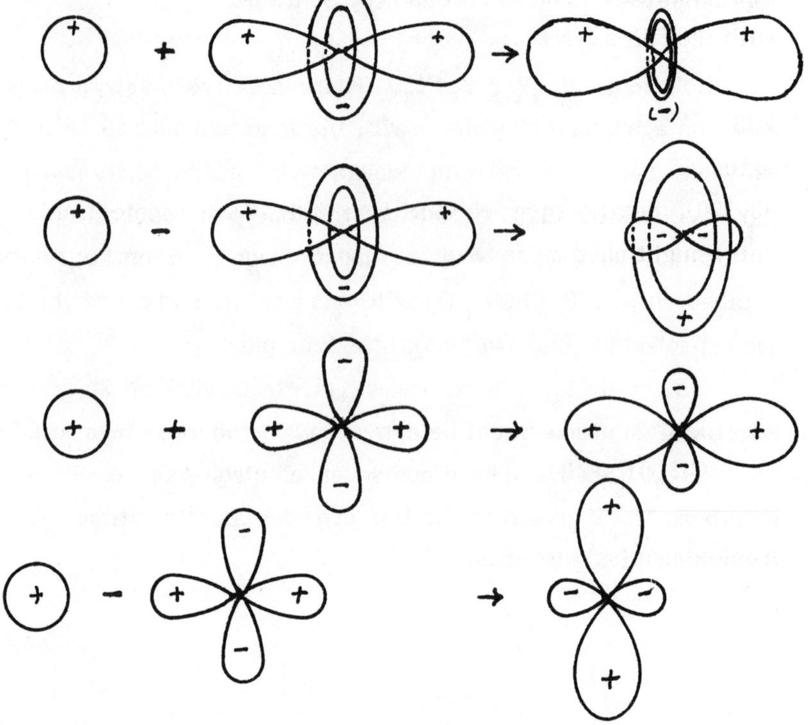

(b) $p_z \pm d_{z^2}$ — Each combination increases one lobe of d_{z^2} and diminishes the other (one combination shown).

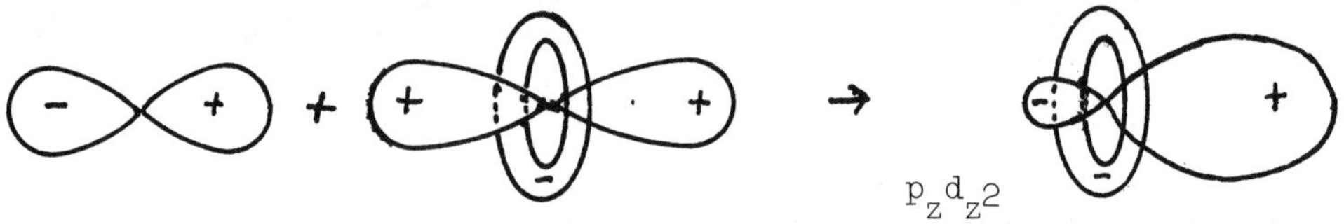

2.12 Problem: The Ru-O-Ru bond angle in $(Cl_5Ru)_2O$ is 180°. What is the state of hybridization of the oxygen? Explain the reasons for the large bond angle. (See R.J. Gillespie, J. Am. Chem. Soc. 1960, 82, 5978.)

2.12   Solution: The $180°$ bond angle at O for $(Cl_5Ru)_2O$ suggests sp hybridization for O with the filled p orbitals on O involved in d-d-π bonding with Ru.

2.13   Problem: Using the general principle of isoelectronic groups and H.A. Bent's extension of isoelectronic groupings (see DMA, p. 60 or H.A. Bent, J. Chem. Educ. 1966, 43, 170), identify the species of line (2) that are "isoelectronic" with those of line (1). Each entry might have more than one "isoelectronic" partner in the other line.

1. $(H_2N)_2CO$ $HONO_2$ OCO   CO $(CN)_2$ ClO_2^+ $Si_3O_9^{6-}$ $TeCl_2$ Bi_9^{5+}

2. H_2CCO $(HO)_2CO$ ONN H_2CNN H_3CNO_2 F_2CO BF B_2O_2 $(CH_3)_2CO$
    CH_2CN^- SO_2 cyclic $(SO_3)_3$ ICl_2^+ Pb_9^{4-}

Isoelectronic groups (read horizontally):

CO	CN^-	NO^+	NN	BF
$-CH_3$	$-\ddot{N}H_2$	$-\ddot{O}H$	$-\ddot{\ddot{F}}\!:$	
$>CH_2$	$>\ddot{N}H$	$>\ddot{O}\!:$		
$\geqslant CH$	$\geqslant N\!:$	$\geqslant \ddot{O}$ (+)		
$=CH_2$	$=\ddot{N}H$	$=\ddot{\ddot{O}}$		

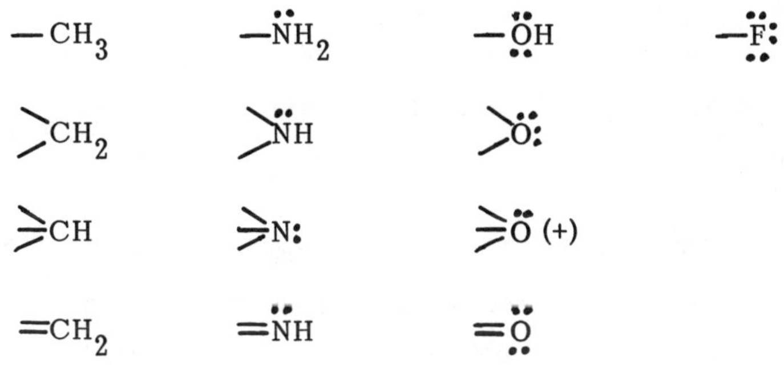

2.13   Solution: For each horizontal line shown the molecules, ions, or the groups underlined are "isoelectronic."

(1)	(2)
$\underline{HO}NO_2$	$\underline{CH_3}NO_2$
$(H_2N)_2CO$	$(\underline{HO})_2CO$, $\underline{F}_2CO$, $(CH_3)_2CO$

(1)	(2)
OCO, HNCO	ONN, H_2CNN, CH_2CN^-, H_2CCO
a b a b	a b a b a b a b
CO	BF
$(CN)_2$	B_2O_2 (2 x 9 electrons)
ClO_2^+	SO_2 (18 electrons)
$Si_3O_9^{6-}$	$(SO_3)_3$ (3 x 24 electrons)
$TeCl_2$	ICl_2^+ (20 electrons)
Bi_9^{5+}	Pb_9^{4-} (40 electrons)

2.14 Problem: Pauling's electroneutrality principle states that electrons are distributed in a molecule in such a way as to make the residual charge on each atom zero or nearly zero, except that hydrogen and the most electropositive metals can acquire partial positive charge and the most electronegative atoms can acquire partial negative charge. On this basis describe the reasonable formal charge distribution in NH_4^+ and in SO_4^{2-}.

2.14 Solution: For NH_3 the formal charge is +1 on N assuming equal sharing of electrons, but we expect the bonds to be polar (N is more electronegative than H), displacing charge toward N so that N becomes neutral and the positive charge is distributed equally among the four H atoms. For SO_4^{2-} we expect the degree of π bonding and the polarity of the bonds to result in the distribution of the negative charge equally (-1/2 each) on the four oxygen atoms. The limiting cases involving zero and four double bonds are shown. Two double bonds make the formal charge on S zero, neglecting bond polarity.

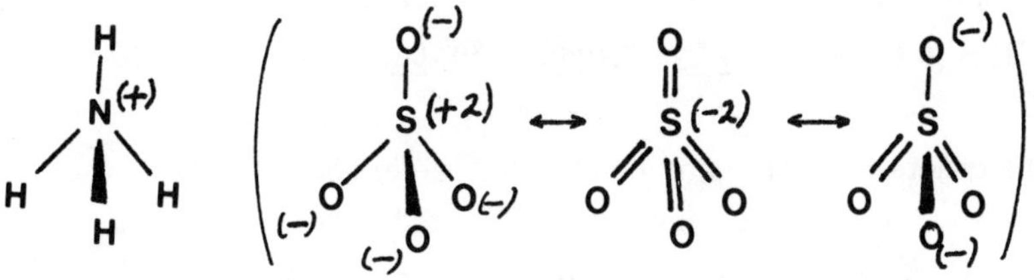

2.15 Problem: The bromine atom in BrF_5 is below the plane of the base of the tetragonal pyramid. Explain.

2.15 Solution: In BrF_5 the important repulsion is between the lone pair and the bonding pairs of the four Br-F bonds in the base of the pyramid, displacing the F atoms upward and leaving the Br below the plane of the base of the pyramid.

2.16 Problem: $H_2C{=}SF_4$ gives the expected isomer with the double-bonded methylene group in an equatorial position. From the orientation of the π bond, would you expect the H atoms to be in the axial plane or the equatorial plane? (See K.O. Christe and H. Oberhammer, Inorg. Chem. 1981, 20, 296.)

2.16 Solution: π bonding can involve either d_{xz} (or d_{yz}), with the π lobes above and below the equatorial plane, or the d_{xy} orbital, with the π lobes in the equatorial plane. Since there is less crowding in the equatorial plane, the latter is expected. The 97° $S\langle^F_F$ angle in the equatorial plane (see above reference) is consistent with this expectation. The σ hybridization for S is dsp^3, using d_{z^2}

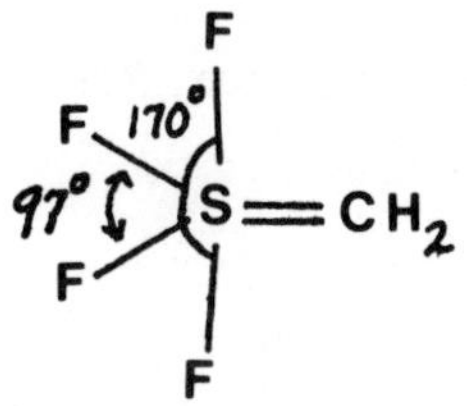

2.17 Problem: Compare the expected bond orders of ClO_4^- and IO_4^-, and of IO_3^-, BrO_3^-, and ClO_3^-. (See R. Nightingale, J. Phys. Chem. 1960, 64, 162.)

2.17 Solution: The orders of decreasing bond order are expected to be:

$$ClO_4^- > IO_4^- \text{ and } ClO_3^- > BrO_3^- > IO_3^-$$

π bonding is more favorable for atoms of nearly the same size (Cl and O). It is less effective with increasing size of the p orbitals (poorer overlap). The size of the p orbitals increases with increasing n (major quantum number). Data given in the reference agree with this order.

2.18 Problem: The C—C bond distance in methyl acetylene is unusually short (146 pm) for a single bond. Show how this short bond can be rationalized in more than one way.

2.18 Solution: Since the strength of a bond involving some kind of s-p hybrid increases with the amount of s character (the lower-energy s orbitals lower the energy of the hybrids), the bond involving the sp hybrid orbitals for C-CH_3 is stronger and shorter than C-CH_3 bonds formed using sp^2 and sp^3 hybrids. Another explanation is that hyperconjugation involving "no bond" structures, $H\text{-}\overset{\bullet\bullet}{\underset{(-)}{C}}\text{=C=C}\begin{smallmatrix}H \\ H,\end{smallmatrix}$ (+)H increases the bond order of the C-CH_3 bond.

2.19 Problem: Calculate the dipole moment to be expected for the ionic structure H^+Cl^- using the same internuclear separation as for the HCl molecule. Calculate the dipole moment for HCl assuming 19% ionic character.

2.19 Solution: The bond distance for HCl is 127.4 pm, assuming +1 and -1 charges separated by 127 pm,

$$\mu = qd = 4.80 \times 10^{-10} \text{ esu} \times 127 \text{ pm} = 6.10 \text{ D}$$

This is for 100% ionic character, if the bond is only 19% ionic, the calculated dipole moment is 1.16 D, 0.19 x 6.10 = 1.16 D. The actual dipole moment for HCl is 1.03 D, corresponding to about 17% ionic character.

2.20 Problem: Calculate the heats of formation (from electronegativities) for H_2S, H_2O, SCl_2, NF_3, and NCl_3. Calculate the heats of formation of these compounds from the bond energies for comparison.

$$-\Delta H_f = 96.49n(\chi_A - \chi_B)^2$$

where n is the number of bonds.

2.20 Solution:

H_2S	$-\Delta H_f = 2(96.5)(2.6 - 2.2)^2 = 0.32(96.5) = -31$ kJ/mole
H_2O	$-\Delta H_f = 2(96.5)(3.4 - 2.2)^2 = 2.88\ (96.5) = -278$ kJ/mole
SCl_2	$-\Delta H_f = 2(96.5)(3.2 - 2.6)^2 = -69$ kJ/mole
NF_3	$-\Delta H_f = 3(96.5)(4.0 - 3.0)^2 = -290$ kJ/mole

correcting for the extraordinary stability of the $N\equiv N$ bond, requiring more energy (231 kJ/N) to break the bond,

$$-\Delta H_f = -290 + 231 = -59 \text{ kJ/mole}$$

H_2S	$1/8S_8 + H_2$	$\rightarrow$	H_2S	
	$+ 226 + 432$		$- 2 \times 363$	$\Delta H_f = -68$ kJ/mole
H_2O	$H_2 + 1/2O_2$	$\rightarrow$	H_2O	
	$+432 + 1/2 \times 494$		$- 2 \times 459$	$\Delta H_f = -239$
SCl_2	$1/8S_8 + Cl_2$	$\rightarrow$	SCl_2	
	$+ 226 + 239$		$- 2 \times 255$	$\Delta H_f = -45$
NF_3	$1/2\ N_2 + 3/2\ F_2$	$\rightarrow$	NF_3	
	$+ 1/2 \times 942 + 3/2 \times 155$		$- 3 \times 283$	$\Delta H_f = -146$

The values from bond energies are more reliable since the bond energies are obtained more directly from thermodynamic data. The agreement is poorest for NF, indicating that in this case the correction made for the N≡N bond is overdone.

2.21 Problem: Calculate the electronegativity differences from the bond energies for H-Cl, H-S, and S-Cl. Using Allred's electronegativity of H, compare the electronegativity values that can be obtained for S and Cl with those tabulated.

Bond energies			
H-Cl	428 kJ/mole	H_2	432 kJ/mole
H-S	363	Cl_2	239
S-Cl	255	$(1/8)S_8$	226

$$D(A\text{-}B) = 1/2\ [D(A\text{-}A) + D(B\text{-}B)] + 96.5(\chi_A - \chi_B)^2$$

$$\chi_H = 2.2$$

2.21 Solution:

HCl $\quad 428 = 1/2(432 + 239) + 96.5(\Delta\chi)^2$

$\Delta\chi = 1.0$

HS $\quad 363 = 1/2(432 + 226) + 96.5(\Delta\chi)^2$

$\Delta\chi = 0.6$

SCl $\quad 255 = 1/2(226 + 239) + 96.5(\Delta\chi)^2$

$\Delta\chi = 0.5$

Using $\chi_H = 2.2$, $\chi_{Cl} = 3.2$, $\chi_S = 2.8$

Accepted values $\chi_{Cl} = 3.2$, $\chi_S = 2.6$

2.22 Problem: Allred and Rochow (J. Inorg. Nucl. Chem. 1958, 5, 264-8, 269-88) have proposed the following empirical equation for the calculation of electronegtivities:

$$\chi = 0.359\ \frac{Z_{eff}}{r^2} + 0.744$$

where Z_{eff} is calculated using Slater's rules (see DMA p. 32) and r is the covalent radius (in A). Using $r_{As} = 1.22$ A and $r_{Br} = 1.14$ A, calculate the electronegativities of As and Br.

Little and Jones used this equation to calculate a complete set of electronegativities (J. Chem. Educ. 1960, 37, 231). What are the advantages and disadvantages of this method of obtaining electronegativities?

2.22 Solution: For As and Br, n = 4 for the valence shell. There are 10 electrons below n = 3, giving complete shielding; 18 in the shell below n = 4, shielding by 0.85 for each electron, and the shielding of one electron by another in the valence shell is 0.35.

		As	$(1s^2 2s^2 2p^6)$	$(3s^2 3p^6 3d^{10})$	$(4s^2 4p^3)$	
Z_{eff}	=	33	-10(1.00)	-18(0.85)	-5(0.35)	= 5.95

$$\chi_{As} = 0.359 \frac{5.95}{(1.22)^2} + 0.744 = 2.2$$

		Br	$(1s^2 2s^2 2p^6)$	$(3s^2 3p^6 3d^{10})$	$(4s^2 4p^5)$	
Z_{eff}	=	35	-10(1.00)	-18(0.85)	-7(0.35)	=7.25

$$\chi_{Br} = 0.359 \frac{7.25}{(1.14)^2} + 0.744 = 2.7$$

This approach permits the calculation of electronegativities from easily accessible information. It might be applied to ions. The equation cannot be expected to "fit" elements in all parts of the periodic table equally well.

2.23 Problem: Compare the electronegativity of the N orbital involved in the N-H bond in each of the following species:

NH_3 NH_4^+ NH_2^-  $\overset{+}{N}$-H $CH_3C{\equiv}NH^+$

2.23 Solution: Electronegativity, as defined by Pauling, is the ability of an atom in a molecule to attract electrons. We expect the electronegativity to increase with increasing s character in the hybrid orbital, and with an increase in the positive formal charge. These considerations lead to an electronegativity order of N in the above series of

$$\chi_{CH_3CNH^+} > \chi_{C_5H_5NH^+} > \chi_{NH_4^+} > \chi_{NH_3} > \chi_{NH_2^-}$$

2.24 Problem: What is the expected bond order of the SeSe bonds in the cyclic Se_4^{2+} species?

2.24 Solution: The contributing structures to the resonance hybrid which are expected to be of major importance would be:

$$\left\{ \begin{array}{cc} \ddot{S}e - \ddot{S}e^{2+} \\ \| \quad | \\ Se - \ddot{S}e \end{array} \quad \begin{array}{cc} \ddot{S}e = Se^{2+} \\ | \quad | \\ \ddot{S}e - \ddot{S}e \end{array} \quad \begin{array}{cc} \ddot{S}e - \ddot{S}e^{2+} \\ | \quad \| \\ \ddot{S}e - Se \end{array} \quad \begin{array}{cc} \ddot{S}e - \ddot{S}e^{2+} \\ | \quad | \\ \ddot{S}e = Se \end{array} \right\}$$

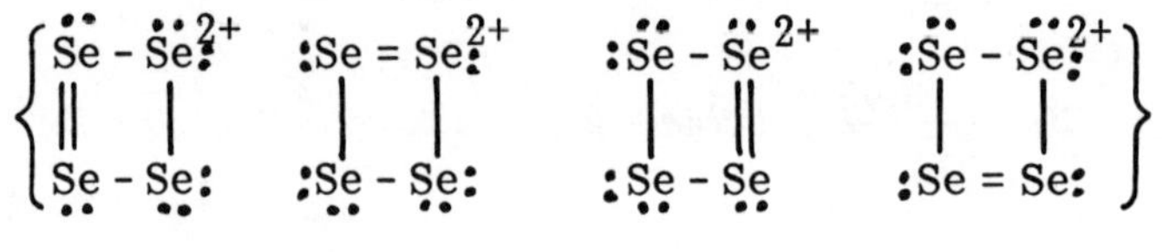

From the above we expect the Se-Se bond order to be 1 1/4, that is, between any pair of Se atoms, there is one double bond in a sum of four contributing structures.

2.25 Problem: Which of the possible isomers of $(CH_3)_2PF_3$ is the one encountered? Why is there a site preference for the methyl group?

2.25 Solution: In $(CH_3)_2PF_3$ the methyl groups occupy the equatorial positions. In such cases the less electronegative substituents occupy equatorial positions in order to minimize bonding pair repulsion. There is greater displacement of electron density in the bonds toward the more electronegative substituent, decreasing the bonding pair-bonding pair repulsion. It is the bond angle between the methyl groups that is greater (124°) than the 120° angle in the trigonal plane of a regular trigonal bipyramid. The two axial F substituents bend away slightly from the CH_3 groups.

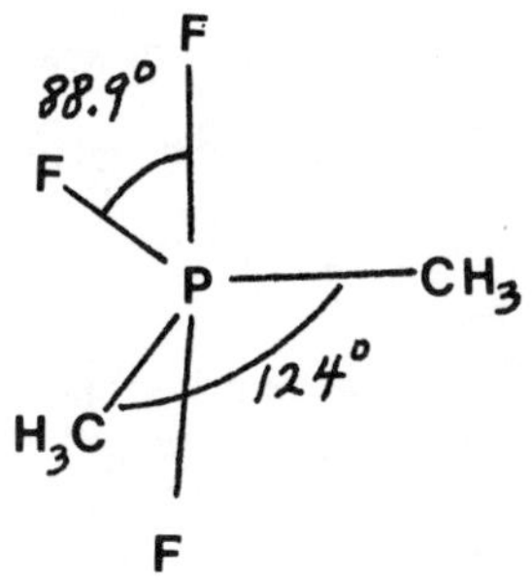

2.26 Problem: There are striking differences between the physical properties of the second period elements C and N and the corresponding elements of the third period, Si and P. Similarly, there are important differences between the oxides and oxoacids (just consider those with the highest oxidation state). What bonding tendencies account for the differences and why do they occur?

2.26 Solution: Carbon exists as diamond with each C bonded to four others by single bonds and as graphite where each C is bonded to three others in a planar hexagonal network — the latter occurs because of delocalized π bonding. Silicon gives only the single bonded structure.

Nitrogen occurs as triple bonded $N \equiv N$ while P_4 (only single bonds) is the stable form.

The oxides and oxoacids are:

CO_2	H_2CO_3	N_2O_5	HNO_3
SiO_2	H_4SiO_4	P_4O_{10}	H_3PO_4

In CO_2, carbon forms two sigma bonds and two π bonds. In H_2CO_3, N_2O_5, and HNO_3, C or N form only three σ bonds and a π bond. In all of the Si and P oxides and oxoacids considered here Si or P form four σ bonds (sp^3 hybridization). Any π bonding must involve the d orbitals of the Si or P and occurs only to the extent needed to achieve low charges on all atoms.

Pi bonding occurs to a much greater extent for C and N than for Si and P because of the size of the atoms and of the orbitals. The sidewise π overlap is most favorable for small atoms (short X-X distance). Pi bonding is also unfavorable for large, diffuse orbitals because of poor overlap.

2.27 Problem: Explain why the valence bond structure of carbon monoxide is best described as $:C\overset{\leftarrow}{\equiv}O:$, featuring a dative bond from O to the less electronegative C.

2.27 Solution: The ground state electron configuration of C is $2s^2 2p^2$. We consider that the s orbital and the empty p orbital form two sp hybrids, one directed toward O and one directed 180° opposite. This contains the C lone pair, leaving empty the sp hybrid pointing toward O. Hence, the C-O σ hybrid is formed by overlap of the empty sp hybrid with a filled O σ orbital. The two unhybridized singly occupied C 2p orbitals form π bonds by overlap with singly occupied O 2p orbitals.

III Symmetry

3.1 Problem: Use matrices to show that $\sigma_1\sigma_2 = C_4$ and $\sigma_2\sigma_1 = C_4^3$ for the kaleidoscope group in the figure below. Take the intersection of σ_1 and σ_2 as the z axis (that is, the axis perpendicular to the plane of the paper) and let the x axis be the intersection of σ_1 and the plane of the paper.

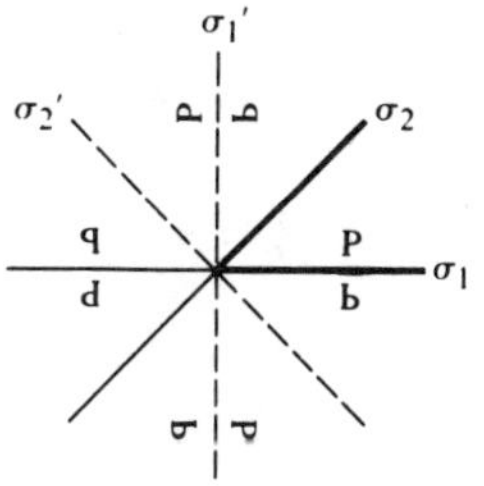

Reflections in a kaleidoscope.

3.1 Solution: The mirror plane σ_1 carries **P** into **b**, or $x_2 = x_1$, $y_2 = -y_1$, and $z_2 = z_1$ (x and z are unchanged, y goes into the negative of itself). The matrix may be set up as:

NEW \ ORIG	x_1	y_1	z_1
x_2	1		
y_2		-1	
z_2			1

or

$$\sigma_1 = \begin{pmatrix} 1 & 0 & 0 \\ 0 & -1 & 0 \\ 0 & 0 & 1 \end{pmatrix} \begin{pmatrix} x \\ y \\ z \end{pmatrix}$$

The mirror plane σ_2 carries **P** into , or $x_2 = y_1$, $y_2 = x_1$, and $z_2 = z_1$.

	x_1	y_1	z_1
x_2		1	
y_2	1		
z_2			1

or $$\sigma_2 = \begin{pmatrix} 0 & 1 & 0 \\ 1 & 0 & 0 \\ 0 & 0 & 1 \end{pmatrix} \begin{pmatrix} x \\ y \\ z \end{pmatrix}$$

The C_4 operation carries **P** into , or $x_2 = y_1$, $y_2 = -x_1$, and $z_2 = z_1$. Thus

	x_1	y_1	z_1
x_2		1	
y_2	-1		
z_2			1

or $$C_4 = \begin{pmatrix} 0 & 1 & 0 \\ -1 & 0 & 0 \\ 0 & 0 & 1 \end{pmatrix} \begin{pmatrix} x \\ y \\ z \end{pmatrix}$$

The $C_4^{\,3}$ operation carries **P** into , or $x_2 = -y_1$, $y_2 = x_1$, and $z_2 = z_1$. Thus

	x_1	y_1	z_1
x_2		-1	
y_2	1		
z_2			1

or $$C_4^3 = \begin{pmatrix} 0 & -1 & 0 \\ 1 & 0 & 0 \\ 0 & 0 & 1 \end{pmatrix} \begin{pmatrix} x \\ y \\ z \end{pmatrix}$$

We are now to show by matrix multiplication that $\sigma_1 \sigma_2 = C_4$ and $\sigma_2 \sigma_1 = C_4^{\,3}$. We will use the product rule for matrix multiplication

$$p_{ik} = \sum_j a_{ij} b_{jk}$$

where p is an element in the ith row and jth column of the product matrix, a and b are elements in matrices A and B and are located in the ith row and jth column and jth row and kth column, respectively. The sum of each p_{ik} element is taken over all elements in the ith row of A with all elements in the kth column of B. Labeling each matrix with subscripts to follow the multiplication more readily, we have

$$\underset{\sigma_1}{\begin{pmatrix} 1_{11} & 0_{12} & 0_{13} \\ 0_{21} & -1_{22} & 0_{23} \\ 0_{31} & 0_{32} & 1_{33} \end{pmatrix}} \underset{\sigma_2}{\begin{pmatrix} 0_{11} & 1_{12} & 0_{13} \\ 1_{21} & 0_{22} & 0_{23} \\ 0_{31} & 0_{32} & 1_{33} \end{pmatrix}} =$$

$$\begin{pmatrix} (1_{11}\cdot 0_{11}+0_{12}\cdot 1_{21}+0_{13}\cdot 0_{31}) & (1_{11}\cdot 1_{12}+0_{12}\cdot 0_{22}+0_{13}\cdot 0_{32}) & (1_{11}\cdot 0_{13}+0_{12}\cdot 0_{23}+0_{13}\cdot 1_{33}) \\ (0_{21}\cdot 0_{11}-1_{22}\cdot 1_{21}+0_{23}\cdot 0_{31}) & (0_{21}\cdot 1_{12}-1_{22}\cdot 0_{22}+0_{23}\cdot 0_{32}) & (0_{21}\cdot 0_{13}-1_{22}\cdot 0_{23}+0_{23}\cdot 1_{33}) \\ (0_{31}\cdot 0_{11}+0_{32}\cdot 1_{21}+1_{33}\cdot 0_{31}) & (0_{31}\cdot 1_{12}+0_{32}\cdot 0_{22}+1_{33}\cdot 0_{32}) & (0_{31}\cdot 0_{13}+0_{32}\cdot 0_{23}+1_{33}\cdot 1_{33}) \end{pmatrix}$$

$$= \begin{pmatrix} 0 & 1 & 0 \\ -1 & 0 & 0 \\ 0 & 0 & 1 \end{pmatrix} \equiv C_4$$

And, with less fanfare

$$\underset{\sigma_2}{\begin{pmatrix} 0 & 1 & 0 \\ 1 & 0 & 0 \\ 0 & 0 & 1 \end{pmatrix}} \underset{\sigma_1}{\begin{pmatrix} 1 & 0 & 0 \\ 0 & -1 & 0 \\ 0 & 0 & 1 \end{pmatrix}} = \underset{C_4^3}{\begin{pmatrix} 0 & -1 & 0 \\ 1 & 0 & 0 \\ 0 & 0 & 1 \end{pmatrix}}$$

Thus, σ_1 and σ_2 do not commute.

3.2 Problem: Select the point group to which each of the species below belongs.

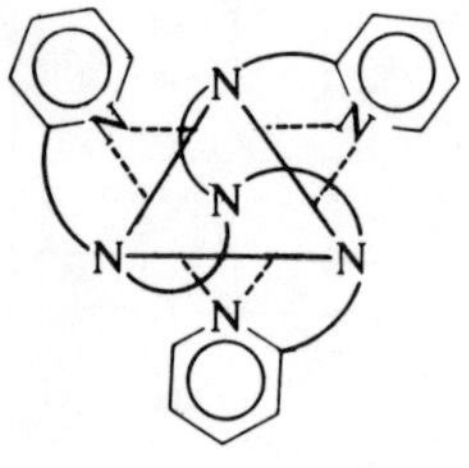

(a)

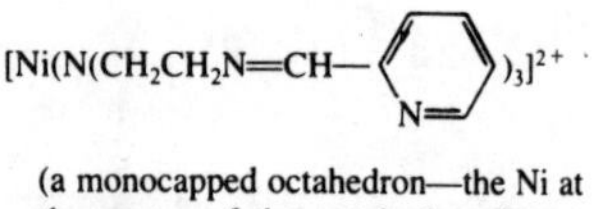

(a monocapped octahedron—the Ni at the center of the octahedron is not shown. From L. J. Wilson and N. J. Rose, *J. Am. Chem. Soc.* 1968, *90*, 6041.

(b) $C_{10}H_6F_2$

(c) C_6H_6

All of the tribromobenzenes

(d)

3.2 **Solution:** The "flow chart" below indicates the minimum number of generators which should be found to identify the point group to which a species belongs.

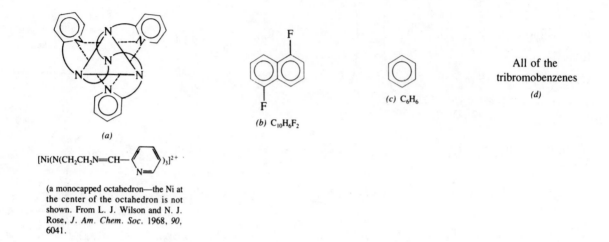

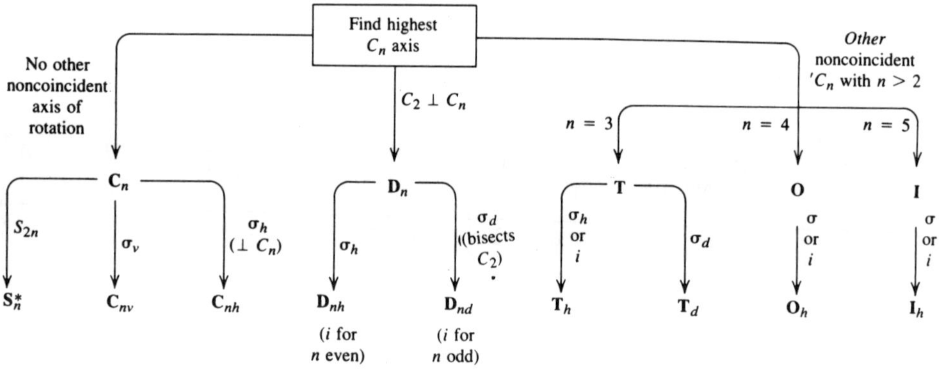

*Special nomenclature:

$\mathbf{S_1} \equiv \mathbf{C_{1v}} \equiv \mathbf{C_{1h}} \equiv \mathbf{C_s}$

$\mathbf{S_n} \equiv \mathbf{C_{nh}}$ for odd n

$\mathbf{S_2} \equiv \mathbf{C_i}$

Flow chart for assignment of molecules to point groups.

(a) The highest C_n is C_3. There are no other C_3 and there are no $C_2 \perp C_3$. The rotational group is therefore C_3. The molecule is chiral—there are no σ, or other S_n, so the point group is C_3.

(b) Highest $C_n = C_2$; no other C_2 present; a $\sigma_h \perp C_2$ is present. Point group is C_{2h}.

(c) Highest $C_n = 6$, $\perp C_2$, σ_h. Point group is D_{6h}.

(d) 1, 2, 3 isomer C_2, no $\perp C_2$, σ_v. Point group is C_{2v}.
1, 2, 4 isomer C_1, σ . Point group is C_s.
1, 3, 5 isomer C_3, $\perp C_2$, σ_h, Point group is D_{3h}.

3.3 Problem: Assign the molecules depicted below to the appropriate point groups.

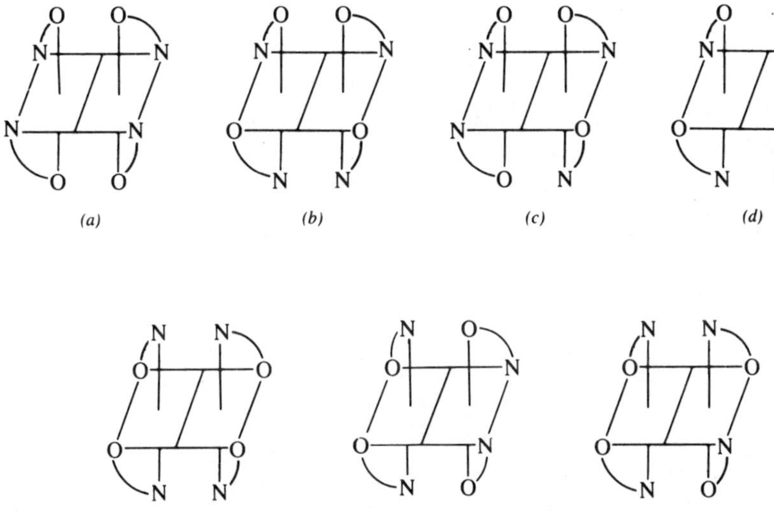

Isomers of $[(gly)_2Co(OH)_2Co(gly)_2]$

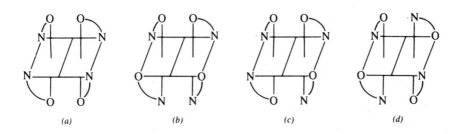

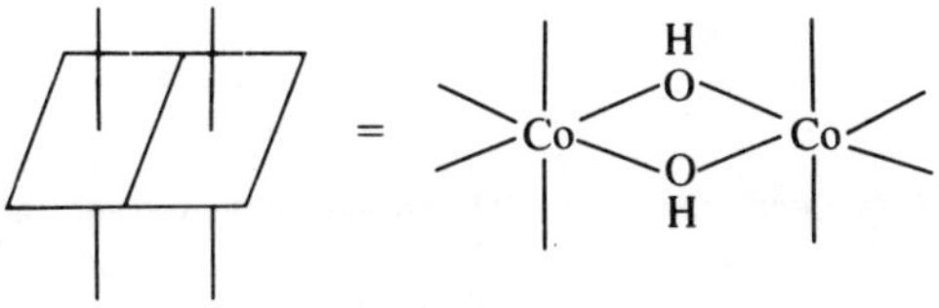

gly = N⌒O = $NH_2CH_2CO_2^-$

3.3 Solution:

(a)	C_2, no $\perp C_2$, σ_h. Point group is	C_{2h}
(b)	only σ.	C_s
(c)	only E.	C_1
(d)	only i.	C_i
(e)	C_2. no $\perp C_2$, σ_h.	C_{2h}
(f)	only C_2.	C_2
(g)	only E.	C_1

3.4 Problem: Assign the species shown below to the appropriate point groups.

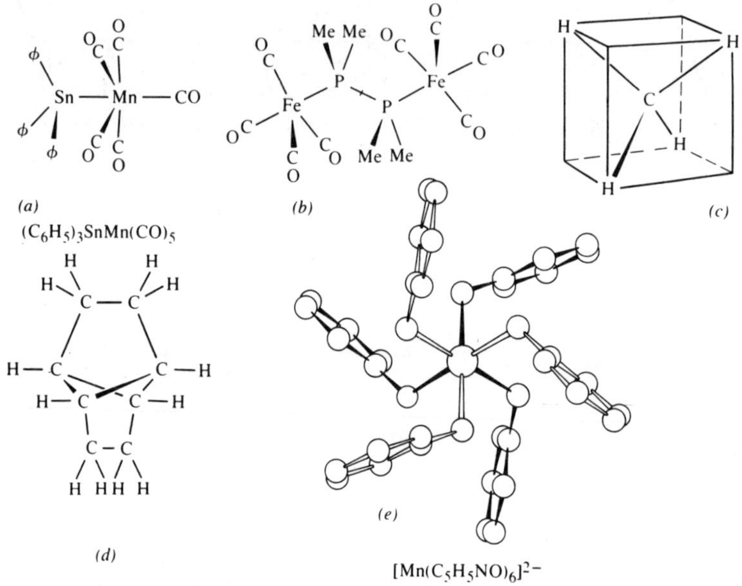

(*a*) is from H. P. Weber and R. F. Bryan, *Acta Cryst.* 1967, *22,* 822. (*d*) after J. F. Chiang and S. H. Bauer, *Trans. Faraday Soc.* 1968, *64,* 2248. (*e*) is reprinted with permission from T. J. Bergendahl and J. S. Wood, *Inorg. Chem.* 1975, *14,* 338; copyright 1975, American Chemical Society.

3.4 Solution:

(a) only σ. Point group is C_s

(b) C_2, σ_h. C_{2h}

(c) C_3, + additional C_3, σ_d. T_d

(d) C_2, $\perp C_2$, σ_d. D_{2d}

(e) C_3, no $\perp C_2$; S_6 S_6

3.5 Problem: A scalene triangle has three unequal sides. A regular scalene tetrahedron may be constructed by folding the pattern obtained by joining the midpoints of the sides of a scalene triangle having acute angles. To what point group does the regular scalene tetrahedron belong?

3.5 Solution: The regular scalene tetrahedron has C_2 axes through each pair of opposite edges. It belongs to a D_2 point group.

3.6 Problem: Assign point groups to the following figures and their conjugates. What are the shapes of the conjugate figures?

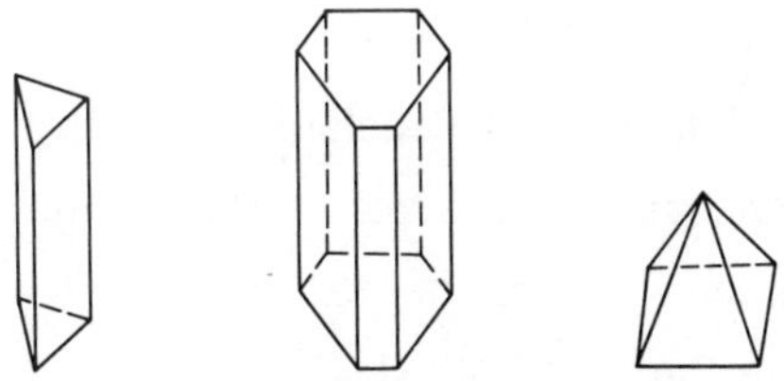

(a) Trigonal prism *(b)* Hexagonal prism with alternate sides of equal area *(c)* Square pyramid

3.6 Solution:

(a) C_3, $C_2 \perp C_3$, σ_h. Point group is D_{3h}

(b) C_3, $C_2 \perp C_3$, σ_h. D_{3h}

(c) C_4, no $\perp C_2$, σ_v. C_{4v}

The conjugate figures are obtained by connecting the centers of the faces and would be:

(a) a trigonal bipyramid.

(b) a hexagonal bipyramid—the hexagon having alternate interior angles of the same size.

(c) a square pyramid.

The conjugates would all have the same point groups as the figures to which they are conjugate.

3.7 Problem: On page 47 there are figures which can be copied and used for the construction of a tetrahedron, an octahedron, and a cube. Identify the point groups for each, considering the shading of the faces.

3.7 Solution: The regular tetrahedron belongs to the T_d group since it has more than one noncoincident C_3 axis and a dihedral mirror plane that carries one C_3 axis into another. The shaded tetrahedron retains all of the proper rotational axes of the regular tetrahedron, but loses all of the improper axes of rotation. Thus, it belongs to the T group--a pure rotational group. In similar fashion we find that the shaded cube has several noncoincident C_4 axes, but no improper axes, and belongs to the O group. The shaded octahedron is somewhat more interesting since the shading reduces the number of both proper and improper symmetry elements. The highest rotational axis is now a C_3 axis, of which there are several, but a center of inversion is retained, making this an example of a T_h group.

3.8 Problem: On pages 47 and 48 there are figures which can be copied and used for the construction of a dodecahedron with trigonal faces, a dodecahedron with pentagonal faces, and an icosahedron. Determine the point groups for the polyhedra taking into account any shading on the faces. Ignoring the shading, which of the polyhedra are conjugate Platonic solids?

3.8 Solution: The trigonal dodecahedron is not a Platonic solid since all faces are not equivalent. It has three mutually perpendicular C_2 axes, and several dihedral mirror planes, so it belongs to the D_{2d} group. The icosahedron with shaded faces has no mirror planes or other S_n axes, but it does have several C_5 axes. It belongs to an I group. The pentagonal dodecahedron also has several C_5 axes, but with no shading it also has an inversion center, thus it belongs to the I_h group. The pentagonal dodecahedron and the regular icosahedron are conjugate solids, that is, one can be formed from the other by connecting the centers of the faces.

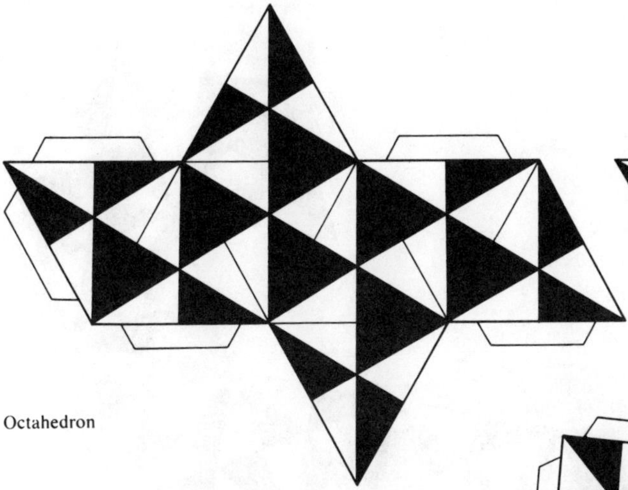

Octahedron

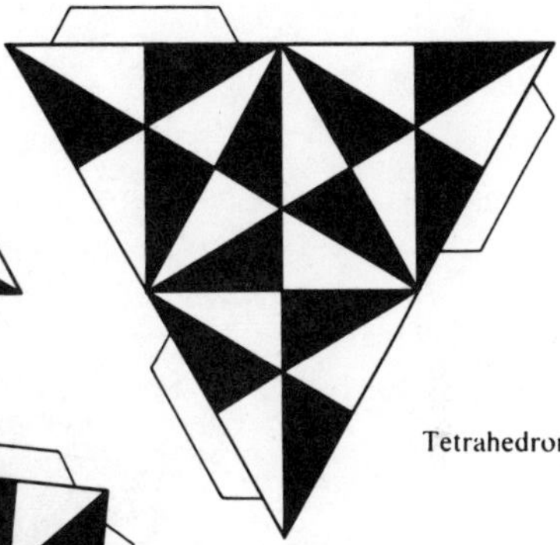

Tetrahedron

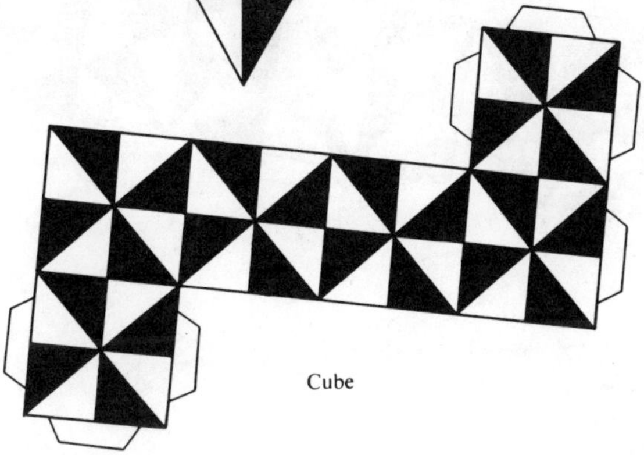

Cube

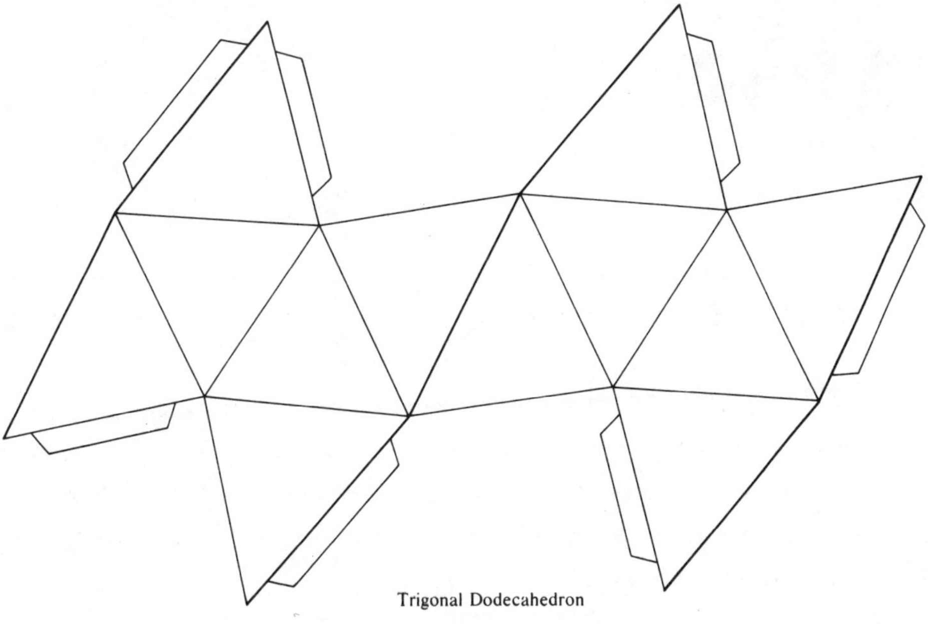

Trigonal Dodecahedron

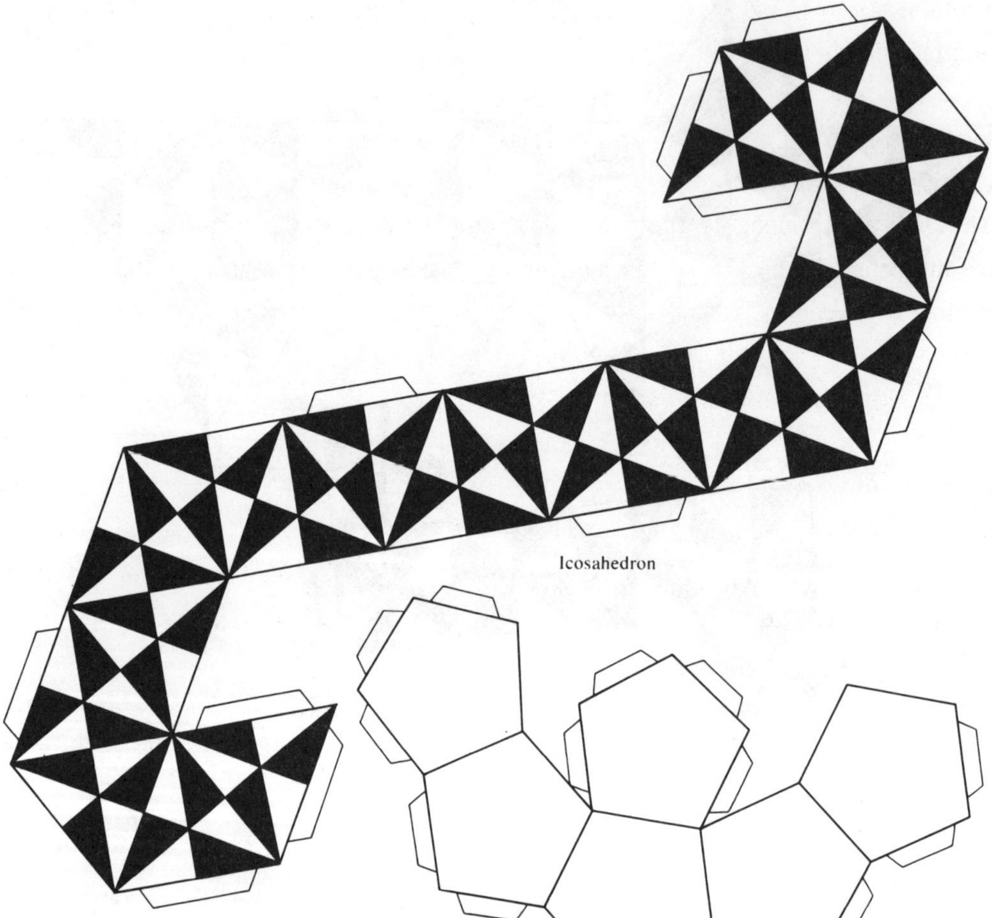

Icosahedron

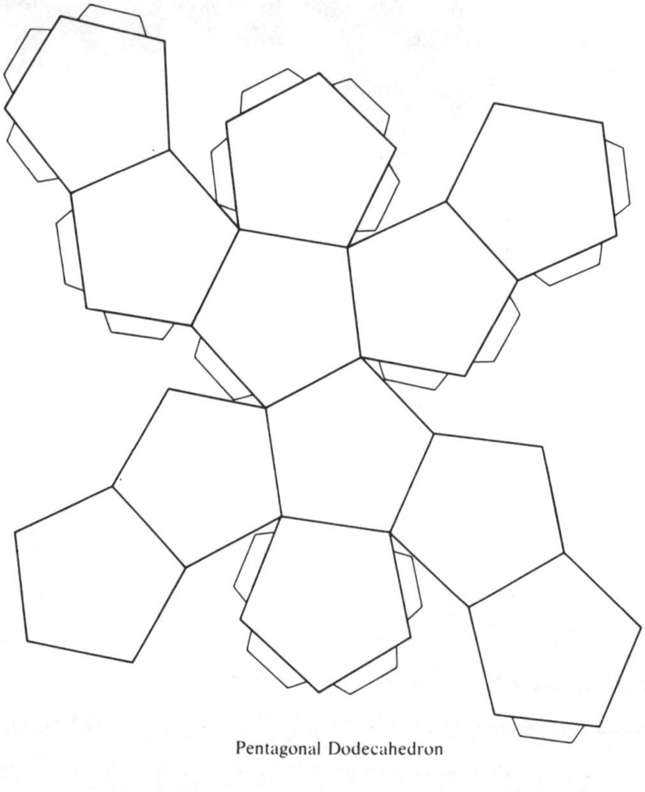

Pentagonal Dodecahedron

3.9 Problem: Determine the order of the D_{3h} group, the number of classes of operations, and the dimensions of the irreducible representations.

3.9 Solution: The order of the C_3 subgroup is 3; adding a perpendicular C_2 to give D_3 doubles the order to give 6; adding σ_h again doubles the order to give 12 for D_{3h}.

The $3C_2$ axes belong to the same class and the $3\sigma_v$ belong to the same class, since they are interchanged by C_3, so the number of classes cannot be greater than 9 (without determining if C_3 and C_3^2, and S_3 and S_3^2 are in the same class). For any group, the dimensions of the representation are the same as the character under the identity operation—when these are squared and summed over all irreducible representations the order of the group is obtained. We find on assuming 8 irreducible representations, with one being two-dimensional, $2^2 + 1^2 + 1^2 + 1^2 + 1^2 + 1^2 + 1^2 + 1^2 = 11 \neq$ order of the group. Hence there must be at least 2 two-dimensional representations and 6 irreducible representations, since $2^2 + 2^2 + 1^2 + 1^2 + 1^2 + 1^2 = 12$, or 6 classes, 2 two-dimensional and 4 one-dimensional representations.

3.10 Problem: Indicate how the slash mark on the "vee-bar" shown below would be shifted under each of the operations of the C_{2v} group. Use your results as a guide to construct the multiplication table for the C_{2v} group.

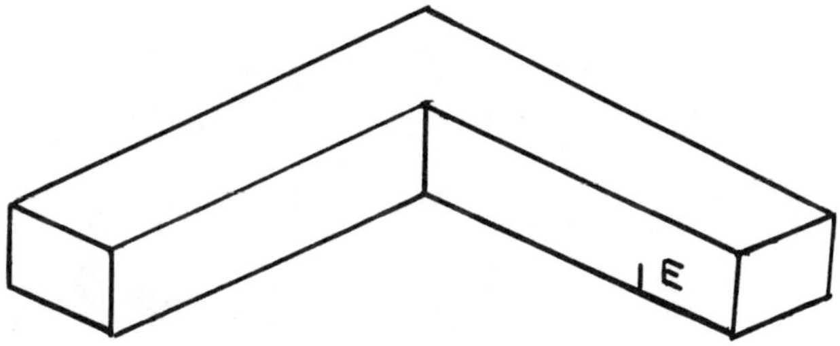

3.10 Solution: Let the z axis be the C_2 axis, let the y axis be in the plane of the "vee." Under the operations of the C_{2v} group, the slash mark is brought to the new position indicated on the figure below. The multiplication table for the C_{2v} group may be constructed by noting the position of the slash mark after two successive operations, and noting the single operation which would carry the slash mark to the same position. The operation carried out first is listed at the top of the multiplication table.

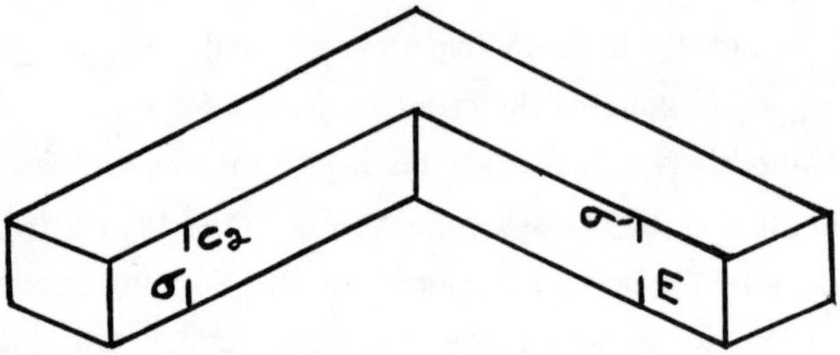

C_{2v}	E	C_2	σ	σ'
E	E	C_2	σ	σ'
C_2	C_2	E	σ'	σ
σ	σ	σ'	E	C_2
σ'	σ'	σ	C_2	E

3.11 Problem: Determine the independent operations of the C_{3h} group and construct the group multiplication table. Determine whether C_3 and C_3^2 are in the same class.

3.11 Solution: The operations of C_{3h} are E, C_3, C_3^2, σ_h, S_3, and S_3^5. Even powers of an S_n operation generate rotations, while odd powers yield further improper rotations.

$$S_3 = \sigma_h C_3$$

$$S_3^2 = (\sigma_h C_3)(\sigma_h C_3) = C_3^2$$

$$S_3^3 = (\sigma_h C_3)C_3^2 = \sigma_h$$

$$S_3^4 = C_3^2 C_3^2 = C_3$$

$$S_3^5 = (\sigma_h C_3)C_3 = \sigma_h C_3^2$$

$$S_3^6 = E$$

C_{3h}	$E=S_3^6$	S_3^1	S_3^2	S_3^3	S_3^4	S_3^5
E	E	S_3	S_3^2	S_3^3	S_3^4	S_3^5
S_3	S_3	S_3^2	S_3^3	S_3^4	S_3^5	E
S_3^2	S_3^2	S_3^3	S_3^4	S_3^5	E	S_3
S_3^3	S_3^3	S_3^4	S_3^5	E	S_3	S_3^2
S_3^4	S_3^4	S_3^5	E	S_3	S_3^2	S_3^3
S_3^5	S_3^5	E	S_3	S_3^2	S_3^3	S_3^4

In groups having a single generating operation (Abelian groups), all operations are in separate classes. Thus, the C_3 and C_3^2 (or C_3^-) operations are not interchanged by σ_h, or any other operation of the group.

3.12 Problem: Test the mutual orthogonality of the representations of the D_{3h} group.

D_{3h}	E	$2C_3$	$3C_2$	σ_h	$2S_3$	$3\sigma_v$		
A_1'	1	1	1	1	1	1		x^2+y^2,z^2
A_2'	1	1	-1	1	1	-1	R_z	
E'	2	-1	0	2	-1	0	(x,y)	(x^2-y^2,xy)
A_1''	1	1	1	-1	-1	-1		
A_2''	1	1	-1	-1	-1	1	z	
E"	2	-1	0	-2	1	0	(R_x,R_y)	(xy,yz)

3.12 Solution: Orthogonality requires that for each pair of symmetry species given in the character table the sum of the product of the characters under each of the operations of the group should be zero. We will illustrate this test for orthogonality for the A_2'' and E" symmetry species above.

$$\begin{matrix} E & & 2C_3 & & 3C_2 & & \sigma_h & & 2S_3 & & 3\sigma_v & & \\ (1)(2) & + & 2(1)(-1) & + & 3(-1)(0) & + & (-1)(-2) & + & 2(-1)(1) & + & 3(1)(0) & = & 0 \end{matrix}$$

3.13 Problem: Derive the C_{3v} character table.

3.13 Solution: The C_{3v} point group has the operations E, $2C_3$, and $3\sigma_v$. All of the σ_v are in the same class since the C_3 operation carries one σ_v into another successively. Also, both C_3 are in the same class since reflection in a vertical mirror plane reverses the sense of the rotation. Since there are three classes of operations, there will be three distinct irreducible representations. Since the order of the group is six, one of the irreducible representations will be two-dimensional and the other two will be one-dimensional. Examining the transformation matrices for x, y, and z for one operation in each class (see Problem 3.1 for instructions on setting up the transformation matrices) we have

$$\underset{E}{\begin{pmatrix} 1 & 0 & 0 \\ 0 & 1 & 0 \\ 0 & 0 & 1 \end{pmatrix}} \qquad \underset{C_3}{\begin{pmatrix} -1/2 & \sqrt{3}/2 & 0 \\ \sqrt{3}/2 & -1/2 & 0 \\ 0 & 0 & 1 \end{pmatrix}} \qquad \underset{(\sigma_v)_{xz}}{\begin{pmatrix} 1 & 0 & 0 \\ 0 & -1 & 0 \\ 0 & 0 & 1 \end{pmatrix}}$$

From the above it is readily apparent that z is invariant under all of the operations and gives an irreducible representation having a single unit element for each matrix of the representation. This representation is called the totally symmetric representation and given the Mulliken symbol A (or A_1 or A') and such a representation occurs for every group. The remaining two-dimensional representation is termed an E representation, and is the one to which x and y belong. The sum of the diagonal elements of these matrices gives the character of the representation. The characters for the remaining one-dimensional representation can be obtained from the orthogonality relationships. The character table is shown below.

C_{3v}	E	$2C_3$	$3\sigma_v$
A_1	1	1	1
A_2	1	1	-1
E	2	-1	0

3.14 Problem: Derive the character table for the C_3 group, expressing the characters as simple and complex numbers.

3.14 Solution: Both C_n and S_n groups are cyclic (or Abelian) groups whose representations, and characters, may be developed by taking a single generator which may take as its value the n roots of 1 where n is the order of the group. (See DMA Section 3.5.6 or F. Theobald, J. Chem. Educ. 1982, 59, 277.) Successive powers of the value taken by the generator give the characters of the remaining elements of the group. The values of the n roots of 1 are easily found by geometric construction locating n points regularly on a circle with coordinates chosen as the real number line and the imaginary number line, illustrated below for n = 3.

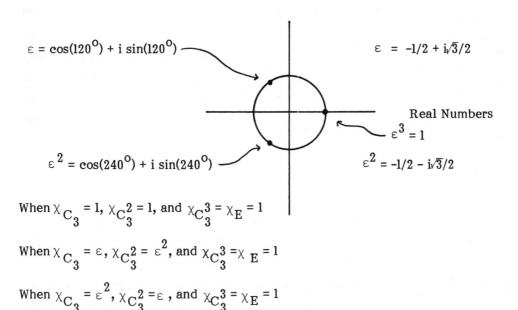

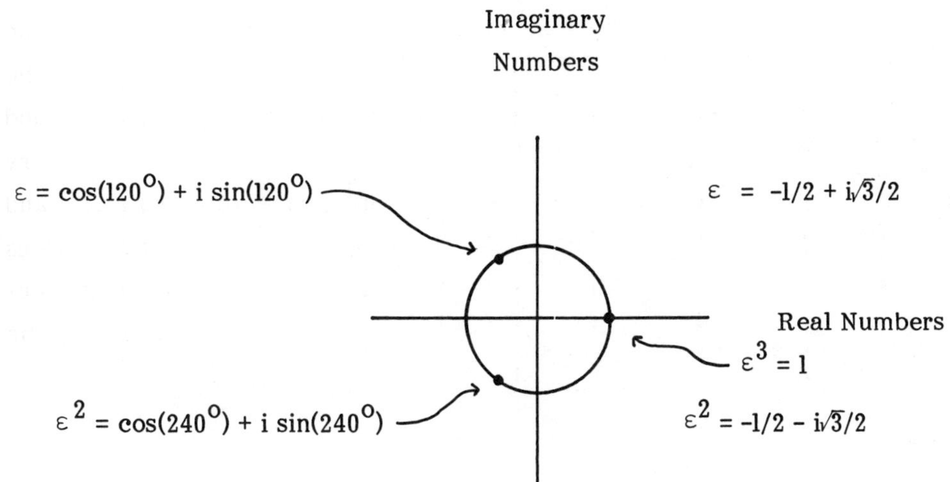

When $\chi_{C_3} = 1$, $\chi_{C_3^2} = 1$, and $\chi_{C_3^3} = \chi_E = 1$

When $\chi_{C_3} = \varepsilon$, $\chi_{C_3^2} = \varepsilon^2$, and $\chi_{C_3^3} = \chi_E = 1$

When $\chi_{C_3} = \varepsilon^2$, $\chi_{C_3^2} = \varepsilon$, and $\chi_{C_3^3} = \chi_E = 1$

The character table for the C_3 group is thus:

C_3	E	C_3	C_3^2
A	1	1	1
E	$\begin{Bmatrix} 1 \\ 1 \end{Bmatrix}$	$(-1/2 + i\sqrt{3}/2)$ $(-1/2 - i\sqrt{3}/2)$	$(-1/2 - i\sqrt{3}/2)$ $(-1/2 + i\sqrt{3}/2)$

3.15 **Problem:** Take the following direct products and find the irreducible representations contained.

(a) $E_g + E_u$ in D_{3d}

(b) $E_g \times T_{1g}$ in O_h

(c) $E_g \times T_{2g}$ in O_h

(d) $E_u \times E_u$ in O_h

The character table below will be needed.

O_h	E	$8C_3$	$6C_2$	$6C_4$	$3C_2(=C_4^2)$	i	$6S_4$	$8S_6$	$3\sigma_h$	$6\sigma_d$		
A_{1g}	1	1	1	1	1	1	1	1	1	1		$x^2 + y^2 + z^2$
A_{2g}	1	1	−1	−1	1	1	−1	1	1	−1		
E_g	2	−1	0	0	2	2	0	−1	2	0		$(2z^2 - x^2 - y^2, x^2 - y^2)$
T_{1g}	3	0	−1	1	−1	3	1	0	−1	−1	(R_x, R_y, R_z)	
T_{2g}	3	0	1	−1	−1	3	−1	0	−1	1		(xz, yz, xy)
A_{1u}	1	1	1	1	1	−1	−1	−1	−1	−1		
A_{2u}	1	1	−1	−1	1	−1	1	−1	−1	1		
E_u	2	−1	0	0	2	−2	0	1	−2	0		
T_{1u}	3	0	−1	1	−1	−3	−1	0	1	1	(x, y, z)	
T_{2u}	3	0	1	−1	−1	−3	1	0	1	−1		

3.15 Solution: The D_{3d} group is a subgroup of the O_h group. Its elements are E, $2C_3$, $3C_2$, i, $2S_6$ and $3\sigma_d$. With fewer classes of symmetry elements than the O_h group, it has fewer irreducible representations—it has no T representations. The D_3 group is a subgroup of both the D_{3d} and O_h groups. Since the D_3 group has no inversion operation, its symmetry species do not carry a g or u subscript. The D_3 character table is set off in the O_h character table above by dashed lines—the g and u subscripts should be omitted for D_3. In the following solution we make use of the fact that g x g = g and u x u = g whereas g x u = u. Accordingly we need examine only products from the rotation groups alone.

(a) To find the irreducible representations contained in the direct product of E_g and E_u under D_{3h} symmetry we first obtain the characters for the reducible representation by multiplying together the characters under each class.

D_3	E	$2C_3$	$3C_2$
E	2	-1	0
E	2	-1	0
Γ_R	4	1	0

In this case inspection shows that the product is made up of $1A_1 + 1A_2 + 1E$. These will all be ungerade (g x u = u), that is $1A_{1u}$, $1A_{2u}$, and $1E_u$.

(b) In similar fashion

O	E	$8C_3$	$6C_2$	$6C_4$	$3C_2(=C_4^2)$
E	2	-1	0	0	2
T_1	3	0	-1	1	-1
Γ_R	6	0	0	0	-2 $\rightarrow T_1 + T_2$

or $\qquad E_g \times T_{1g} \rightarrow T_{1g} + T_{2g}$

(c) Same results as in (b)

(d) Same results as in (a) except all products are gerade (u x u = g).

In more complex cases it may be necessary to use the decomposition formula

$$n_\Gamma = \frac{1}{h} \sum_g n_g \chi_R \chi_\Gamma$$

where n_Γ is the number of times a particular irreducible representation is contained in a reducible representation, h is the order of the group, n_g is the number of operations in the class g, χ_R is the character of the reducible representation and χ_Γ that of the irreducible representation under the operations of class g, and the summation is taken over all classes. Using this formula we reduce the set of characters in part (b) as follows:

$$\begin{array}{lllllll} & E & 8C_3 & 6C_2 & 6C_4 & 3C_2 & \\ n_{A_1} = \frac{1}{24} & [1 \cdot 6 \cdot 1 + & 8 \cdot 0 \cdot 1 + & 6 \cdot 0 \cdot 1 + & 6 \cdot 0 \cdot 1 + & 3 \cdot -2 \cdot 1] & = 0 \\ n_{A_2} = \frac{1}{24} & [1 \cdot 6 \cdot 1 + & 8 \cdot 0 \cdot 1 + & 6 \cdot 0 \cdot -1 + & 6 \cdot 0 \cdot -1 + & 3 \cdot -2 \cdot 1] & = 0 \\ n_E = \frac{1}{24} & [1 \cdot 6 \cdot 2 + & 8 \cdot 0 \cdot -1 + & 6 \cdot 0 \cdot 0 + & 6 \cdot 0 \cdot 0 + & 3 \cdot -2 \cdot 2] & = 0 \\ n_{T_1} = \frac{1}{24} & [1 \cdot 6 \cdot 3 + & 8 \cdot 0 \cdot 0 + & 6 \cdot 0 \cdot -1 + & 6 \cdot 0 \cdot 1 + & 3 \cdot -2 \cdot -1] & = 1 \\ n_{T_2} = \frac{1}{24} & [1 \cdot 6 \cdot 3 + & 8 \cdot 0 \cdot 0 + & 6 \cdot 0 \cdot 1 + & 6 \cdot 0 \cdot -1 + & 3 \cdot -2 \cdot -1] & = 1 \end{array}$$

3.16 Problem: Give the symmetry labels for the p and d orbitals for $[Co(NH_3)_6]^{3+}$, $[Co(en)_3]^{3+}$, $[Co(edta)]^-$, and <u>trans</u>-$[Cr(Cl)_2(NH_3)_4]^+$.

3.16 Solution: We must first assign the species to the appropriate point group (see Problems 3.2-3.4). We then find how the appropriate function describing the angular part of the wave function transforms in the given point group (see discussion in DMA). Most character tables list these functions at the far right of the table for the appropriate species. We will examine several ways of assigning labels to the orbitals in a given symmetry as an aid to understanding how the information included in character tables is derived.

$[Co(NH_3)_6]^{3+}$ This species has octahedral geometry, and, for freely rotating NH_3 ligands, would belong to the O_h point group. From the O_h character table given on page 54 we find that x, y, and z transform as T_{1u}; $2z^2 - x^2 - y^2$ and $x^2 - y^2$ as E_g; and xy, xz and yz as T_{2g}. The symmetry labels become the orbital labels, lower case letters being used—t_{1u}, t_{2g} and e_g.

$[Co(en)_3]^{3+}$ This species belongs to D_3, a subgroup of O_h. We can obtain the symmetry species for the p orbitals in D_3 symmetry by reducing the characters of the T_{1u} symmetry species for those classes that belong to D_3.

D_3	E	$2C_3$	$3C_2$		
Γ_p	3	0	-1	→	$E + A_2$

By examining the behavior of the p_z orbital under D_3 symmetry (it goes into itself under E or C_3 and inverts under C_2) we see that it belongs to A_2 symmetry. The p_x and p_y orbitals must belong to E symmetry species. We obtain the characters of the reducible representation of the d orbitals by adding the appropriate characters for the E_g and T_{2g} representation in O_h symmetry. Reduction under D_3 symmetry gives the appropriate labels for D_3.

D_3	E	$2C_3$	$3C_2$
Γ_d	(2+3)	(-1+0)	(0+1)

$$n_{A_1} = \frac{1}{6} [1 \cdot 5 \cdot 1 + 2 \cdot -1 \cdot 1 + 3 \cdot 1 \cdot 1] = 1$$

$$n_{A_2} = \frac{1}{6} [1 \cdot 5 \cdot 1 + 2 \cdot -1 \cdot 1 + 3 \cdot 1 \cdot -1] = 0$$

$$n_E = \frac{1}{6} [1 \cdot 5 \cdot 2 + 2 \cdot -1 \cdot -1 + 3 \cdot -1 \cdot 0] = 2$$

The d_{z^2} orbital is invariant under the D_3 operations and belongs to the totally symmetric A_1 species. The d_{xy}, $d_{x^2-y^2}$ and d_{xz}, d_{yz} form two sets of E symmetry species.

$[Co(edta)]^-$ belongs to the C_2 point group, a subgroup of the O_h group. There are only two symmetry species for this Abelian group of order two, the symmetric A species and the antisymmetric B species.

C_2	E	C_2
A	1	1
B	1	-1

Letting the z axis be the C_2 axis, we examine each orbital for symmetric or antisymmetric behavior under the C_2 operation.

	Behavior under C_2	Symmetry Species
p_z	symmetric—goes into itself	A
p_y	antisymmetric—goes into its negative	B
p_x	antisymmetric—goes into its negative	B

d_{z^2}	symmetric—goes into itself	A
$d_{x^2-y^2}$	symmetric—goes into itself	A
d_{xy}	symmetric—goes into itself	A
d_{xz}	antisymmetric—goes into its negative	B
d_{yz}	antisymmetric—goes into its negative	B

<u>trans</u>-$[Cr(Cl)_2(NH_3)_4]^+$ belongs to the D_{4h} group, a subgroup of the O_h group. The correlation of symmetry species between the groups is not as simple as with the previous subgroups of O_h discussed in this problem. Thus, we will work directly with the D_4 character table given below.

D_4	E	$2C_4$	$C_2(=C_4^2)$	$2C_2'$	$2C_2''$
A_1	1	1	1	1	1
A_2	1	1	1	-1	-1
B_1	1	-1	1	1	-1
B_2	1	-1	1	-1	1
E	2	0	-2	0	0

Here the C_4 axis is chosen to coincide with the z axis and the $2C_2'$ axes are chosen to coincide with the x and y axes. Let us assign symmetry labels to the p and d orbitals in D_{4h} symmetry. We know that the p orbitals will be ungerade and the d orbitals gerade, so we can use the D_4 character table to obtain the desired information. Looking at the transformation of each orbital individually we find

D_4	E	C_4	C_4^2	C_2'	C_2''
p_x	1	0	-1	1	0
p_y	1	0	-1	-1	0

Here we have arbitrarily chosen a single C_2' coincident with the x axis. The zero under C_4 indicates that the orbital goes into neither itself nor its negative. We find that neither the p_x nor the p_y orbital transforms as a symmetry species belonging to D_4 but together they transform as an E species. Accordingly p_x and p_y are labeled as e_g orbitals in D_{4h} symmetry. The remaining orbitals transform as follows

D_4	E	C_4	C_4^2	C_2'	C_2''		
p_z	1	1	1	-1	-1	→	a_{2u}
d_{z^2}	1	1	1	1	1	→	a_{1g}
$d_{x^2-y^2}$	1	-1	1	1	-1	→	b_{1g}
d_{xy}	1	-1	1	-1	1	→	b_{2g}
d_{xz}	1	0	-1	-1	0	→	e_g
d_{yz}	1	0	-1	1	0		

The reader interested in obtaining the labels of species of atomic orbitals and spectroscopic states by descent from spherical symmetry should consult R.L. DeKock, A.J. Kromminga, and T.S. Zwier, J. Chem. Educ. 1979, 56, 510.

3.17 Problem: The representations for the same orbitals in centrosymmetric groups and closely related subgroups usually differ only in the dropping of the g and u subscripts. Explain why the p orbitals are T_{1u} in O_h and T_2 in T_d.

3.17 Solution: The T_1 or T_2 distinction for O (or O_h) is made using the major axis, C_4, since both have characters of zero under the S_6 operations. Since C_4 is missing for T_d, the T_1 or T_2 distinction is made using S_4.

In a similar vein we find the A_2 labeling in O_h is based on the behavior under the S_6 operation. On descending to the D_{4h} group the S_6 disappears so the C_4 operation defines the A and B labeling and the A_2 species in O_h becomes a B_1 species when the symmetry is lowered to D_{4h}.

3.18 Problem: Instead of using Cartesian coordinates x and y, the position of a point in the xy plane can be specified by the polar coordinates r and ϕ.

(a) What is the mathematical relation between x, y, and r, ϕ?

(b) Rotation around the z axis through an angle θ takes the point (x,y) into (x', y'). The new polar coordinates are [r, $(\phi-\theta)$]. Express mathematically the relation between (x', y') and the polar coordinates [r, $(\phi-\theta)$]. Convert the relationship into one between (x', y') and (x,y).

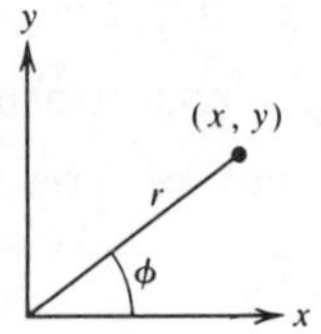

(c) Express the results of (b) in matrix notation, thereby obtaining the standard rotation matrix.

3.18 Solution:

(a) $r = \sqrt{x^2+y^2}$

$\tan\phi = \dfrac{y}{x} \qquad x = r\cos\phi \qquad y = r\sin\phi$

(b) $x' = r\cos\phi' = r\cos(\phi-\theta)$

$y' = r\sin\phi' = r\sin(\phi-\theta)$

From the trigonometric relationships for cos and sin of $\phi - \theta$ we have

$x' = r[\cos\phi\cos\theta + \sin\phi\sin\theta]$

or $x' = x\cos\theta + y\sin\theta$

$y' = r[\sin\phi\cos\theta - \cos\phi\sin\theta]$

$y' = y\cos\theta - x\sin\theta$ or $-x\sin\theta + y\cos\theta$

(c) Expressing the above in matrix form we have

$$\begin{pmatrix} x' \\ y' \end{pmatrix} \quad \begin{pmatrix} \cos\theta & \sin\theta \\ -\sin\theta & \cos\theta \end{pmatrix} \begin{pmatrix} x \\ y \end{pmatrix}$$

3.19 Problem: Nuclei are termed chemically equivalent if permuted by a C_n operation, enantiotopically equivalent if permuted only by an S_n operation, and magnetically equivalent if permuted by an operation which leaves all other magnetically active nuclei invariant. Indicate the types of equivalent sets of halogen nuclei found in the following species.

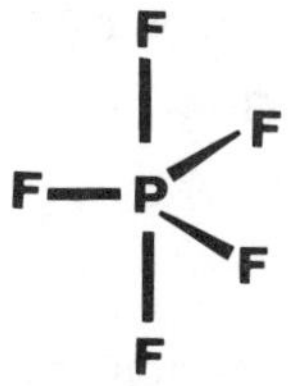

(a)

(b)

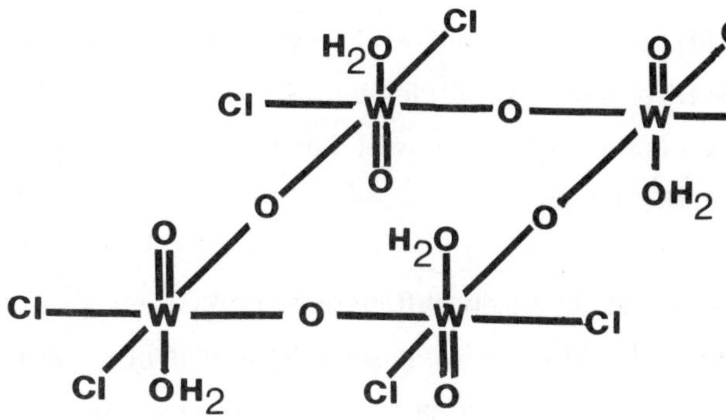

(c)

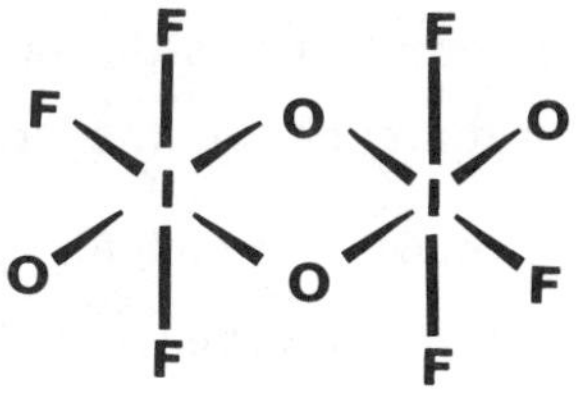

(d)

3.19 Solution: We first assign the species to the appropriate point group. Proceeding as in Problem 3.2, we obtain D_{3h} for PF_5, O_h for SF_6, D_{2d} for the W tetramer, and C_{2h} for the I dimer. We must examine the permutations possible under each operation of the group (see D.H. McDaniel, J. Phys. Chem. 1981 85, 471), with the following results:

(a) In PF_5, the equatorial F are chemically and magnetically equivalent (C_3), the axial F are chemically equivalent (C_2) and magnetically equivalent (σ_h). The magnetic equivalence of the axial F would be removed in a chiral environment, that is, in an optically active solvent.

(b) In SF_6 all F are chemically equivalent ($6C_4$, etc.) and magnetically equivalent.

(c) Here the $3C_2$ operations show that there are two sets of chemically equivalent Cl (four Cl having parallel bond directions in each set); the σ_d's or S_4's scramble these two sets giving the result that all Cl atoms are enantiotopically equivalent; all Cl atoms are magnetically unique in a chiral environment, but equivalent in an achiral environment.

(d) There are three sets of F atoms which give chemically equivalent pairs under the C_2 operation; under the i and σ_h we find one set of four enantiotopically equivalent F atoms and one set of two F; all F atoms are magnetically unique.

3.20 Problem: If A and B are symmetry elements belonging to the same class, then $X^{-1}AX = B$ for some group member X and its inverse X^{-1}. Show that if all operations of a group commute, the group will have only one-dimensional representations. Give some examples of such groups.

3.20 Solution: If all operations commute, then $X^{-1}AX = X^{-1}XA = A$ for all operations of the group and thus each symmetry element is in a class by itself. Since there are as many different representations as there are classes, the number of representations must equal the number of symmetry elements, that is, the order of the group. Also, since the character under the identity operation for an irreducible representation is the same as the dimension of the matrix representation, and $\sum_{\Gamma} [\chi(E)]^2 = h$, each of the irreducible representations must be one-dimensional representations. Groups for which all operations commute are termed Abelian groups. All cyclic groups, groups generated by the successive application of one symmetry operation (such as S_n in the case of S_n and C_{nh} groups or C_n for C_n groups) are Abelian groups, but the converse is not necessarily true. Thus, the D_2 group is an Abelian group, but is not a cyclic group. In cases where an E representation is given in the character table for an Abelian group, it may be noted that it consists of a pair of one-dimensional representations.

3.21 Problem: (a) For a reducible representation, demonstrate that the character under any operation R is $\chi(R) = \sum_{\Gamma} n_{\Gamma}\chi_{\Gamma}(R)$ where the sum is over the Γ irreducible representations of the group, n_{Γ} is the number of times the Γth irreducible representation is contained in the reducible one and $\chi_{\Gamma}(R)$ is the character of the Γth irreducible representation under the operation R. (b) Using the result of (a) and the orthogonality of irreducible representations, show that $n_{\Gamma} = \frac{1}{h}\sum_{\Gamma} \chi(R)\chi_{\Gamma}(R)$. This formula allows you to determine the number of times the Γth irreducible representation is contained in any reducible representation.

3.21 Solution: (a) The demonstration follows from the rules for matrix addition and multiplication of a matrix by a constant. For matrix addition:

$$\begin{pmatrix} a_{11} & a_{12} \cdots & a_{1n} \\ a_{21} & a_{22} \ddots & \\ & & a_{nn} \end{pmatrix} + \begin{pmatrix} b_{11} & b_{12} \cdots & b_{1n} \\ b_{21} & b_{22} \ddots & \\ & & b_{nn} \end{pmatrix} = \begin{pmatrix} a_{11} + b_{11} & a_{12} + b_{12} \cdots & a_{1n} + b_{1b} \\ a_{21} + b_{21} & a_{22} + b_{22} \ddots & \\ & & a_{nn} + b_{nn} \end{pmatrix}$$

and

$$p \begin{pmatrix} a_{11} & a_{12} \cdots & a_{1n} \\ a_{21} & a_{22} \ddots & \\ & & a_{nn} \end{pmatrix} = \begin{pmatrix} pa_{11} & pa_{12} \cdots & pa_{1n} \\ pa_{21} & pa_{22} \ddots & \\ & & pa_{nn} \end{pmatrix}$$

We illustrate with a sum of three matrices which can be generalized to any number. Suppose the matrix for the operation R in the reducible representation is

$$\begin{pmatrix} d_{11} & d_{12} \cdots & d_{1n} \\ d_{21} & d_{22} \ddots & \\ d_{n1} & & d_{nn} \end{pmatrix} = n_a \begin{pmatrix} a_{11} & a_{12} \cdots & a_{1n} \\ a_{21} & a_{22} \ddots & \\ a_{n1} & & a_{nn} \end{pmatrix} + n_b \begin{pmatrix} b_{11} & b_{12} \cdots & b_{1n} \\ b_{21} & b_{22} \ddots & \\ b_{n1} & & b_{nn} \end{pmatrix} + n_c \begin{pmatrix} c_{11} & c_{12} \cdots & c_{1n} \\ c_{21} & c_{22} \ddots & \\ c_{n1} & & c_{nn} \end{pmatrix}$$

$$= \begin{pmatrix} n_a a_{11} + n_b b_{11} + n_c c_{11} & n_a a_{12} + n_b b_{12} + n_c c_{12} & \cdots\cdots \\ n_a a_{21} + n_b b_{21} + n_c c_{21} & n_a a_{22} + n_b b_{22} + n_c c_{22} \ddots & \\ \cdots\cdots & & n_a a_{nn} + n_b b_{nn} + n_c c_{nn} \end{pmatrix}$$

The character of the matrix for the reducible representation is

$$d_{11} + d_{22} + d_{33} + \ldots + d_{nn} = \sum_{j=1}^{n} d_{jj} = \chi(R) =$$

$$\sum_{j=1}^{n} n_a a_{jj} + n_b b_{jj} + n_c c_{jj} = n_a \sum_{j=1}^{n} a_{jj}$$

$$+ n_b \sum_{j=1}^{n} b_{jj} + n_c \sum_{j=1}^{n} c_{jj} = n_a \chi_a(R) + n_b \chi_b(R) +$$

$$n_c \chi_c(R) = \sum n_\Gamma \chi(R)$$

This demonstration assumes all the matrices to be of the same dimension and so is not a general proof.

(b) Since we know for each matrix in the reducible representation

$$\chi(R) = \sum_{\Gamma} n_{\Gamma} \chi_{\Gamma}(R),$$

We can multiply by $\chi_{\Gamma}'(R)$ and sum over all operations R in the group.

$$\sum_{R} \chi(R) \chi_{\Gamma}'(R) = \sum_{\Gamma} \sum_{R} n_{\Gamma} \chi_{\Gamma}(R) \chi_{\Gamma}'(R)$$

But, $\sum_{R} \chi_{\Gamma}(R) \chi_{\Gamma}'(R) = 0$ if $\Gamma \neq \Gamma'$

$= h$ if $\Gamma = \Gamma'$

Then $\sum_{R} \chi(R) \chi_{\Gamma}(R) = \sum_{\Gamma} n_{\Gamma} h = n_{\Gamma} h$

So, $n_{\Gamma} = \frac{1}{h} \sum_{R} \chi(R) \chi_{\Gamma}(R)$

IV Molecular Orbital Theory

4.1 Problem: Give the bond order and the number of unpaired electrons for: Be_2^+, B_2^+, C_2^+, O_2, O_2^+, O_2^-, O_2^{2-}.

4.1 Solution: The molecular orbital energy level diagram for diatomic molecules as a function of atomic number is shown below.

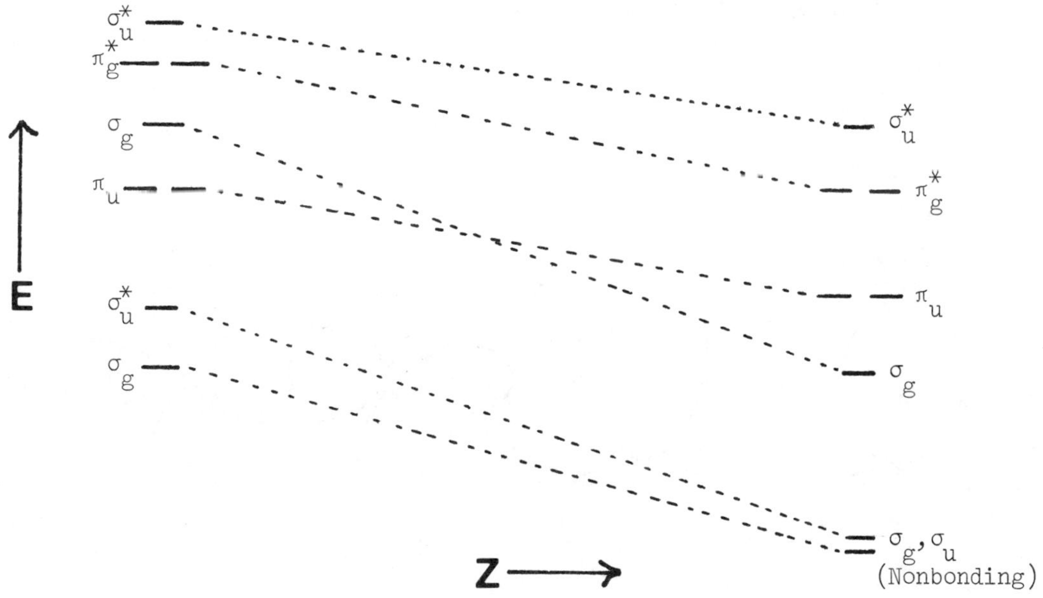

Configuration interaction (or, if you prefer, hybridization) mixes the σ_g orbitals of the lighter diatomic molecules as shown on the left--this ordering is shown through N_2. Electrons populate the orbitals in accordance with Hund's rules of maximum multiplicity; the bond order is obtained by subtracting the number of antibonding electrons from the number of bonding electrons and dividing by two. The answers for our problem are thus:

	Be_2^+	B_2^+	C_2^+	O_2	O_2^+	O_2^-	O_2^{2-}
Bond order	0.5	0.5	1.5	2	2.5	1.5	1
Number of unpaired electrons	1	1	1	2	1	1	0

4.2 Problem: Sketch sigma bonding orbitals that result from the combination of the following orbitals on separate atoms: p_z and d_{z^2}, s and p_z, $d_{x^2-y^2}$ and $d_{x^2-y^2}$ (let the bond be along the x axis in the latter case).

4.2 Solution:

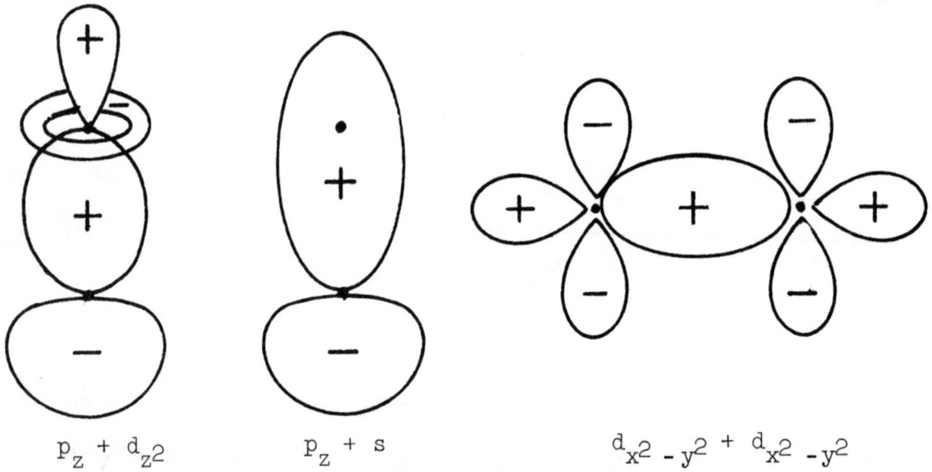

$p_z + d_{z^2}$ $p_z + s$ $d_{x^2-y^2} + d_{x^2-y^2}$

4.3 Problem: Sketch π bonding orbitals that result from combination of the following orbitals on separate atoms: p_x and p_x, p_x and d_{xz}, d_{xz} and d_{xz}.

4.3 Solution:

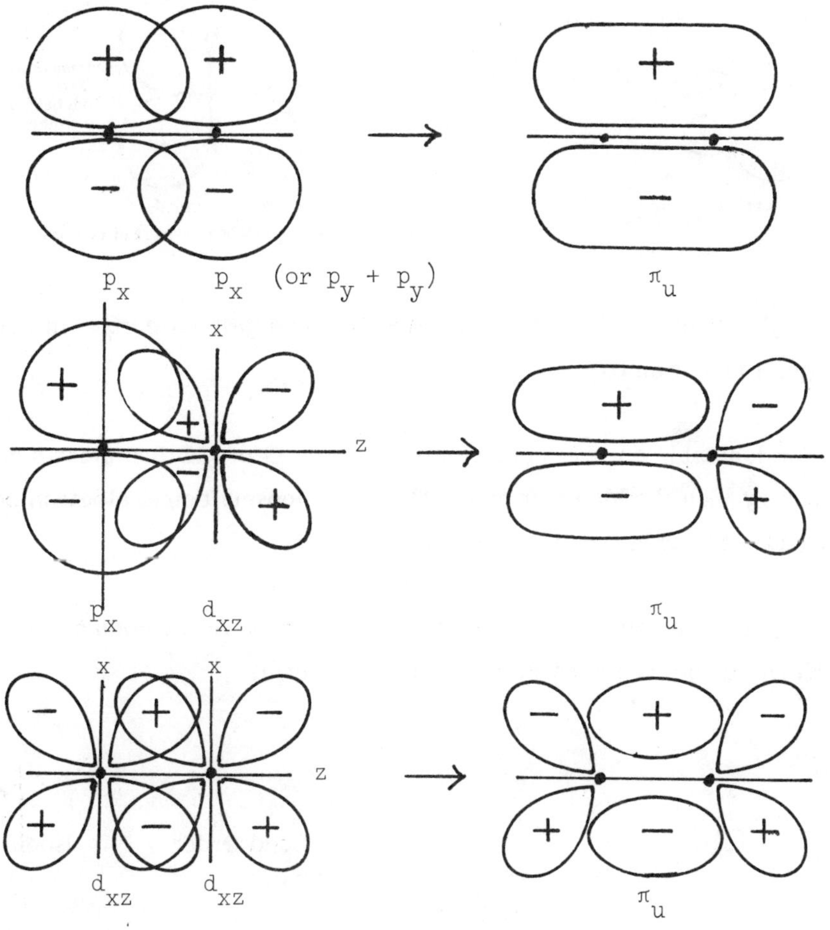

4.4 Problem: Sketch a delta bond for an X—Y molecule, identifying the atomic orbitals and showing the signs of the amplitudes of the wave functions (signs of the lobes).

4.4 Solution:

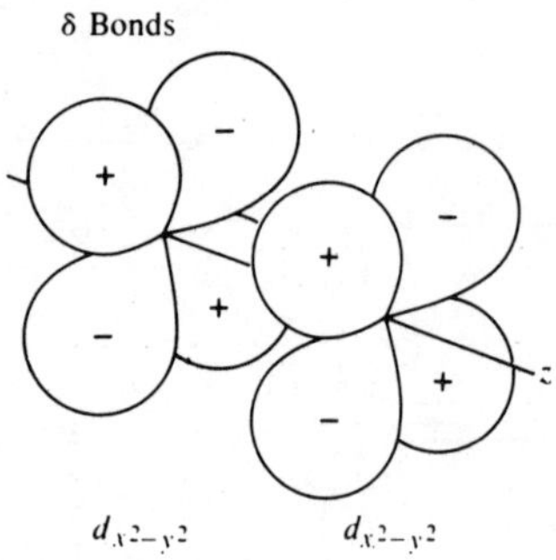

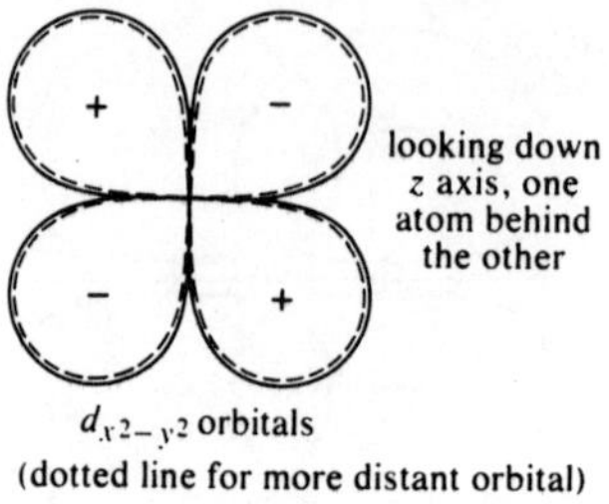

4.5 Problem: Explain why the CO ligand donates the lone pair on carbon in bonding to metals, rather than the pair on oxygen.

4.5 Solution: The lone pair on the more electronegative O of CO has more s character and is lower in energy than the lone pair on C. The more loosely bound electron pair on C is the pair donated to metals in forming carbonyl complexes.

4.6 Problem: Write the molecular orbital configurations and give the bond orders of NO^+, NO, and NO^-. Which of these species should be paramagnetic?

4.6 Solution:

		Bond Order	Isoelectronic with
NO^+	$\pi_x^2\pi_y^2\sigma_{p_z}^2$	3 Diagmagnetic	N_2
NO	$\pi_x^2\pi_y^2\sigma_{p_z}^2\pi_x^{*1}$	2.5 Paramagnetic	O_2^+
NO^-	$\pi_x^2\pi_y^2\sigma_{p_z}^2\pi_x^{*1}\pi_y^{*1}$	2.0 Paramagnetic	O_2

4.7 Problem: (a) Draw an energy diagram for the molecular orbitals that would arise in a diatomic molecule from the combination of unhybridized d orbitals. Label the atomic orbitals being combined and the resulting molecular orbitals. Let the z axis lie along the bond. (b) Sketch the shape of these orbitals.

4.7 Solution:

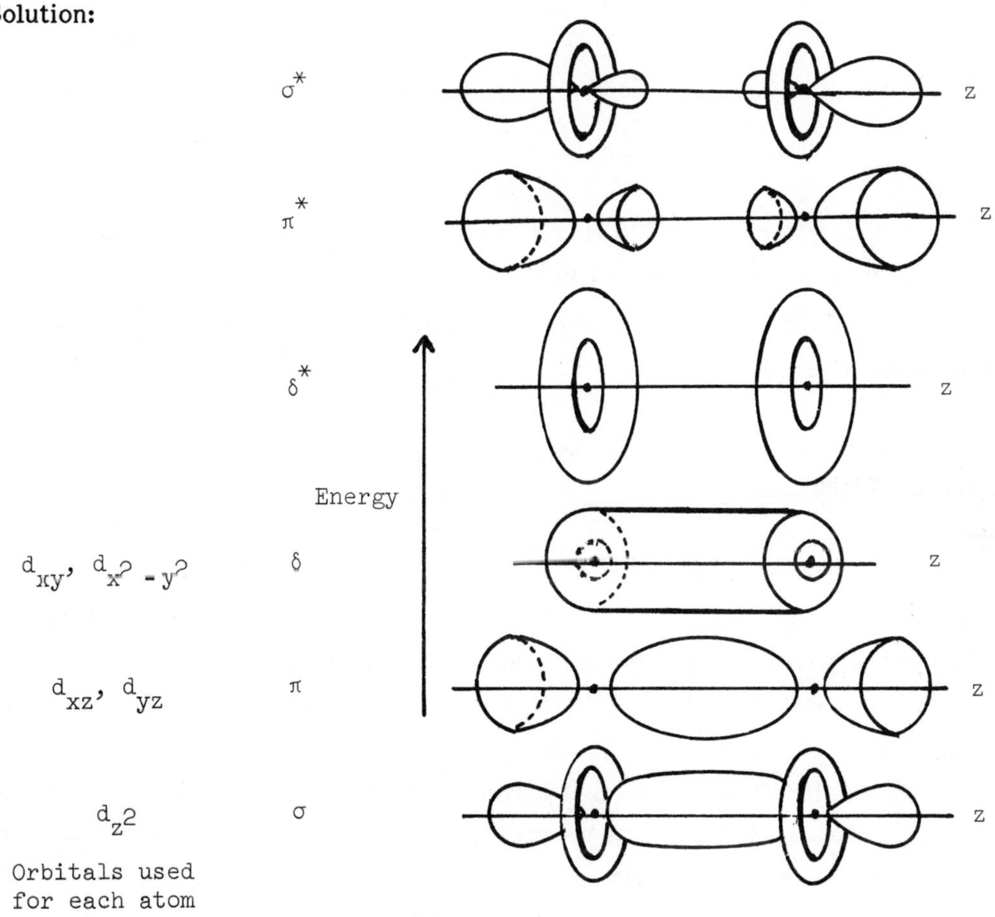

The σ interaction should be most favorable (lowest in energy) because of ideal, directed overlap. Overlap is poorer for π interaction and poorer still for the parallel lobes giving δ bonding. The antibonding orbitals increase in energy in the reverse order, $\delta^* < \pi^* < \sigma^*$.

4.8 Problem: Account for the differences in dissociation energies and bond lengths on addition and removal of an electron from (a) O_2, (b) N_2, (c) CO, (d) NO.

	O_2^+	O_2	O_2^-	N_2^+	N_2	N_2^-
Bond energy	642.9	493.6	395.0	840.7	941.7	765 kJ/mole
Bond length	111.6	120.8	(135)	109.8	111.6	119 pm
	CO^+	CO	CO^-	NO^+	NO	NO^-
Bond energy	804.5	1070.2	784	1046.9	626.9	487.8 kJ/mole
Bond length	111.5	112.8	—	106.3	115.1	125.8 pm

4.8 Solution: In general, we find that removing an electron from a diatomic species tends to decrease interelectronic repulsion and thus to shorten the bond length. This factor may be opposed if the electron being removed is being taken from a bonding orbital—as is the case on going from N_2 to N_2^+—and little effect is observed. Shorter bond lengths mean stronger interactions among all overlapping orbitals, not just the set involving the electron being removed or added. The electron being removed or added will come from the highest occupied molecular orbital (HOMO) or go into the lowest unoccupied molecular orbital (LUMO). These frontier orbitals control the chemistry of these species and are well worth looking at in detail.

(a) With O_2 both the HOMO and the LUMO are identical, that is the π_g^* orbitals. Since these orbitals are antibonding, removal of an electron increases the bond strength whereas addition of an electron decreases the bond strength, the former effect being slightly larger.

(b) With N_2 the HOMO is a σ_g which is only slightly bonding. In valence bond terminology we would say there is strong mixing between the lone pairs and the sigma bonding orbital. Removal of an electron from this orbital decreases the bond energy somewhat and has little effect on the bond length. The LUMO is a π_g^* orbital. Adding an electron here is equivalent to removing an electron from a π_u bonding orbital, and results in a decrease in bond energy and an increase in bond length.

(c) CO is isoelectronic with N_2 and might be expected to behave in a similar fashion. It does but the effects are larger. The σ orbital is somewhat more bonding in CO (compare ionization energies of CO and C to those of N_2 and N). Another complication is that in comparing dissociation energies we must consider the energy of the products. CO^+ will dissociate to C^+ and O, but CO^- will dissociate to C and O^-. Accordingly the antibonding character of the π* orbital in CO must be gauged by comparing the electron affinity of CO with that of O. The data presented here allow the EA of CO to be calculated if we know that of O, and it turns out that it costs approximately 125 kJ/mole to force electrons onto CO.

(d) With NO we again have a case where the HOMO and LUMO are the same, the π* orbitals. Here the asymmetry in the dissociation products of NO^+ and NO^- is particularly noticeable. The high ionization energy of N leads to a high dissociation energy of NO^+, so we find that removing the π* electron of NO increases the dissociation energy much more than adding a π* electron decreases the dissociation energy.

4.9 Problem: The bond dissociation energy of C_2 (599 kJ/mole) decreases slightly on forming C_2^+ (513 kJ/mole) and increases greatly on forming C_2^- (818 kJ/mole). Why is the change much greater for the addition of an electron?

4.9 Solution: $C_2\ \sigma_{2s}^2\sigma_{2s}^{*2}\pi^4\sigma_{2p}^0$. Both the π and σ_{2p} orbitals are bonding orbitals, so the direction of change in bond energies are those anticipated. The bond length changes mitigate the energy changes expected from orbital population, bond shortening due to ionization having a greater effect than bond lengthening on forming C_2^- (See Problem 4.8).

4.10 Problem: The Lewis structure for CO_2 shows four lone pairs of electrons on the oxygens. There are two nonbonding pairs in the MO description. How can you reconcile the two descriptions?

4.10 Solution: The Lewis structure of CO_2 ($\ddot{\underset{..}{O}}$=C=$\ddot{\underset{..}{O}}$) shows electron pairs on individual atoms or shared between two atoms. In the MO description the two e_{1g} electron pairs are localized on the oxygens. The lower energy a_{1g} and a_{1u} electron pairs are nonbonding and largely localized on oxygens. The higher energy a_{1g} and a_{1u} electron pairs form the bonds.

4.11 Problem: The preferred electron dot structure for NO_2 shows five lone pairs on the oxygens and one unshared electron on N. Show how this description is consistent with the MO description.

4.11 Solution: The bonding orbitals for NO_2 are a_1 and b_2 (σ) and b_1 (π). The $\cdot N(=\ddot{O}:)(-\ddot{O}:)$ very low energy a_1 and b_2 (from s) orbitals are largely localized on the oxygens because of their low energy. The nonbonding a_2, a_1, and b_2 orbitals (from p) are localized on the oxygens. These are the orbitals for the five O "lone pairs". The a_1 nonbonding orbital using the oxygen p_z orbitals is localized largely on N to accommodate the unshared electron. See the energy level diagram in problem 4.14 for NO_2^-.

Molecular orbitals for NO_2.

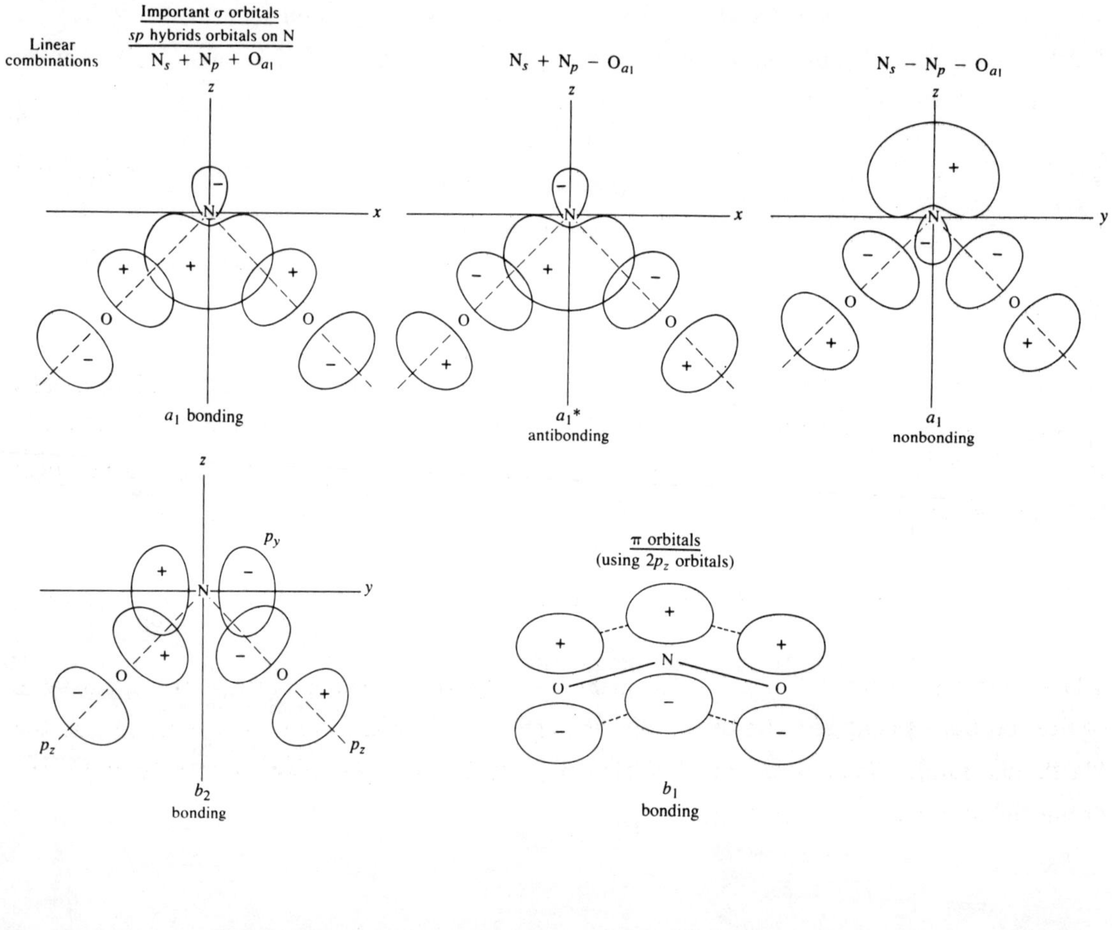

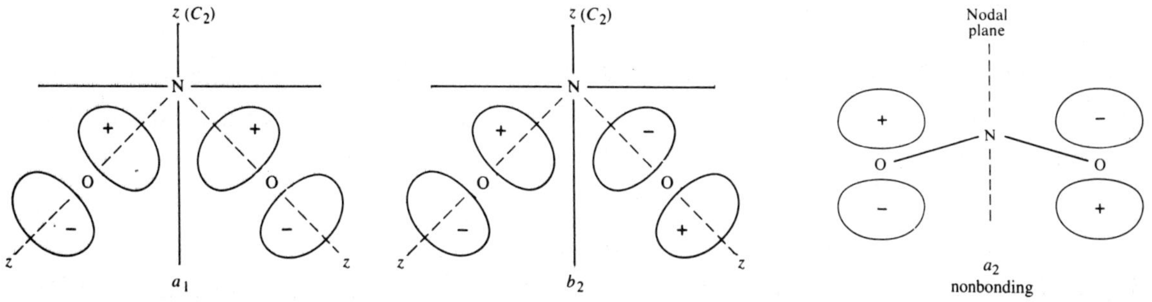

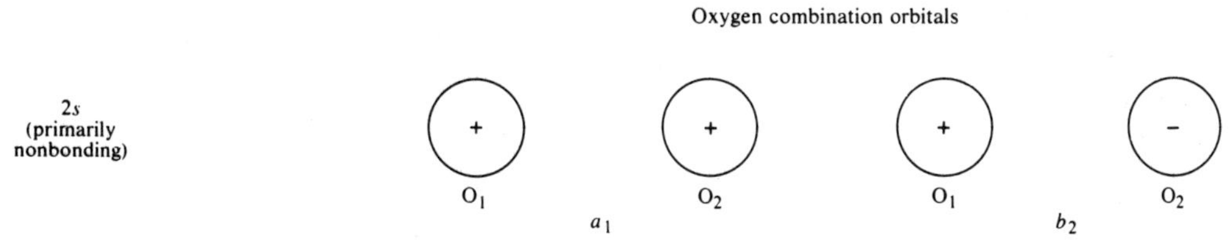

4.12 Problem: In forming NO_2^+ from NO_2, from which molecular orbital is the electron lost? On which atom is this electron localized? To which orbital is an electron added in forming NO_2^-? What are the consequences, in terms of molecular shape, of the loss and gain of an electron?

4.12 Solution: In forming NO_2^+ from NO_2 the electron is lost from the nonbonding HOMO a_1 orbital, largely localized on N (see Problem 4.11). To form NO_2^- an electron is added to this orbital. NO_2^+ is linear (CO_2 structure) and NO_2^- is angular with a smaller bond angle because the directed nonbonding a_1 orbital has two electrons, rather than one as in NO_2.

4.13 Problem: Why is the first excited state of BeH_2 bent, whereas that of BH_2 is linear?

4.13 Solution: H:Be:H is linear. A promoted electron must go into a third, directed orbital on Be, causing it to be bent. BH_2 is bent with an electron in the b_2 orbital (see the Walsh diagram). Promotion of this electron would favor a linear molecule with the promoted electron in the nonbonding e_{1u}.

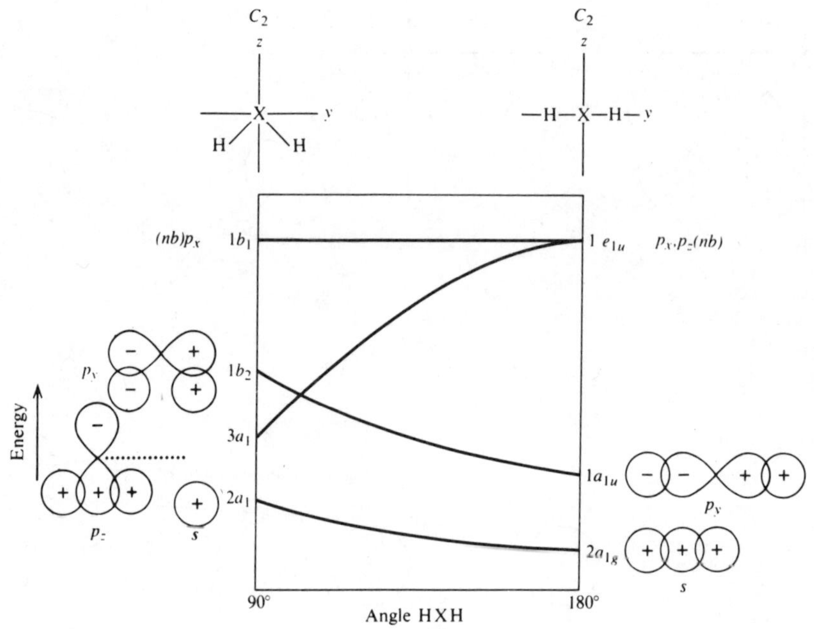

Walsh correlation diagram for XH_2.

4.14 Problem: Apply the group theoretical treatment to obtain the bonding description for NO_2^-.

4.14 Solution: The NO_2^- ion has C_{2v} symmetry.

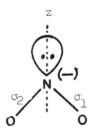

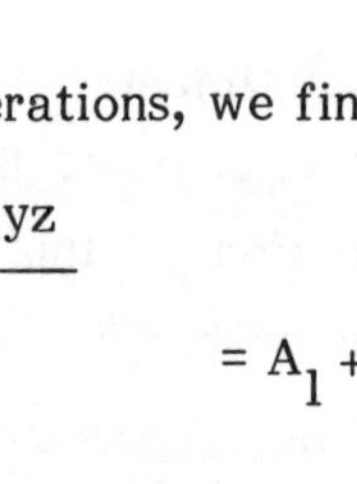

Examining the σ orbitals unchanged by the group operations, we find they transform as $A_1 + B_2$:

	E	C_2	σ_{xz}	σ_{yz}	
Γ_σ	2	0	0	2	$= A_1 + B_2$

For N, s and p_z (represented in the sketches for Problem 4.11 as the sp hybrid) belong to a_1 and can combine with the oxygen a_1 group orbital to give bonding (σ), nonbonding, and antibonding MO's. The N p_y orbital (b_2) combines with the oxygen b_2 group orbital to form the second σ bond. The oxygen p_y orbitals transform in the same way as σ_1 and σ_2, but these are nonbonding because of poor overlap with the N orbitals. The p_x orbitals of the oxygens belong to A_2 and B_1:

	E	C_2	σ_{xz}	σ_{yz}	
Γ_π	2	0	0	-2	$= A_2 + B_1$

The a_2 group orbital is nonbonding since there is no a_2 orbital for N. The π bonding MO is b_1. The low energy oxygen s orbitals are nonbonding. Sketches of the orbitals are given in Problem 4.11 and the energy level diagram is presented on the next page.

4.15 Problem: Apply the group theoretical treatment to obtain the MO bonding description for σ bonding in $PF_5(D_{3h})$. Note that the two axial sigma bonds are treated as one set and the three sigma bonds in the equatorial plane are in a separate set.

4.15 Solution: The σ orbitals are in two sets—2 axial and 3 equatorial. Those within a set are interchanged by symmetry operations of the D_{3h} group, but those of one set are not interchanged with those of the other set by any operation of the group. The D_{3h} character table is given in Problem 3.12.

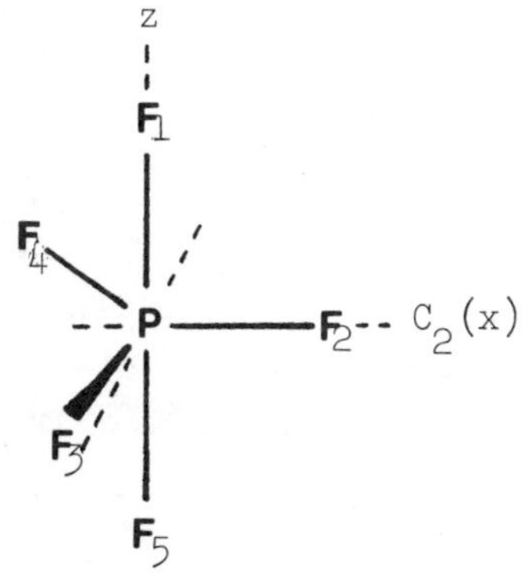

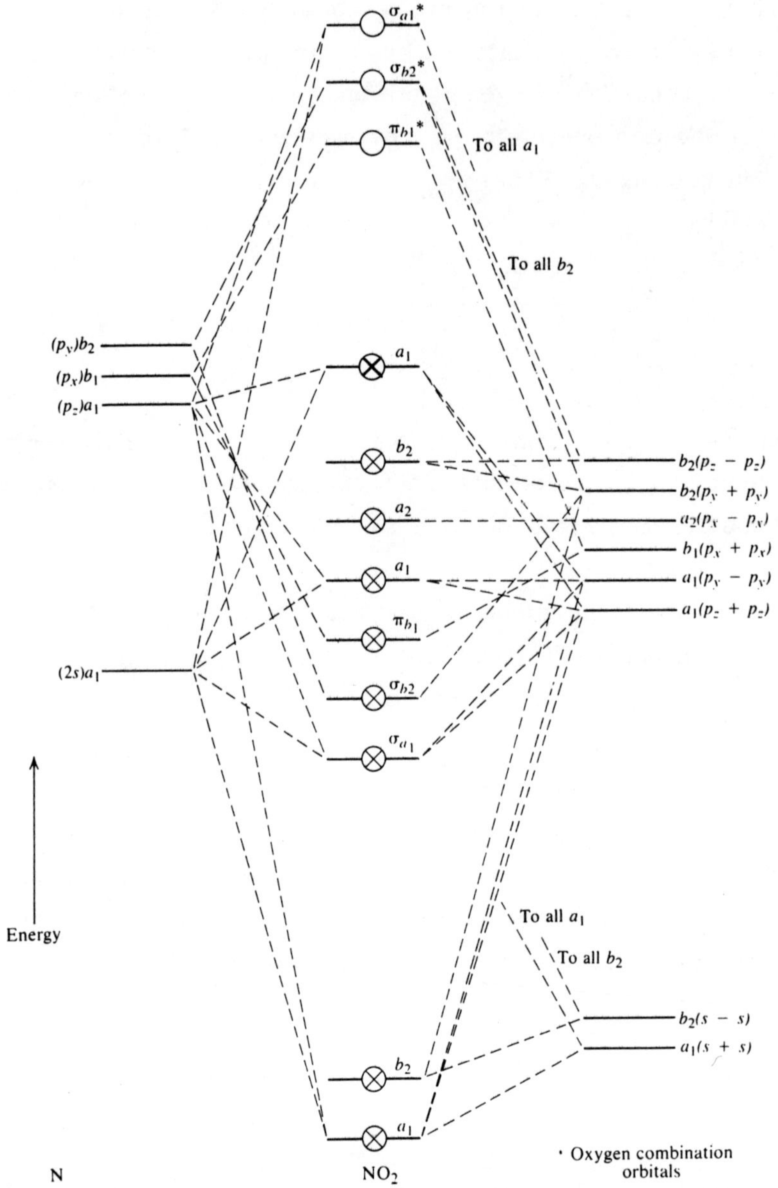

Qualitative molecular orbital energy-level diagram for NO_2^-.

The σ orbitals unchanged by group operations are:

D_{3h}	E	C_3	C_2	σ_h	S_3	σ_v	
$\Gamma_{\sigma_{ax}}$	2	2	0	0	0	2	$= A_1' + A_2''$
$\Gamma_{\sigma_{eq}}$	3	0	1	3	0	1	$= A_1' + E'$

These are the representations for the F group orbitals and the P σ bonding orbitals. The atomic orbitals on P belonging to these representations can be identified from the D_{3h} character table.

a_1'	a_2''	e'
s	p_z	(p_x,p_y)
d_{z^2}		$(d_{xy},d_{x^2-y^2})$

We need both s and d_{z^2} (there are $2a_1'$), $p_z(a_2'')$ and either (p_x,p_y) or $(d_{xy},d_{x^2-y^2})$, or some combination of the two e' pairs. Since the p orbitals are much lower in energy, we shall use (p_x,p_y). We do not need to apply the projection operator method to obtain the axial group orbitals, since one is totally symmetric $(\sigma_1+\sigma_5)$ and the other $(\sigma_1-\sigma_5)$ is (a_2''), changing sign for σ_h, C_2, and S_3. Applying the projection operator method (see DMA p. 156) for the equatorial orbitals, we have:

D_{3h}	E	C_3	C_3'	C_2	C_2'	C_2''	σ_h	S_3	S_3'	σ_v	σ_v'	σ_v''
σ_2	σ_2	σ_3	σ_4	σ_2	σ_4	σ_3	σ_2	σ_3	σ_4	σ_2	σ_4	σ_3

Multiplying by characters for a_1' (all + 1), summing, simplifying, and normalizing we get:

$$a_1' = \frac{1}{\sqrt{3}}(\sigma_2+\sigma_3+\sigma_4)$$

For e', we get: $e' = (2\sigma_2 - \sigma_3 - \sigma_4)$ (one of the e' pair)

Performing the C_3 operation on this orbital gives $-\sigma_2 + 2\sigma_3 - \sigma_4$, but we have only interchanged subscripts. Multiplying this by 2 and adding to e' above, we get:

$$
\begin{array}{ll}
(e')_a & 2\sigma_2 - \sigma_3 - \sigma_4 \\
 & -2\sigma_2 + 4\sigma_3 - 2\sigma_4 \\
\hline
(e')_b & 3\sigma_3 - 3\sigma_4 \text{ or } \sigma_3 - \sigma_4
\end{array}
$$

Normalizing the 2 LCAO,

$$(e')_a = \frac{1}{\sqrt{6}}(2\sigma_2 - \sigma_3 - \sigma_4) \quad (e')_b = \frac{1}{\sqrt{2}}(\sigma_3 - \sigma_4)$$

The equatorial orbitals are the same as the σ orbitals for BF_3. They are sketched in the accompanying figure. The P a_1' orbital for bonding to the axial F is $s + d_z2$, giving maximum overlap along z. For the equatorial F, $s - d_z2$ gives better overlap. The a_2'' group orbital overlaps with p_z. The e' group orbitals overlap with p_x and p_y.

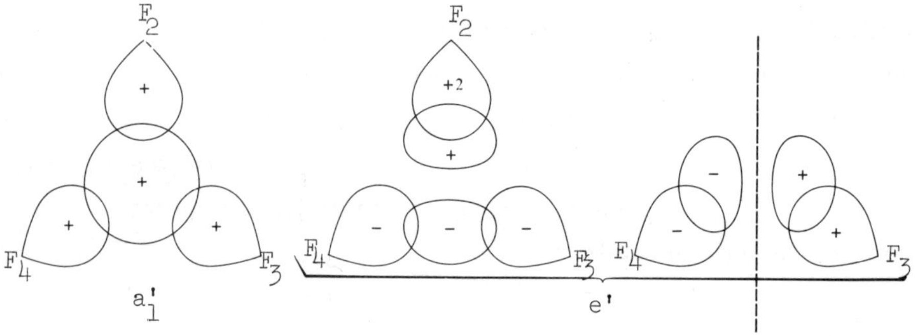

Bonding σ molecular orbitals for the equatorial bonds

4.16 Problem: Apply the group theoretical treatment to obtain the MO bonding description for sigma bonding for $SF_6(O_h)$.

4.16 Solution: Sigma bonding in SF_6 (O_h). We can look at the transformation properties of the central atom orbitals to identify their representations, but we know that the s orbital is totally symmetric (A_{1g}), the p orbitals are T_{1u}, and the d orbitals are E_g and T_{2g} (character table, Problem 3.15). We obtain the representations of the LGO's by examining the effects of all classes of operations on the individual ligand orbitals. The result for the vectors numbered as shown is given in the table for the O group. We can add the subscripts g or u by examining the effects of i and/or planes of symmetry. The reducible representation for the σ vectors reduces to $A_1 + E + T_1$ or, for O_h, $A_{1g} + E_g + T_{1u}$. Using the corresponding sulfur orbitals as templates, we can sketch the LGO's as shown. The sulfur t_{2g} orbitals are nonbonding.

The systematic way to obtain the detailed description of the LGO's and their LCAO wave functions is to apply the projection operator approach (see DMA p. 156). Here we tabulate all of the symmetry operations and write down the orbital obtained as a result of performing each operation on this orbital. For example, choosing σ_1, this is still σ_1 after the identity operation, it becomes σ_2 after one C_3 operation, σ_3 after another C_3 operation, etc.

Next we multiply each orbital in the table by the character for the corresponding symmetry operation, for the A_1 representation (all +1), and sum. In this case we get $a_1 = 4\sigma_1 + 4\sigma_2 + 4\sigma_3 + 4\sigma_4 + 4\sigma_5 + 4\sigma_6$, or simplifying, $a_1 = \sigma_1 + \sigma_2 + \sigma_3 + \sigma_4 + \sigma_5 + \sigma_6$. This is normalized to give the LCAO wave function by dividing by the square root of the sum of the squares of the coefficients.

$$\psi_{a_1} = \frac{1}{\sqrt{6}}(\sigma_1 + \sigma_2 + \sigma_3 + \sigma_4 + \sigma_5 + \sigma_6).$$

Multiplying the orbitals in the table by the characters for the representation E and summing gives:

$$e(1) = 4\sigma_1 + 4\sigma_6 - 2\sigma_2 - 2\sigma_3 - 2\sigma_4 - 2\sigma_5$$

or

$$2\sigma_1 + 2\sigma_6 - \sigma_2 - \sigma_3 - \sigma_4 - \sigma_5$$

Orientation of the ligand σ orbitals for SF_6

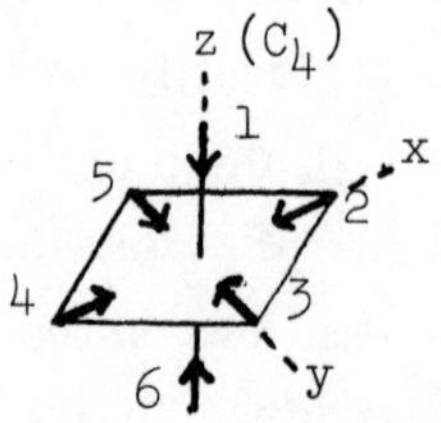

Effects of the Symmetry Operations of the O Point Group on the Ligand Orbitals of an Octahedral Complex

O	E	C_4^z	$C_2^z(C_4^2)$	C_3	C_2	i (for O_h)
σ_1	1	1	1	0	0	0
σ_2	1	0	0	0	0	0
σ_3	1	0	0	0	0	0
σ_4	1	0	0	0	0	0
σ_5	1	0	0	0	0	0
σ_6	1	1	1	0	0	0
Γ_σ	6	2	2	0	0	0 = A_1+E+T_1

Results of Applying All Symmetry Operations to σ_1

E	$C_4(1)$	$C_4(2)$	$C_4(3)$	$C_4(4)$	$C_4(5)$	$C_4(6)$	$C_2(1)$	$C_2(2)$	$C_2(3)$
σ_1	σ_1	σ_5	σ_2	σ_3	σ_4	σ_1	σ_1	σ_6	σ_6

$C_3(1)$	$C_3(2)$	$C_3(3)$	$C_3(4)$	$C_3(5)$	$C_3(6)$	$C_3(7)$	$C_3(8)$
σ_2	σ_3	σ_4	σ_5	σ_5	σ_2	σ_3	σ_4

$C_2'(1)$	$C_2'(2)$	$C_2'(3)$	$C_2'(4)$	$C_2'(5)$	$C_2'(6)$
σ_6	σ_6	σ_2	σ_3	σ_5	σ_4

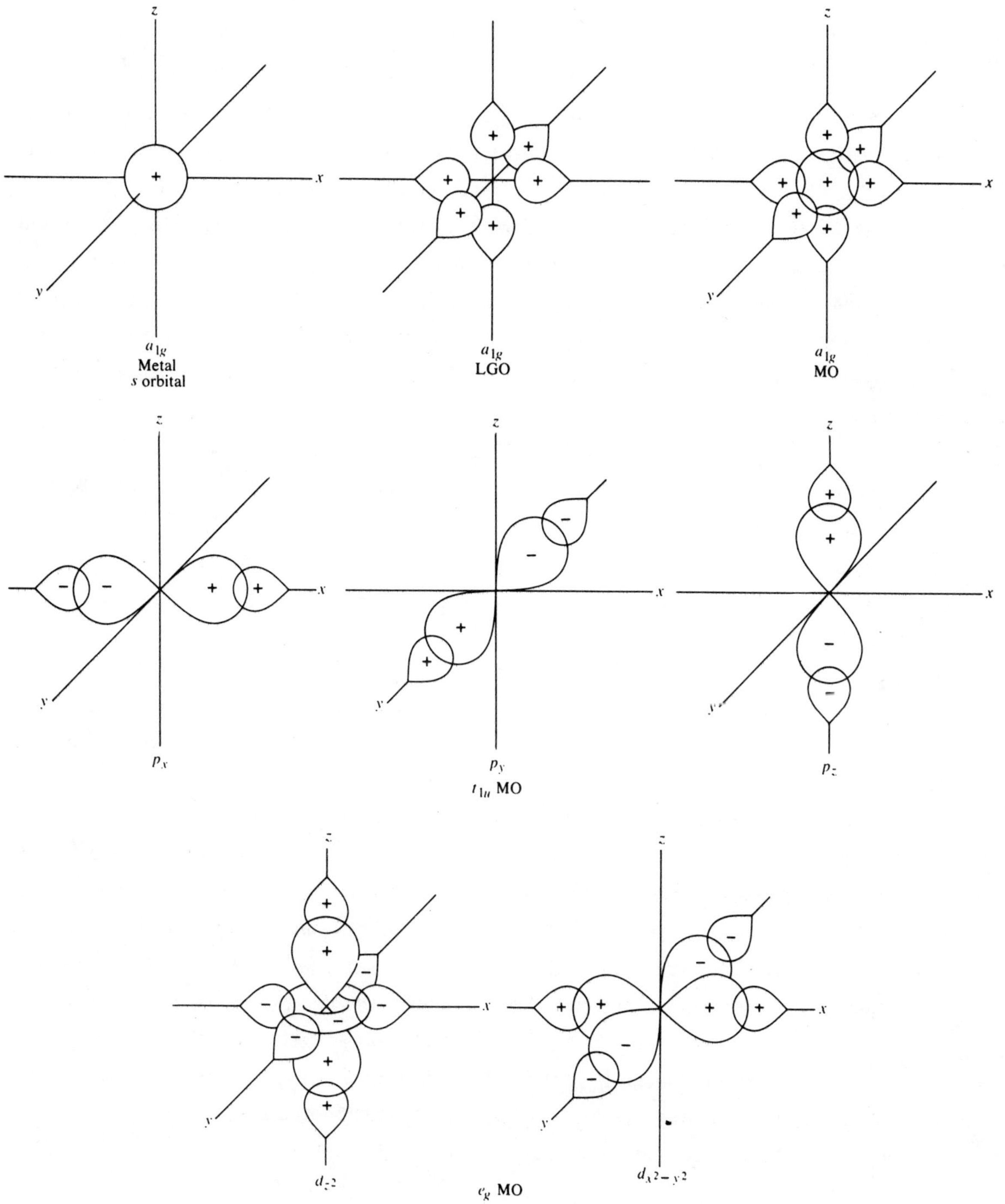

Combination of atomic orbitals and LGO to form sigma bonds in an octahedral complex.

and $\psi_{e(1)} = \frac{1}{\sqrt{12}} (2\sigma_1 + 2\sigma_6 - \sigma_2 - \sigma_3 - \sigma_4 - \sigma_5)$. This is one of the pair of e LGO's. This one matches perfectly with the sulfur d_{z^2} orbital. We can get the other one by performing an operation which interchanges ligands along z with those in the xy plane. Using a C_2 axis such that $\sigma_1 \rightarrow \sigma_2$ and $\sigma_6 \rightarrow \sigma_4$, we get $2\sigma_2 + 2\sigma_4 - \sigma_1 - \sigma_3 - \sigma_5 - \sigma_6$.

This is equivalent to the LGO already obtained, it corresponds to a different numbering scheme. Adding it to the one obtained just gives another equivalent combination. If we keep in mind that the other sulfur e orbital is $d_{x^2-y^2}$, we recognize that we want an LGO entirely contained in the xy plane with σ_1 and σ_6 not participating. Multiplying the second LGO obtained above by 2 and adding to the first LGO, we get the second e orbital.

$$3\sigma_2 + 3\sigma_4 - 3\sigma_3 - 3\sigma_5 \text{ and } \psi_{e(2)} = \frac{1}{2} (\sigma_2 - \sigma_3 + \sigma_4 - \sigma_5).$$

The t_1 LGO can be obtained similarly, recognizing that for each we need a pair of orbitals of opposite sign along one axis.

$$t_1(1) = \frac{1}{\sqrt{2}} (\sigma_1 - \sigma_6)$$

$$t_1(2) = \frac{1}{\sqrt{2}} (\sigma_3 - \sigma_5) \text{ and } t_1(3) = \frac{1}{\sqrt{2}} (\sigma_2 - \sigma_4)$$

In this case we can get $t_1(2)$ and $t_1(3)$ by operating on $t_1(1)$ with C_4 along x and y, respectively, giving us the independent LGO's directly.

The S-F bond length (156 pm) in SF_6 is shorter than expected for a single bond. We have used the equivalent of sp^3d^2 hybridization, with the possibility of some degree of S-F p-d π bonding to account for the short bond length. There is controversy concerning the participation of outer d orbitals in bonding. Zare* prefers a molecular orbital description using p orbitals on S for 3-center bonding as for XeF_2 (see DMA p. 165). In SF_6 there are three equivalent 3-center, 4-electron bonds. Although the bond order is 1/2 for each S-F bond of the 3-center bond, there is an ionic contribution. Presumably this is sufficiently important to enhance the strength of the S-F bond, with a covalent bond order of 1/2, to account for the bond length somewhat shorter than that expected for a single bond.

*T. Kiang and R.N. Zare, *J. Am. Chem. Soc.* 1980, **102**, 4024.

4.17 Problem: The H_3^+ ion is the simplest case of a three-center bond. The three H atoms are in a triangular arrangement (D_{3h}). Sketch the three MO's or (LCAO) and give a qualitative energy level diagram.

4.17 Solution: There are three LCAO. The bonding combination is the sum of the three ls orbitals. There are just two other LCAO and these must involve a change in sign and, hence, a nodal plane. From the D_{3h} character table (see Problem 3.12) we see that the only representations that change sign upon rotation by C_3 are E' and E". The remaining two MO's are degenerate so each involves one nodal plane. The representation must be E' since for s orbitals there is no change in sign for the σ_h operation. The two equivalent LCAO correspond to the σ group orbitals for BF_3 (see DMA p. 157), except that here we are using only s orbitals.

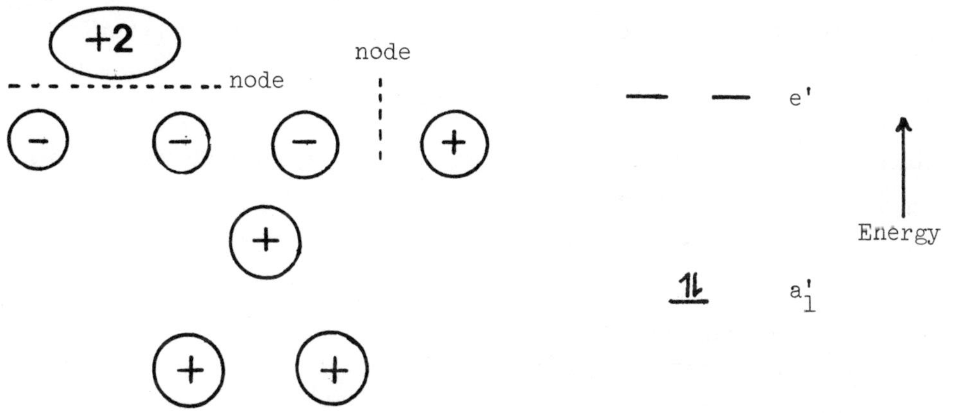

We can find the representations for the MO's more systematically by seeing how many AO's are left unchanged by each symmetry operation of the group.

D_{3h}	E	C_3	C_2	σ_h	S_3	σ_v
Γ_H	3	0	1	3	0	1

Γ_H can be reduced by inspection to give A' + E'. We can apply the projection operator method exactly as for the σ LCAO for BF_3 (see DMA p. 154f) to obtain the result given above pictorially.

4.18 Problem: XeF_4 is square planar (D_{4h}). The usual valence bond description assumes participation of the Xe d orbitals. Assuming that the d orbitals are too high in energy for bonding and that the low energy 5s electrons are nonbonding, give a qualitative bonding description using only the Xe p_x and p_y orbitals for bonding.

4.18 Solution: Along the x axis Xe uses the p_x orbital for bonding to 2F. There are three LCAO, the bonding orbital where the signs match perfectly, the antibonding orbital where the lobes touching have opposite signs, and a nonbonding orbital that does not involve participation of the Xe orbital. There is an identical three-center bond along the y axis. Since each three-center MO involves one electron pair for the two Xe–F bonds, each Xe–F bond has a bond order of 0.5. (See DMA p. 162f for a description of similar bonding in XeF_2 and K.A.R. Mitchell, "The Use of Outer d Orbitals in Bonding," Chem. Rev. 1969, 69, 157.)

4.19 Problem: Diatomic molecules can be assigned term symbols by analogy with atoms. Angular momentum (λ) is quantized around the internuclear axis and the individual quantum numbers are given as follows:

Molecular orbital	λ	m_ℓ
σ	0	0
π	1	± 1
δ	2	± 2

Proceeding by analogy to Russell-Saunders coupling and using Hund's rule, give term symbols for the ground state of H_2, B_2 and O_2. The symbols used are the representations of the $D_{\infty h}$ group, Σ, Π, and Δ, corresponding to Λ (for L) = 0, 1, and 2, respectively. (For an approach using group theory, see D.I. Ford, J. Chem. Educ. 1972, 49, 336.)

4.19 Solution: For H_2 the electron configuration is σ_g^2. Since both electrons are in the same orbital we must have the quantum numbers $m_\ell = 0$, $m_s = -1/2$ and $m_\ell = 0$, $m_s = 1/2$.

Hence $M_L = 0 + 0$ and $\Lambda = 0$; $M_S = 1/2 - 1/2 = 0$ and $S = 0$. The ground state is $^1\Sigma$. Since the wavefunction for each electron is g, the total wavefunction is also g, giving $^1\Sigma_g$.

For B_2 the electron configuration is $\sigma_g^2\ \sigma_u^{*2}\ \pi_u^2$. From the Pauli principle, the following sets of quantum numbers are possible for the π_u electrons. (Filled shells contribute $M_L = 0$, $M_S = 0$.)

	$m_s = +1/2$		$m_s = -1/2$		$M_L = \Sigma m_\ell$	$M_S = \Sigma m_s$
m_ℓ =	+1	-1	+1	-1		
	↑	↑			0	1
	↑		↓		2	0
	↑			↓	0	0
		↑	↓		0	0
		↑		↓	-2	0
			↓	↓	0	-1

The number of individual microstates corresponding to various M_L and M_S values is:

M_L \ M_S	-1	0	1
2		1	
0	1	2	1
-2		1	

Hence, the possible values of Λ are 2 and 1 and S = 1,0. Terms are $^1\Sigma$, $^1\Delta$, $^3\Sigma$. Because both electrons are u (u x u = g), the possible terms become $^1\Sigma_g$, $^1\Delta_g$, and $^3\Sigma_g$. From Hund's rule, the state of maximum spin multiplicity, $^3\Sigma_g$, is the predicted ground state.

For O_2 the electron configuration is $\sigma_g^2\ \sigma_u^{*2}\ \sigma_g^2\ \pi_u^4\ \pi_g^{*2}$. The same array of microstates as for B_2 will be generated and the ground state is again $^3\Sigma_g$.

4.20 Problem: Develop a qualitative molecular orbital energy level diagram for acetylene. Label the orbitals.

4.20 Solution: Both C atoms are equivalent; group orbitals may be constructed as the sum and the difference of the atomic orbitals, or preferably, of the sp hybridized orbitals. Symmetry labels may be assigned readily by considering the appearance of a cross-section perpendicular to the line or bonding and the atomic orbital to which it corresponds (see DMA p. 134). Group orbitals for the H atoms are formed from the sum and difference of their ls orbitals. Molecular orbitals result from the combination of the H group orbitals with the C group orbitals of the same symmetry. The resulting molecular orbital energy diagram is given below.

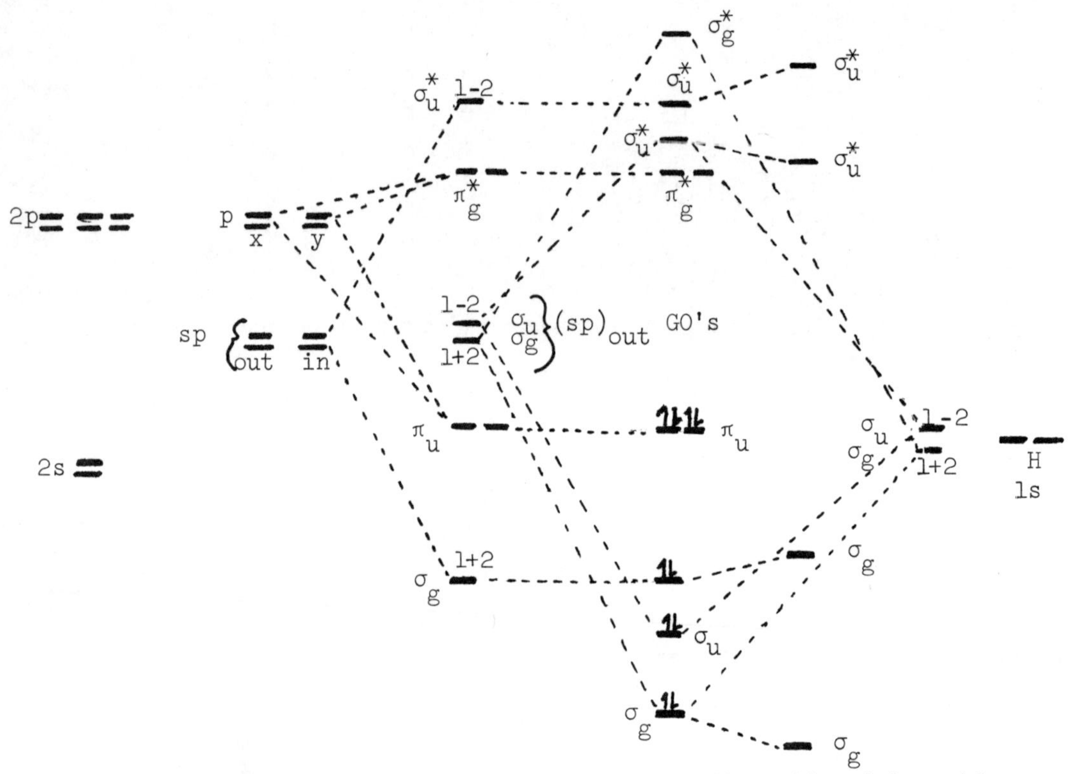

Configuration interaction spreads levels of identical symmetry ($2\sigma_g$ and $2\sigma_u$)

4.21 Problem: Given below is a Walsh diagram for AH_3 molecules correlating the molecular orbital energy levels of pyramidal geometry with those of planar geometry. Compare the predictions of geometry based on the Walsh diagram with those based on the Gillespie VSEPR approach and on the Pauling hybridization approach for AH_3 molecules having 6, 7, 8, and 10 valence electrons.

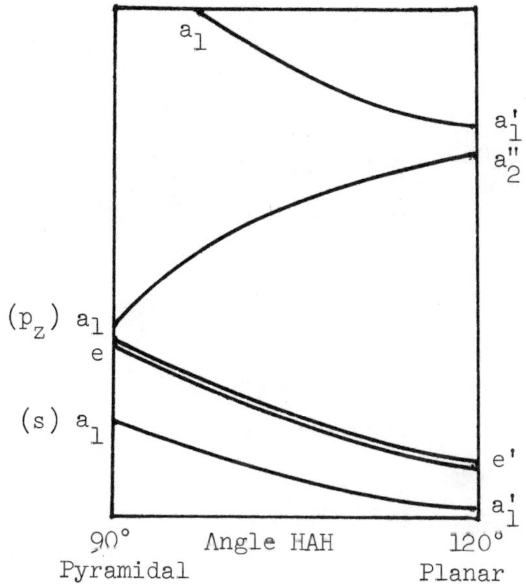

4.21 Solution: The geometry predicted from the Walsh diagram will be the one of lowest energy for the particular electron count. The VSEPR and the hybridization approach are discussed in DMA Chapter 2, and touched on in Problems 2.1, 2.5, 2.9···. Resulting geometries predicted by these various approaches are given below.

	WALSH	GILLESPIE	PAULING
6 electrons	trigonal planar	trigonal planar	sp^2 trigonal planar
7 electrons	HAH angle ~ 115° pyramidal	shallow pyramid (less repulsion from single electron than bond pairs)	sp^3 pyramidal (with more s character in bonding orbitals giving HAH angle greater than 109°)
8 electrons	HAH angle 90° pyramidal	pyramidal with HAH angle less than 109° due to electron pair-bond pair repulsion being greater than bond pair-bond pair repulsion	sp^3 pyramidal
10 electrons	trigonal planar	"T"-shaped	must include 5 orbitals from A, sp^3d pseudo-trigonal bipyramidal

V Hydrogen Bonding and Weak Interactions

5.1 Problem: List the substances in each of the following groups in order of increasing boiling points. (Hint: first group the substances according to the type of interaction involved.)

(a) LiF, LiBr, CCl_4, NH_3, CH_4, SiC, CsI

(b) Xe, NaCl, NO, CaO, BrF, Al_2O_3, SiF_4

5.1 Solution:

(a) LiF, LiBr, CsI are ionic salts with both melting points and boiling points expected to decrease as the distance between the centers of the ions increases. However, we find that radius ratio effects lead to high anion-anion repulsion in LiBr and consequent lower values of the melting point and boiling point than anticipated (see Section 6.2.2 of DMA). SiC is a giant covalent molecule which remains solid at higher temperatures than any of the other compounds listed. It decomposes at temperatures above 2200°C. CH_4, NH_3, and CCl_4 contain discrete molecular units with covalent bonding. Only NH_3 has a non-zero dipole moment and the possibility of hydrogen bonding. We should know from experience that CH_4 and NH_3 are gases at room temperature whereas CCl_4 is a liquid. CCl_4 exposes more surface electrons and accordingly displays much greater London forces than CH_4 and NH_3. CH_4 with no dipole moment, and no surface electrons has the lowest boiling point in this set. Combining groups, the net order of boiling points is thus:

$$CH_4 < NH_3 < CCl_4 < LiBr < CsI < LiF < SiC$$

(b) NaCl, CaO, and Al_2O_3 are salt-like solids with attraction between ions increasing with increasing charge on the ions. NO, BrF, and SiF_4 are molecular substances, all of which are gases at room temperature. As a first approximation we might expect the boiling point of NO to be approximately an average of the boiling points of N_2 and O_2, increased somewhat by the dipole moment ($\Delta\chi = 0.5$). The actual boiling point is -151.8^{o}C. For BrF the dipolar contribution should be much more significant ($\Delta\chi = 1.2$), and the boiling point of BrF is 20°C compared to that of -34°C for Cl_2 (the "molecular average" of Br_2 and F_2).

That SiF_4 exists as a substance with discrete molecules, rather than as a network solid (cf SiC_2, AlF_3) is not a matter of any change from ionic to covalent bonding but rather that the coordination number of Si is satisfied by the number of halogens corresponding to a formula unit. The fluorine atoms presented to the outside world by SiF_4 are not very polarizable and the intermolecular attractions are weak (sub.p. -95.7^{o}C). The low intermolecular interactions in the silicon tetrahalide have been the provocation for much speculation by chemists—(see J.H. Hildebrand, J. Chem. Phys. 1947, 15, 727).

The boiling point of Xe is expected to be higher than that of any of the noble gases of lower atomic weight. We may recall that Ar is obtained from fractional distillation of liquid air and that Ar boils at a temperature between that of O_2 and N_2. Congeners in the 3rd, 4th, and 5th periods differ by 30 to 40 degrees in boiling points (MeCl, -24°C; MeBr, 3.5^{o}C; MeI, 42.4^{o}C) so we might expect Xe to boil 70 to 80 degrees above the range for liquid air. The actual boiling point is -109.3^{o}C. The order of increasing boiling points is thus: $NO < Xe < SiF_4 < BrF < NaCl < CaO < Al_2O_3$.

5.2 Problem: Which of the following mixtures would be expected to have maximum boiling points and which to have minimum boiling points?

(a) Methyl acetate and chloroform

(b) C_6H_{12} and C_2H_5OH

5.2 Solution:

(a) Methyl acetate and chloroform give a maximum boiling mixture. Neither component can form hydrogen bonds individually, but $HCCl_3$ can serve as a hydrogen donor and $CH_3CO_2CH_3$ can serve as a hydrogen acceptor. Accordingly, the interaction between the different molecules is greater than that between like molecules, in this case leading to large negative deviations from Raoult's law and resulting azeotropic behavior.

(b) Ethanol is a highly self-associated liquid as a result of hydrogen bonding. Hexane disrupts the hydrogen bonds in ethanol, causing the mixture to form a minimum boiling azeotrope.

5.3 Problem: Although ΔH_{vap} for HF is lower than that for H_2O, HF forms stronger H-bonds. Explain.

5.3 Solution: Each HF molecule forms two H-bonds (one as an H-donor, one as an H-acceptor) whereas each H_2O molecule forms four H-bonds. Hydrogen fluoride vapor at the normal boiling point of HF contains polymeric units up to $(HF)_6$ species—thus not all of the hydrogen bonds in HF are broken on vaporization, whereas water vapor at the boiling point of water consists essentially of monomeric H_2O molecules. Thus, four H-bonds are broken for every H_2O molecule vaporized, whereas less than two H-bonds are broken per HF vaporized.

5.4 Problem: When no chemical reaction occurs, the solubility of a gas in a liquid is proportional to the magnitude of the van der Waals interaction energy of the gas molecules. Indicate the relative solubility of O_2, N_2, Ar, and He in water. Why do deep-sea divers use a mixture of He and O_2 instead of air?

5.4 Solution: The van der Waals interaction leading to solution of these gases is expected to be the geometric mean of the van der Waals interaction between gas molecules and the van der Waals interaction between solvent molecules. The van der Waals interaction for the gases will parallel their boiling points (the higher boiling having the stronger van der Waals interaction). The boiling point order and hence that of increasing solubility is:

$$\mathrm{He} < N_2 < \mathrm{Ar} < O_2.$$

Helium is used instead of N_2 in the "air" for deep-sea divers because of its much lower solubility in blood—thereby preventing the "bends", a painful condition arising from the dissolution of N_2 as the pressure decreases during ascent.

5.5 Problem: Estimate the dispersion energy for H_2O from its "ideal" boiling point and compare your estimate to Coulson's estimate of 12 kJ/mole in H-bonded water.

5.5 Solution: If water were not hydrogen bonded, it would boil at ~ -100°C or about 170K. The dispersion energy is approximately 10 (b.p. K) J/mole or for H_2O would be estimated at 1700 J/mole or 1.7 kJ/mole. This estimate is much less than Coulson's estimate; the 1.7 kJ is an estimate for the interaction at a normal van der Waals contact distance in an idealized nonhydrogen-bonded water. Both the attractive and repulsive terms are much greater at the shorter distance of the H-bond. The repulsive term ($\propto 1/r^{12}$) increases faster than the attractive term ($\propto 1/r^6$) with the shortening of the intermolecular distance, leading to a net repulsive contribution from van der Waals interactions to the H-bond energy in H_2O.

5.6 Problem: Select from among the following radii and bond distances to calculate the S...S distance for SF_6 molecules in contact in the solid

Covalent Bond Length		Crystal Radii		van der Waals Radii	
		S(VI)	43 pm	S	180 pm
S-F (in SF_6)	156 pm	F^-	119	F	147

5.6 Solution: The distance S—F...F—S is the sum of S—F covalent distance, the F...F van der Waals distance and the F—S covalent distance, that is, 156 + 2(147) + 156 = 606 pm.

5.7 Problem: Calculate the expected O...O separation in a system containing O—H O if no hydrogen bonding were to occur and the O—H O atoms are linearly arranged. Compare your results with the nearest O—O distance, 275 pm, in ice $\underline{I_h}$. Use any tables of appropriate radii.

5.7 Solution: The van der Waals radius of O is 152 pm. Thus, O...O contact for non-bonded atoms would be 304 pm. To calculate the O—H...O distance expected for no H-bonding, take the O—H covalent distance and add the H...O van der Waals distance.

$$r_{O\text{-}H(covalent)} + r_{H(vdW)} + r_{O(vdW)}$$
$$95.7 + 120 + 152 = 368 \text{ pm}$$

The observed distance of 275 pm is much less than the O—H...O distance calculated for a van der Waals contact distance. Such shortening of the van der Waals contact distance is typical for H-bonded molecules.

5.8 Problem: The intensity of an infrared absorption is proportional to the change in the dipole moment occurring during the vibration. The asymmetric stretching vibration in IHI^- is far more intense than that found for the stretching vibration of HI. Offer a reasonable explanation. Does your explanation also hold for the intensities of other H-bonded species?

5.8 Solution: In the asymmetric stretching vibration of IHI^-, the negative charge may be viewed as moving from one iodine to the other for only a small dispacement of the hydrogen ($^-$I...HI going to IH...I$^-$). Since $\partial\mu/\partial q$ is large, the intensity of the absorption is large. ($\partial\mu/\partial q$ is the change in dipole moment with displacement along the vibrational coordinate.) For HI, μ is very small since $\Delta\chi$ is small, and μ changes little during vibration, hence the IR absorption for this stretch is weak. Similar arguments hold for other hydrogen bonded species, but the $\partial\mu/\partial q$ may not be as striking in other cases.

5.9 Problem: Indicate whether positive, negative, or no deviation from Raoult's law might be expected for the following binary systems:

(a) HCl—$(CH_3)_2O$ (c) $HCCl_3$—CCl_4

(b) H_2O—C_8H_{18} (d) $HCCl_3$—$(C_2H_5)_3N$

5.9 Solution: If a hydrogen-bonded system can be formed in the mixture, but not in the individual components, negative deviation from Raoult's law will occur--that is, the mixture will have a lower vapor pressure than that predicted by Raoult's law. This is the situation in cases (a) and (d). If a hydrogen-bonded system is disrupted on forming a mixture, positive deviation from Raoult's law will occur as in case (b) where the positive deviation is so extreme that phase separation occurs. Case (c) exhibits no H-bonding tendency in either the mixture or the components alone. Since van der Waals interactions are expected to be similar, the solution should approach ideal behavior.

5.10 Problem: The heat released in the reaction $Br^-(g) + HCl(g) \rightarrow BrHCl^-(g)$ has been found to be 38 kJ. By devising a suitable thermochemical cycle, and using available data, evaluate the expected heat released in the reaction $Cl^-(g) + HBr(g) \rightarrow ClHBr^-(g)$.

5.10 Solution:

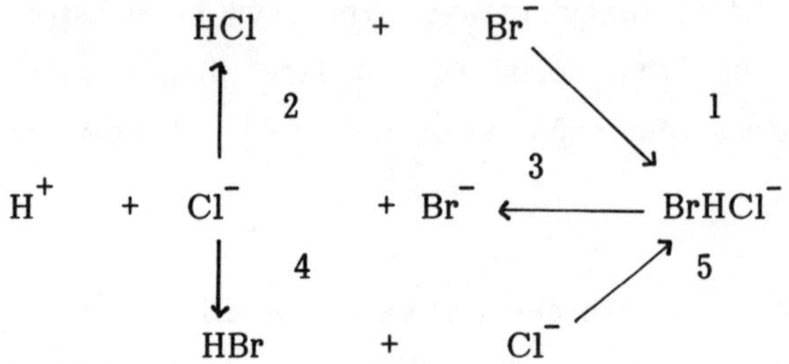

In the above cycles (all species considered as gas phase species) we are given only $\Delta H_1 = -38kJ$; $\Delta H_2 = -(D_{H-Cl} + IE_H - EA_{Cl})$ and likewise $\Delta H_4 = -(D_{HBr} + IE_H - EA_{Br})$. The sum of enthalpy changes around any cycle is zero, therefore

$$\Delta H_1 + \Delta H_2 + \Delta H_3 = 0 = \Delta H_3 + \Delta H_4 + \Delta H_5$$

$$-38kJ + EA_{Cl} - D_{HCl} = EA_{Br} - D_{HBr} + \Delta H_5$$

$$\Delta H_5 = -38 + 348.8 - 427.78 - 324.6 + 362.6 = -79kJ/mole$$

The heat released in step 5 is 79kJ/mole.

Which reaction, (1) or (5), should be considered as defining the H-bond energy in $BrHCl^-$? McDaniel and Vallee (Inorg. Chem. 1962, 2, 1000) propose that the H-bond energy be defined by the reaction in which the proton in the reactant already resides on the weaker acid, otherwise the "H-bond energy" will include a proton transfer energy as well as the hydrogen bond energy.

5.11 Problem: Explain why comparing a normal melting point with a melting point taken with the compound under water sometimes helps distinguish between intra- and intermolecular hydrogen bonding. (See E.D. Amstutz, J.J. Chessick, and I.M. Hunsberger, Science 1950, 111, 305.)

5.11 Solution: See reference cited. When water competes with intermolecular hydrogen bonds, the intermolecular attractions will be greatly weakened and the melting point decreased by 20 or 30 degrees.

5.12 Problem: Why does CsCl(s) react with HCl(g) at low temperatures whereas NaCl(s) does not?

5.12 Solution: In order to form MHX_2 from MX and HX, the energy released in formation of the XHX^- species must be greater than that required to expand the MX lattice to accommodate the XHX^- ion. This condition is a general one for formation of molecular adducts of solids, such as tetrahaloborate salts from MX and BX_3, etc.

5.13 Problem: It has been suggested that the higher melting point of p-methylpyridine-N-oxide compared with the o- and m-isomer may stem from hydrogen bonding. How might this be tested experimentally? Be specific. Mention the data to be gathered, with what the data might be compared, and how this would serve to establish hydrogen bonding or its absence.

5.13 Solution: Numerous approaches are possible. The most unambiguous experiment would be the determination of the crystal structure of the p-isomer and a comparison of the C—H...O distance with that predicted for van der Waals interaction alone (see Problem 5.7). A simpler experiment would be to examine the C—H stretching region of the IR spectrum for solutions in CCl_4 as a function of the concentration of the p-isomer. A concentration dependence would be expected only for a system in which the CH_3- group is participating in H-bonding.

5.14 Problem: Explain why the anhydrous acid $HICl_4$ cannot be isolated, but the crystalline hydrate $HICl_4 \cdot 4H_2O$ may be obtained from ICl_3 in aqueous HCl.

5.14 Solution: $HICl_4$ has no good site for the proton. In the hydrate the proton is found in the $H_9O_4^+$ species.

5.15 Problem: Most substances expand when they are heated, leading to decreasing density with increasing temperature. However, $H_2O(\ell)$ has a maximum density at 4^oC.

How can you explain this?

5.15 Solution: In the structure of ice, water molecules are oriented so as to form the maximum number of H-bonds (four per O). This leaves considerable open space in the lattice. As ice melts some H-bonds are broken and the open cage structure collapses, giving a decrease in volume. After $4^{o}C$, the normal volume expansion, with rising T, takes over and density decreases.

5.16 Problem: In a homologous series of compounds the boiling points and melting points often increase with molecular weight. How do you account for this observation? However, replacing a CH_3 group (MW 15) with an OH group (MW 17) or an F group (MW 20) usually has a dramatic effect on the melting or boiling point even though the molecular weight changes only slightly. Explain.

5.16 Solution: The number of electrons in a molecule is roughly proportional to its molecular weight, and the van der Waals interaction is roughly proportional to the number of electrons per molecule. In a homologous series of compounds, the nature of the specific interactions will remain approximately constant, but van der Waals interactions will increase with molecular weight, leading to an increase in melting and boiling points. The series CH_3, OH, and F have the same number of electrons, but the polarizability decreases with increasing atomic number. Thus, for molecular solids we might expect a decrease in melting points and boiling points as F is substituted for CH_3 and this is observed for boron trimethyl and boron trifluoride (b.p. $-21^{o}C$ and $-99^{o}C$, respectively). But for boric acid we find the OH groups form a hydrogen-bonded network, yielding strong interactions. Boric acid decomposes on heating to give the oxide. The methyl derivatives of the elements are generally molecular compounds, whereas the fluoro derivatives frequently form networks by sharing bridging F, or as the electronegativity increases, ionic solids may form. Thus we find NF_3 has a lower m.p. and b.p. than $(CH_3)_3N$, but the reverse is true for the Sb and Bi derivatives.

5.17 Problem: If a table of van der Waals radii is unavailable, what other source might be used for reasonable estimates of van der Waals radii for nonmetals? Why are these values reasonable?

5.17 Solution: Use ionic radii for anions as reasonable values for van der Waals radii. The values are similar since in each case an approaching atom encounters a completed octet of electrons and no bond is formed.

5.18 Problem: Compounds of the type $[\underline{i}\text{-}(C_5H_{11})_4N]Cl{\cdot}yH_2O$ and $[\underline{i}\text{-}(C_5H_{11})_4]_2CrO_4$ $2yH_2O$ (where y is approximately 40) are found to be isomorphous. It is unusual to find isomorphous compounds of the same cation with anions of different charge. What special features of these highly hydrated compounds might be responsible?

5.18 Solution: The structures of these very highly hydrated compounds are dominated by the large hydrogen-bonded cages. The ions fill cavities in the cages without altering the structure. See G.A. Jeffrey and R.K. McMillan, Prog. Inorg. Chem. 1967, 8, 43.

VI Inorganic Solids

6.1 Problem: Variations in hardness for ionic substance correlate well with what one thermodynamic property?

6.1 Solution: Hardness is determined by the resistance to deformation of the lattice and this correlates well with lattice energy.

6.2 Problem: Which of each of the following pairs might be expected to be more ionic?

(a) $CaCl_2$ or $MgCl_2$ (c) NaCl or CuCl (similar radii)

(b) NaCl or $CaCl_2$ (similar radii) (d) $TiCl_3$ or $TiCl_4$

6.2 Solution: Covalent character increases with polarization (charge displacement).

(a) $CaCl_2$ is expected to be more ionic than $MgCl_2$ because the larger Ca^{2+} cation is less polarizing.

(b) NaCl is expected to be more ionic than $CaCl_2$ because the lower cation charge of Na^+ causes less polarization of the Cl^-.

(c) NaCl is more ionic than CuCl because Cu^+ is an 18-electron cation and causes more polarization than for Na^+ (8-electron shell) since the d electrons are less efficient in screening the nuclear charge.

(d) $TiCl_3$ is more ionic than $TiCl_4$ (a liquid) because Ti^{3+} has lower charge and larger radius than Ti^{IV} — both favor less polarization.

6.3 Problem: For each of the following pairs indicate which substance is expected to be:

(a) More covalent (Fajans' rules):

$MgCl_2$ or $BeCl_2$	$SnCl_2$ or $SnCl_4$
$CaCl_2$ or $ZnCl_2$	$CdCl_2$ or CdI_2
$CaCl_2$ or $CdCl_2$	ZnO or ZnS
$TiCl_3$ or $TiCl_4$	NaF or CaO

(b) Harder:

NaF or NaBr

Al_2O_3 or Ga_2O_3

MgF_2 or TiO_2

6.3 Solution: (a) Covalent character increases with increasing polarization. $BeCl_2$ is more covalent than $MgCl_2$ because of the smaller radius of Be^{2+}. $ZnCl_2$ is more covalent than $CaCl_2$ because of the smaller radius of Zn^{2+} and the fact that Zn^{2+} has an 18-electron configuration. $CdCl_2$ is more covalent than $CaCl_2$ because Cd^{2+} has an 18-electron configuration. $TiCl_4$ (a liquid) is more covalent than $TiCl_3$ because of the smaller radius and higher charge of Ti^{IV}. $SnCl_4$ is more covalent than $TiCl_3$ because of the smaller radius and higher charge of Sn^{IV}. CdI_2 is more covalent than $CdCl_2$ because the larger I^- is more easily polarized. CaO is more covalent than NaF because the higher ionic charges increase polarization.

(b) NaF is harder than NaBr because the lattice energy is greater for small ions. Al_2O_3 is harder than Ga_2O_3. The cation radii are almost the same, but Ga^{3+} is an 18-electron ion, increasing polarization and "softening" the lattice. TiO_2 is harder than MgF_2 because of the higher ionic charges, increasing the lattice energy.

6.4 Problem: Which of the following are not possible close-packing schemes?

(a)	ABCABC...	(d)	ABCBC...
(b)	ABAC...	(e)	ABBA...
(c)	ABABC...	(f)	ABCCAB...

6.4 Solution: If the positions of the spheres in the first layer of a close-packed arrangement are designated A, those of adjacent layers can only be B or C, with the spheres of the adjacent layer in the indentations of the first layer. Arrangements (e) and (f) violate this requirement, repeating the same positions (BB and CC). These are not close-packed arrangements.

6.5 Problem: The repeating unit for hcp is P_AP_B (or 2P) and for ccp it is $P_AP_BP_C$ (or

3P). Between any two close-packed layers there are two layers of tetrahedral sites and there is one layer of octahedral sites in the sequence $(P_AT_BO_CT_AP_BT_AO_CT_B)P_A$... for hcp and $(P_AT_BO_CT_AP_BT_CO_AT_BP_CT_AO_BT_C)P_A$... for ccp. For NiAs, the larger ions (As) are hcp with Ni in all O sites, described as $(P_AO_CP_BO_C)P_A$... or 4PO. The wurtzite structure (4PT) of ZnS has open channels along the packing direction, but the zinc blende structure (6PT) does not. Explain. (See references in Problem 6.7).

6.5 Solution: For 4PT (hcp) the P and T sites are at A and B positions, with C sites vacant, forming open channels. For 6PT (ccp) the P and T sites occur at A, B, and C positions, leaving no open channels.

6.6 Problem: Many MX type compounds with C.N. 6 have the NaCl (6PO) structure, whereas few have the NiAs (4PO) structure. What features of the NiAs structure limit its occurrence? What characteristics of the compound are necessary for the NiAs structure? What unusual physical property of NiAs results from its structure? (See references in Problem 6.7).

6.6 Solution: NiAs has the 4PO structure. The As are hcp, occupying A and B positions only, with Ni in the octahedral sites lined up along C positions. There is no shielding between adjacent Ni atoms. This is tolerable only if there is a considerable amount of polarization (covalence) to diminish repulsion. The more ionic MX compounds adopt the NaCl (6PO) structure. There is metallic conduction along one direction of the NiAs crystal because of the Ni "wires" running in this direction.

6.7 Problem: Give the PTOT notation (See DMA p. 202 or S.M. Ho and B.E. Douglas, J. Chem. Educ. 1969, 46, 208) for NaCl, ZnS (both structures), CaF_2, and TiO_2.

6.7 Solution: NaCl has a ccp arrangement of Cl^- with Na^+ in all octahedral sites, 6PO. Actually, the cation and anion sites are equivalent, but the larger ions are generally regarded as occupying the packing (close-packed) positions. The repeating unit (6 layers) is $P_AO_CP_BO_AP_CO_B$.

ZnS gives two structures involving hcp and ccp arrangements of S^{2-}. The C.N. of

Zn^{2+} is four, placing Zn^{2+} in tetrahedral sites. There are two tetrahedral layers between packing layers. The Zn^{2+} ions fill one of these T layers.

Wurtzite, 4PT (hcp S^{2-}) (P_ATP_BT)

Zinc blende, 6PT (ccp S^{2-}) ($P_ATP_BTP_CT$)

In CaF_2 the C.N. of Ca^{2+} is eight and that of F^- is four. The larger Ca^{2+} can be regarded as ccp with the F^- filling all tetrahedral sites in both T layers, giving a PTT arrangement. For a ccp arrangement there are A, B, and C positions for P, so we have $P_ATTP_BTTP_CTT$ or nine layers in the repeating unit, 9PTT.

In the rutile (TiO_2) structure Ti^{IV} has C.N. 6 and O^{2-} has C.N. 3. The Ti^{IV} ions are in octahedral sites and the O^{2-} ions are close packed at A and B positions (hcp). Since there are as many octahedral sites as packing positions in a close-packed structure, but only half as many Ti^{IV} as O^{2-}, only half of the octahedral sites are filled. The notation is $4PO_{1/2}$ (for $P_AO_{1/2}P_BO_{1/2}$). The C.N. is 3 for O^{2-} because of the vacant octahedral sites.

6.8 Problem: What do the following pairs or groups of structures have in common and how do they differ?

(a) 6PO, 6PT, and 4PT.

(b) $9PT_{1/2}T_{1/2}$ and 6PT.

(c) 9PTT and $4PO_{1/2}$.

(d) 2(3/2)PPO and $4PO_{1/2}$.

6.8 Solution: (a) All are MX (1:1) compounds. 6PO and 6PT are ccp (1/2 x 6 = 3 packing layers or ABC for the repeating unit) and 4PT is hcp (1/2 x 4 = 2 packing layers or AB for the repeating unit). In 6PO the cations fill all O sites. In 6PT and 4PT the cations fill all T sites in one T layer, the other T layer is vacant.

(b) Both are MX (1:1) compounds using P and 1/2 of the T sites and both are ccp (3P layers per repeating unit). In $9PT_{1/2}T_{1/2}$, 1/2 of the T sites in each T layer are used. In 6PT all T sites in one layer are used.

(c) Both are MX_2 or M_2X compounds since there are twice as many sites of one kind occupied as of the other kind. 9PTT is ccp (1/3 x 9 = 3 P layers per repeating unit) with (for MX_2) X in all T sites. The C.N. of M is 8, and the C.N. of X is 4. For $PO_{1/2}$ (for MX_2) M is in 1/2 of the octahedral sites. The C.N of M is 6 and the C.N. of X is 3 (rutile). For each case the roles are reversed for M_2X compounds.

(d) For 2(3/2)PPO there are 2 x 3/2 = 3 layers in the repeating unit, P_AP_BO. This is hcp.

For $4PO_{1/2}$ there are 4 layers in the repeating unit, $P_AO_{1/2}P_BO_{1/2}$, which is also hcp. In each case half of the octahedral sites are occupied — all of them in alternate layers for PPO (a layer structure) and half of them in each layer for $PO_{1/2}$ (rutile).

6.9 Problem: Why are layer structures such as those of $CdCl_2$ and CdI_2 usually not encountered for metal fluorides or compounds of the most active metals?

6.9 Solution: Since layer structures have adjacent layers of anions (Cs_2O is an unusual case in having adjacent ccp layers of cations) the repulsion would prohibit the structure unless there were a considerable amount of polarization (covalence). Metal fluorides and compounds of the most active metals are found in the more typical "ionic" structures such as those of NaCl and CaF_2.

6.10 Problem: Rubidium chloride assumes the CsCl structure at high pressures. Calculate the Rb—Cl distance in the CsCl structure from that for the NaCl structure (from ionic radii for C.N.6, r_{Rb^+} = 166 pm, r_{Cl^-} = 167 pm). Compare with the Rb—Cl distance from radii for C.N. 8. (r_{Br^-} = 175 pm for C.N. 8, calculate r_{Cl^-} for C.N. 8).

$$\frac{r_8}{r_6} = \left(\frac{8A_6}{6A_8}\right)^{1/(n-1)}$$

where r_8 and r_6 are the radii (or internuclear distances) for C.N. 8 and 6, respectively, n is the Born exponent (n = 10 for Rb^+ and 9 for Cl^-, or we can use an average value (9.5) for RbCl). The Madelung constant (A) is 1.75 for NaCl and 1.76 for CsCl.

6.10 Solution: The internuclear distance expected for C.N. 6 is 166 + 167 = 333 pm. For C.N. 8 we calculate

$$d_8 = 333 \left(\frac{8A_6}{6A_8}\right)^{1/(9.5-1)} = 333 \left(\frac{4x1.75}{3x1.76}\right)^{0.118} = 343 \text{ pm}$$

$$r_{Cl^-} \text{ (C.N. 8)} = 167 \left(\frac{8x1.75}{6x1.76}\right)^{1/(9-1)} = 173 \text{ pm}$$, giving d_{RbCl} = 348 pm as the sum of the C.N. 8 radii.

6.11 Problem: Calculate the cation/anion radius ratio (by using plane geometry) for a triangular arrangement of anions in which the cation is in contact with the anions but does not push them apart.

6.11 Solution: The lines joining the centers of the anions form an equilateral (60°) triangle

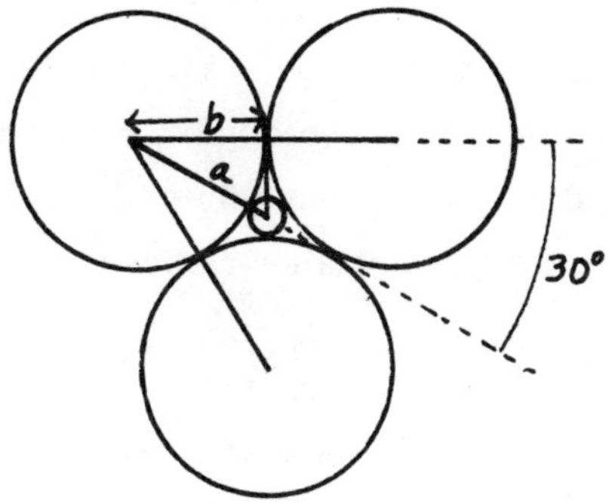

$$\cos 30^{\circ} = \frac{b}{a} = \frac{r_X}{r_X + r_M}$$

$$r_X = 0.866 r_X + 0.866\ r_M$$

$$\frac{r_M}{r_X} = 0.155$$

6.12 Problem: Estimate the density of MgO (6PO) and ZnS (6PT) using radii to determine the cell dimensions and the number of formula units per unit cell.

6.12 Solution: For MgO (6PO or NaCl structure) the cubic cell edge equals

$$2r_{Mg^{2+}} + 2r_{O^{2-}} = 424 \text{ pm} = a$$

The volume of the unit cell = $a^3 = 7.62 \times 10^{-23}$ cm^3

Since there are 4 MgO per unit cell, we have $\frac{4 \times 40.3}{6.02 \times 10^{23}}$ g/unit cell.

The density is given by dividing the mass of the unit cell by the volume of the cell:

$$\frac{161.2}{6.02 \times 10^{23} \times 7.62 \times 10^{-23}} = 3.51 \text{ g/cm}^3.$$ Handbook gives 3.58 g/cm^3.

For ZnS (6PT — the S^{2-} are ccp or face-centered cubic with the Zn^{2+} in T sites within the cell) the diagonal of a face of the cubic cell is $4r_{S^{2-}} = 680$ pm (using the radius for C.N. 6). This is the hypotenuse of a right triangle with sides equal to a, or

$(680)^2 = 2\ a^2$ $\quad$ a = 482 pm $\quad$ $a^3 = 1.12 \times 10^{-22}$ cm^3

4 ZnS/unit cell or $\dfrac{390}{6.02 \times 10^{23}}$ g/unit cell

$$\frac{390}{6.02 \times 10^{23} \times 11.2 \times 10^{-23}} = 5.78 \text{ g/cm}^3$$

The handbook gives 4.10 g/cm^3 for the density.

The cell length for zinc blende is 541 pm. Thus, the effective radius for S^{2-} is 191 pm from this right triangle. ZnS is appreciably covalent so the ionic radius is a poorer value to use than the van der Waals radius (180 pm). Correcting the ionic radius for C.N. 6 to get a value for C.N. 4 gives a smaller radius (163 pm) and poorer agreement with the observed density.

6.13 Problem: $NaSbF_6$ has the NaCl (6PO) structure. The density is 4.37 g/cm^3. Calculate the radius of SbF_6^- using the radius of Na^+, the density, the formula weight, and the number of formula units per unit cell. (r_{Na^+} = 116 pm)

6.13 Solution: For the NaCl structure there are 4 formula units per unit cell and the cell length is $2r_{Na^+} + 2r_{SbF_6^-}$.

Mass of a unit cell = 4 x 259/6.02 x 10^{23} = 1.72 x 10^{-21} g/unit cell. The density is the mass of the unit cell divided by its volume, or

$$4.37 = \text{density} = \frac{1.72 \times 10^{-21}\text{g}}{a^3 \text{ cm}^3} \qquad a^3 = \frac{1.72 \times 10^{-21}}{4.37} = 3.94 \times 10^{-22} \text{ cm}^3$$

$$a = 7.33 \times 10^{-8} \text{ cm} = 733 \text{ pm} = 2r_{Na^+} + 2r_{SbF_6^-} = 2(116) + 2r_{SbF_6^-}$$

$$r_{SbF_6^-} = \frac{501}{2} = 250 \text{ pm}$$

6.14 Problem: Compare the thermochemical data for the alkali halides. What factors are important in establishing the order of increasing heats of formation within each group of halides (fluorides, chlorides, etc.)?

Thermochemical Data for Alkali Metal Halides

S	IE			F	Cl	Br	I
			EA	328.0	348.8	324.6	295.4
S	IE		1/2 D	77.3	119.6	95.1	74.4
160.7	520.1	Li	$-\Delta H_f$	616.9	408.3	350.9	270.1
			U_o	1047	860	802	730
107.8	495.8	Na	$-\Delta H_f$	575.4	411.1	361.4	288
			U_o	928	786	735	671
89.2	418.7	K	$-\Delta H_f$	568.6	435.9	393.8	327.9
			U_o	826	715	672	615
82.0	402.9	Rb	$-\Delta H_f$	549.4	430.5	389	328
			U_o	784	686	644	592
76.7	375.6	Cs	$-\Delta H_f$	554.7	442.8	394	337
			U_o	756	666	617	568

All data in kJ/mole at 298 K from DMA p. 225.
Electron affinity (EA), lattice energy (U_o), and heat of formation (ΔH_f) are thermodynamically negative (energy released).
Sublimation energy (S), ionization energy (IE), and dissociation energy (1/2 D) are thermodynamically positive (energy required).

6.14 Solution: For MCl, MBr, and MI ΔH_f becomes more negative with increasing size of M, a trend established by the ionization and sublimation energies of M. For MF the order is reversed because of the much greater decrease in lattice energy from LiF to CsF.

6.15 Problem: Calculate the heat of formation of NaF_2 and the heat of reaction to produce NaF + $1/2F_2$. (Assume rutile structure, A = 4.816, r_{F^-} (C.N. 3) = 116 pm and estimate $r_{Na^{2+}}$).

6.15 Solution: Assume $r_{Na^{2+}} \simeq r_{Mg^{2+}}$ = 86 pm Z = 1 (highest common factor of charges)
n = 7 (Born exponent for Ne configuration)

$$U_o = \frac{(4.80 \times 10^{-10} \sqrt{\text{dyne}}\ \text{cm})^2\ 1^2\ (6.02 \times 10^{23})\ 4.82}{2.02 \times 10^{-8}\ \text{cm}} \left(1 - \frac{1}{7}\right)$$

$$= 2.83 \times 10^{13}\ \text{erg} = 2{,}830\ \text{kJ/mole}$$

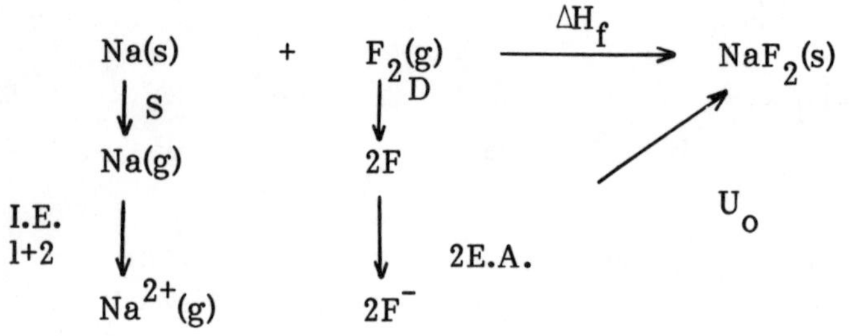

$\Delta H_f = U_o$+D+2E.A.+S+I.E.
= −2830+154−656+108+5058
= 1834

$NaF_2(s)$ NaF(s) + $1/2F_2$ ΔH = −575 − 1834 = −2409 kJ

6.16 Problem: Calculate the heats of formation of Ne^+F^- and Na^+Ne^-, estimating radii of Ne^+ and Ne^-. What factors prohibit the formation of these compounds in spite of favorable lattice energies?

6.16 Solution: Estimate r_{Ne^-} (slightly larger than F^-) and r_{Ne^+} (slightly smaller than Na^+). Hence U_o is approximately that of NaF (928) for NeF or NaNe.

$\Delta H_f(\text{NeF}) = U_o + 1/2D_{F_2}$ + E.A. + I.E. = −930 + 77 − 328 + 2081
= 900 kJ/mole (High I.E. prohibits formation)

Estimate E.A. for Ne as zero

$\Delta H_f(\text{NaNe}) = U_o$ + S + I.E. = −920 + 108 + 496 = −316 kJ/mole

NaNe should be thermodynamically stable from these assumptions, but Ne does not bind an electron.

6.17 Problem: From spectral data the dissociation energy of ClF has been determined to be 253 kJ/mole. The ΔH_f^o of ClF(g) is 50.6 kJ/mole. The dissociation energy of Cl_2 is 239 kJ/mole. Calculate the dissociation energy of F_2.

6.17 Solution:

		ΔH
	$ClF(g) \rightarrow Cl(g) + F(g)$	+253 kJ/mole
(1)	$1/2\ Cl_2(g) \rightarrow Cl(g)$	119.5
(2)	$1/2\ F_2(g) \rightarrow F(g)$	$1/2D_{F_2}$
(3)	$Cl(g) + F(g) \rightarrow ClF(g)$	-253
(4)	$1/2Cl_2(g) + 1/2F_2(g) \rightarrow ClF(g)$	$-50.6 = \Delta H_f$

Since equation (4) is the sum of (1), (2), and (3),

$$-50.6 = -253 + 119.5 + 1/2D_{F_2}$$

$$1/2D_{F_2} = 82.9 \qquad D_{F_2} = 166 \text{ kJ/mole}$$

6.18 Problem: Although the electron affinity of F is lower than that of Cl, F_2 is much more reactive than Cl_2. Account for the higher reactivity of F_2 (a) with respect to the formation of solid halides MX or MX_2 and (b) with respect to the formation of aqueous solutions of MX or MX_2.

6.18 Solution: (a) Less energy is gained for $F + e \rightarrow F^-$ (compared to Cl), but less energy is required to break the F-F bond, and U_o is high for the small F^-, making the formation of metal fluorides very favorable.

(b) The same considerations for the electron affinities and dissociation energies cited in (a) apply. Additional energy is gained from the much higher hydration energy of F^- compared to Cl^-.

6.19 Problem: Given the following crystal radii, explain the differences within each pair.

Crystal radii (C.N. 6)

Na^+	116 pm	S^{2-}	170 pm
Mg^{2+}	86	Cl^-	167

6.19 Solution: Each pair of ions is isoelectronic, so the Pauling univalent radii will be inversely proportional to the effective nuclear charge. Compared to the univalent radius, the crystal radius decreases with increasing ionic charge since a higher ionic charge leads to greater attraction between ions. The two effects operate in the same direction for Na^+ and Mg^{2+} but in opposite directions for S^{2-} and Cl^-. (See DMA pages 216-7 for a more complete discussion of univalent and crystal radii.)

6.20 Problem: If Ge is added to GaAs, the Ge is about equally distributed between the Ga and As sites. Which sites would the Ge prefer if Se is added also? Would GaAs doped with Se be an n-type or a p-type semiconductor?

6.20 Solution: Ge prefers Ga sites if Se is added, since Se occupies As sites, as Se^+ plus an electron, giving an n-type semiconductor.

6.21 Problem: (a) Under what circumstances with regard to relative sizes of ions and degree of nonpolar character are Frenkel and Schottky defects likely?
(b) The phenomenon of "half-melting" or solid electrolytes is related to which type of defect?

6.21 Solution: (a) Schottky defects, involving stoichiometric cation and anion vacancies, are likely for ionic compounds and ions of comparable size. Schottky defects minimize mechanical distortions and charge displacement. Frenkel defects, involving the migration of ions to interstitial sites, are more likely for more covalent compounds and for large (and highly polarizable) anions. Large, polarizable anions make it easier to accommodate the displaced cations and minimize repulsive effects.
(b) Solid electrolytes commonly have Frenkel defects, providing mobile cations. Half-melting refers to a temperature dependent transition, above which the cations are free to migrate.

6.22 Problem: What is the significance of the term "molecular weight" with respect to diamond or SiO_2 (considering a perfect crystal of each)?

6.22 Solution: A crystal is a molecule for diamond (C) and SiO_2. More properly, we should use formula weight, as for ionic compounds, since "molecular weight" has more specific meaning.

6.23 Problem: The triclinic feldspars form an isomorphous series involving replacement of Ca^{2+} for Na^+ from albite, $Na[AlSi_3O_8]$ to anorthite, $Ca[Al_2Si_2O_8]$. How can the series be isomorphous with changing ratio of Si/Al? Why do not K^+ and Ba^{2+} occur in this series?

6.23 Solution: The framework in each case consists of $[(Al,Si)_4O_8]$. The changing proportions of Al and Si are compensated by replacement of Ca^{2+} for Na^+. K^+ and Ba^{2+} are too large to replace Na^+ and Ca^{2+}. They form a separate series of feldspars.

6.24 Problem: Metals that are very malleable (can be beaten or rolled into sheets) and ductile (can be drawn into wire) have the ccp structure. Why are these characteristics favored for ccp rather than hcp?

6.24 Solution: For ccp there are slip planes along any cube diagonal. These occur along one direction only for hcp. Thus, for the ccp structure there are several ways to slide one layer over another without altering the structure.

6.25 Problem: Discuss the possible effects of extreme pressure on a metal (assume bcc at low P) and on a solid nonmetal.

6.25 Solution: The bcc structure is not close packed and only about 90% as dense as ccp and hcp structures. At high pressure the bcc structure would be expected to convert to a more dense close-packed structure (LeChatelier's Principle). Most solid nonmetals have crystal structures that accommodate specific bonding interactions. These are not likely to be altered by pressure effects, at least in some general predictable way. Some nonmetals form molecular crystals (e.g., N_2, O_2, and S_8). In these cases commonly, the molecules are close packed and a structure change with increasing pressure is not expected.

6.26 Problem: Two of the three structures encountered for metals are ccp and hcp. How can the body-centered cubic structure be described in terms of a close-packed structure?

6.26 Solution: The bcc structure is similar to the CsCl structure, but with the same atoms in all positions. It can be described as 3·2PTOT with metal atoms in all P, T, and O sites. The index is written as 3·2 to indicate that there are 6 layers in the repeating unit, based on a ccp arrangement. The sequence is $(P_ATO_CTP_BT)(O_ATP_CTO_BT)$ with six layers in the repeating unit because for ccp P and O positions are equivalent and here they are occupied by the same atoms.

6.27 Problem: Sketch the curves for the distribution of energy states and their electron populations for a metallic conductor, an insulator, and a semiconductor.

6.27 Solution:

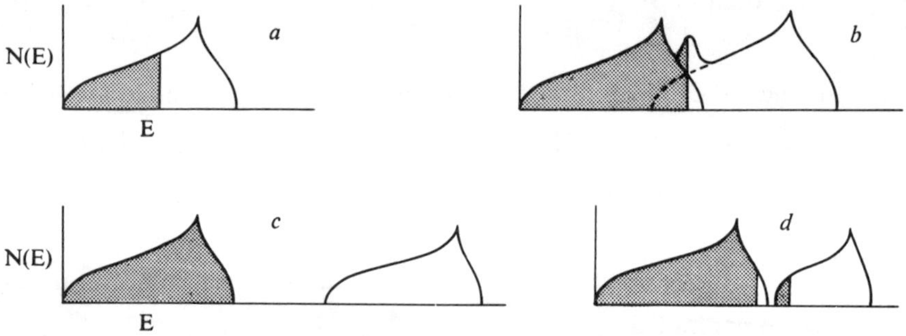

Distribution of energy states. The shading indicates occupied states.

For metals a conducting band is partially filled (a) or a filled band overlaps an empty band (b). The filled and empty bands are well separated for an insulator (c), but the energy levels are close enough for promotion in the case of a semi-conductor (d).

6.28 Problem: Soft metals become "work-hardened." Explain this and how the softness is restored by heating the metal.

6.28 Solution: Working (bending, etc.) disrupts the slip planes, particularly when puckered layers (not corresponding to slip planes) slide over one another. The original structure can be restored by heating (annealing) below the melting point.

6.29 Problem: Consider the sliding of one layer of a crystal over another until an equivalent arrangement is achieved. What are the consequences with respect to hardness, brittleness, and malleability for (1) a metal, (2) an ionic crystal, (3) a covalent crystal, and (4) a molecular crystal. In a crude way, how do these considerations relate to melting point?

6.29 Solution: (1) Metal - all atoms are the same and the close-packed layers or slip planes can slide over one another. Bonding is weakened as the atoms slide from the original P sites, but the same situation is restored by 1 atom displacement. Metals are malleable, particularly those with ccp structures (see Problem 6.24). Hardness and melting points vary with bond strength.

(2) Ionic Crystal - hard but brittle. Great force is required to separate cations from anions, but displacement brings anions into contact and cations into contact, causing the crystal to cleave or shatter.

(3) Covalent Crystal - hard and brittle. Displacement requires the breaking of covalent bonds that are not easily broken or restored.

(4) Molecular Crystal - soft and malleable. There is weak bonding between molecules so they can slide over one another. There is no strong repulsion during deformation.

Ionic crystals and covalent crystals (e.g. diamond and quartz) have high melting points because of strong "bonding." Molecular crystals are low melting because of weak interaction between molecules. Their melting points depend roughly on molecular weights since van der Waals interactions increase with increasing numbers of electrons. Metals vary from low melting points (Hg is a liquid) to extremely high melting points for metals such as W. There are great variations in bond strengths for metals.

6.30 Problem: The Hume-Rothery phases or electron compounds are characterized by electron/atom ratios of 21/14 for bcc, 21/13 for complex cubic, and 21/12 for hcp. What structures are expected for the following:

$CuZn$, Cu_5Zn_8, and $CuZn_3$ (brass phases)
Cu_3Sn, Cu_5Sn, and $Cu_{31}Sn_8$

6.30 Solution: The number of electrons assigned to a metal atom is the group number, except that group VIII metals are assigned zero electrons.

Alloy	Electrons/atoms	
$CuZn$	3/2 = 21/14	bcc
Cu_5Sn	9/6 = 21/14	bcc
Cu_5Zn_8	21/13	complex cubic
$Cu_{31}Sn_8$	63/39 = 21/13	complex cubic
$CuZn_3$	7/4 = 21/12	hcp
Cu_3Sn	7/4 = 21/12	hcp

6.31 Problem: In a solid electrolyte what are the advantages of an hcp array of anions over a ccp array of anions? Assume that the metal ions are in octahedral holes.

6.31 Solution: With the anions in close-packed positions, the octahedral sites are lined up at C positions for hcp, but they are staggered (A, B, and C positions) for ccp. The motion of cations would be less impeded in the C channels for hcp.

6.32 Problem: The compound MX has a complex structure with the following 8 layer repeating sequence:

$$\begin{array}{cccc} A & B & C & B \\ PT_- & PT_+ & PT_- & PT_+ \end{array}$$

All of the indicated T sites are occupied by X atoms. The positions of the packing layer sites are indicated. What positions do the T sites occupy? What is the coordination number of the M atoms of layer A? Of layer B? Of layer C?

6.32 Solution: The T_- tetrahedra point down, the T_+ tetrahedra point up, the bottom packing layer is A. The T sites, from bottom to top, are thus above A, below C, above C, and below A. Since each M atom in an A site has an X atom directly above or below it, its

C.N. is 2; the same is true of the M atoms in C sites. Each M atom in a B site will have three X atoms (in C and A positions) on either side of the B layer, giving a C.N. of 6.

	A	A	A	B	C	C	C	B	A	A
	T_+	P	T_-	P	T_+	P	T_-	P	T_+	P
C.N.		2		6		2		6		

6.33 Problem: The nickel arsenide structure is 4PO, while the cesium iodide structure is $2\cdot\frac{3}{2}$PPO.

(a) Describe each of these structures.

(b) Several pairs of elements (e.g., Ni, Te and Co, Te) form a series of nonstoichiometric compounds in which the limiting composition MX has the NiAs structure and that of limiting composition MX_2 has the cadmium iodide structure. Explain this observation.

6.33 Solution: (a) The NiAs structure consists of an hcp (ABABA...) array of As with octahedral holes filled by Ni. The CdI_2 structure consists of an hcp array of I with half of the octahedral holes filled by Cd. That is, every other layer of octahedral holes is vacant.
(b) When the composition is intermediate between MX and MX_2, the octahedral holes empty in the CdI_2 structure can be occupied partially by M. When all are occupied, the stoichiometry will be MX and the NiAs structure is reached.

6.34 Problem: The $CdCl_2$ structure is a 9PPO structure with Cd occupying the octahedral sites between close packed layers of Cl. $AlCl_3$ may be described as a $9PPO_{2/3}$ structure. Sketch an $AlCl_3$ POP layer showing systematic absences of Al in the O sites. Al_2Br_6 has a $2\cdot6PT_{1/6}T_{1/6}$ structure. What is the coordination number of Al in each of these compounds and what is responsible for the difference in C.N. observed?

6.34 Solution: Our planar unit cell should be chosen in such a fashion that 2/3 times the number of octahedral sites (C positions) contained is equal to a whole number. Further, to preserve the hexagonal symmetry of the close packed POP layer, the $\underline{a}$ distance should equal the $\underline{b}$ distance (note however, that the hexagonal symmtery need not be preserved by the partially occupied sites). These restrictions lead us to start with a planar unit cell having nine octahedral sites, suggesting the structure shown below (planar unit cell shown with darker lines and "bonds" to the Cl atoms in the A and B packing layer sites). The larger Br atom decreases the cation-to-anion ratio, resulting in a change in C.N. from 6 to 4 in going from $AlCl_3$ to Al_2Br_6.

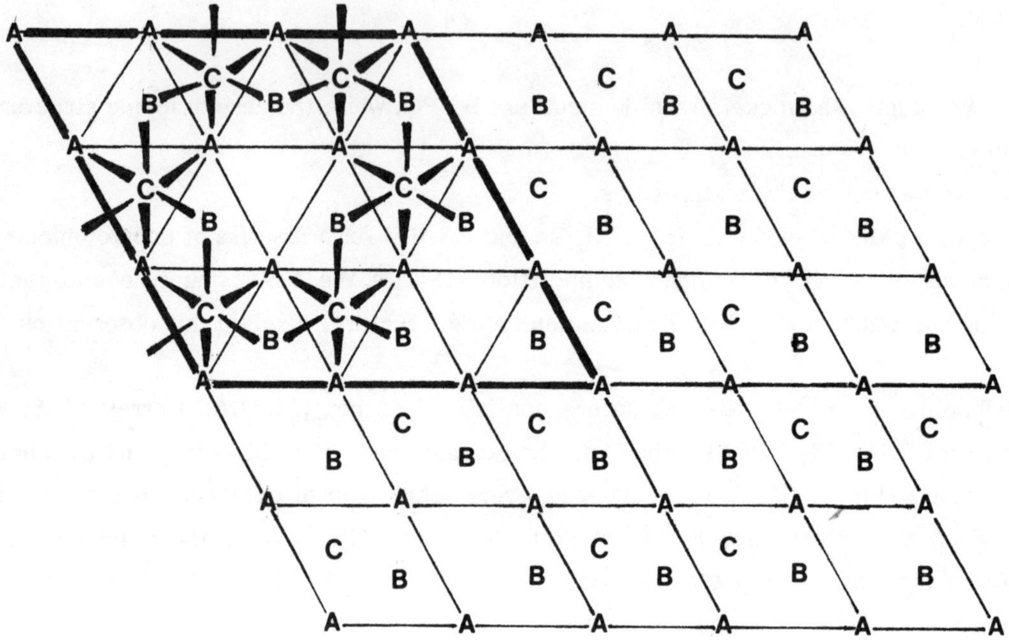

VII Coordination Compounds—Bonding and Spectra

7.1 Problem: Which of the following complexes obey the rule of 18 (EAN rule)?

(a) $Cu(NH_3)_4^{2+}$, $Cu(en)_3^{2+}$, $Cu(CN)_4^{3-}$

(b) $Ni(NH_3)_6^{2+}$, $Ni(CN)_4^{2-}$, $Ni(CO)_4$

(c) $Co(NH_3)_6^{3+}$, $CoCl_4^{2-}$

(d) $Fe(CN)_6^{3-}$, $Fe(CN)_6^{4-}$, $Fe(CO)_5$

(e) $Cr(NH_3)_6^{3+}$, $Cr(CO)_6$

7.1 Solution: To obey the rule of 18 (EAN), the sum of the number of electrons in the metal valence shell [(n-1)dnsnp] and those donated by the ligands should be 18, filling the low-energy metal orbitals.

(a)	$Cu(NH_3)_4^{2+}$	9 + 8 = 17 e	$Cu(en)_3^{2+}$	9 + 12 = 21 e
	$Cu(CN)_4^{3-}$	10 + 8 = <u>18 e</u>		
(b)	$Ni(NH_3)_6^{2+}$	8 + 12 = 20 e	$Ni(CN)_4^{2-}$	8 + 8 = 16 e
	$Ni(CO)_4$	10 + 8 = <u>18 e</u>		
(c)	$Co(NH_3)_6^{3+}$	6 + 12 = <u>18 e</u>	$CoCl_4^{2-}$	7 + 8 = 15 e
(d)	$Fe(CN)_6^{3-}$	5 + 12 = 17 e	$Fe(CN)_6^{4-}$	6 + 12 = <u>18 e</u>
	$Fe(CO)_5$	8 + 10 = <u>18 e</u>		
(e)	$Cr(NH_3)_6^{3+}$	3 + 12 = 15 e	$Cr(CO)_6$	6 + 12 = <u>18 e</u>

7.2 Problem: Determine the number of unpaired electrons and the LFSE for each of the following.

(a) $Fe(CN)_6^{4-}$

(b) $Fe(H_2O)_6^{3+}$

(c) $Co(NH_3)_6^{3+}$

(d) $Cr(NH_3)_6^{3+}$

(e) $Ru(NH_3)_6^{3+}$

(f) $PtCl_6^{2-}$

(g) $CoCl_4^{2-}$ (tetrahedral)

7.2 Solution: For an octahedral field, each t_{2g} electron is stabilized by -4 Dq and each e_g electron is destabilized by +6 Dq. For a tetrahedral field t_2 electrons are destabilized by +4 Dq and e electrons are stabilized by -6 Dq, but Dq is smaller for the tetrahedral case.

		Unpaired electrons	LFSE			Unpaired electrons	LFSE
(a)	t_{2g}^6	0	-24 Dq	(e)	t_{2g}^5	1	-20 Dq
(b)	$t_{2g}^3e_g^2$	5	0 Dq	(f)	t_{2g}^6	0	-24 Dq
(c)	t_{2g}^6	0	-24 Dq	(g)	$e^4t_2^3$	3	-12 Dq
(d)	t_{2g}^3	3	-12 Dq				

7.3 Problem: Discuss briefly the factors working for and against the maximum spin state of d electrons in transition metal complexes.

7.3 Solution: Maximum (high) spin is favored by maximum charge delocalization and exchange energy. The low-spin case is favored by maximum LFSE.

7.4 Problem: Explain why square planar complexes of transition metals are limited (other than those of planar ligands such as porphyrins) to those of (a) d^7, d^8, and d^9 ions and (b) very strong field ligands which can serve as π acceptors.

7.4 Solution: (a) These are the cases for maximum LFSE for the square planar case. The most important feature of the energy level diagram shown is one orbital of very high energy, so that d^8 complexes are diamagnetic.

(b) Only very strong field ligands, particularly π acceptors, give sufficiently strong fields to cause spin pairing for d^7 and d^8.

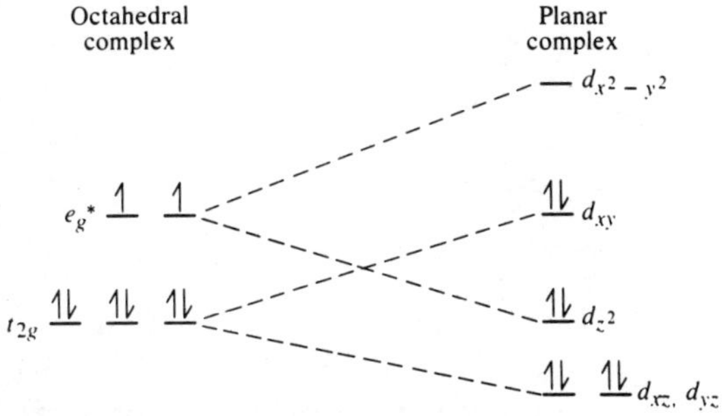

Energy levels for square planar nickel(II) complexes.

7.5 Problem: For which d^n configurations would no Jahn-Teller splitting be expected for the tetrahedral case (ignore possible low-spin cases).

7.5 Solution: d^2, e^2; d^5, $e^2t_2^3$; and d^{10}, $e^4t_2^6$

Distortion of the tetrahedron removes the degeneracy of the d orbitals as shown in Problem 7.6. Without spin pairing, no gain in energy will result for d^2 since the d_{z^2} and $d_{x^2-y^2}$ orbitals will have one electron each. No gain in energy results for high-spin d^5 or for d^{10} since each orbital is singly occupied (d^5) or filled (d^{10}).

7.6 Problem: Give the orbital occupancy (identify the orbitals) for the Jahn-Teller splitting expected for tetrahedral complexes with high-spin d^3 and d^4 configurations. Indicate the nature of the distortions expected.

7.6 Solution: For d^3 the tetrahedron should be elongated along z. This causes the orbitals in the xy plane derived from e and t_2 to have lower energy (the ligands are farther from the xy plane). One t_2 orbital (d_{xy}) has lower energy. For d^4 the tetrahedron should be flattened along z. This raises the energy of the orbitals in the xy plane and lowers the others. Two electrons are accommodated in the two lower energy orbitals from t_2.

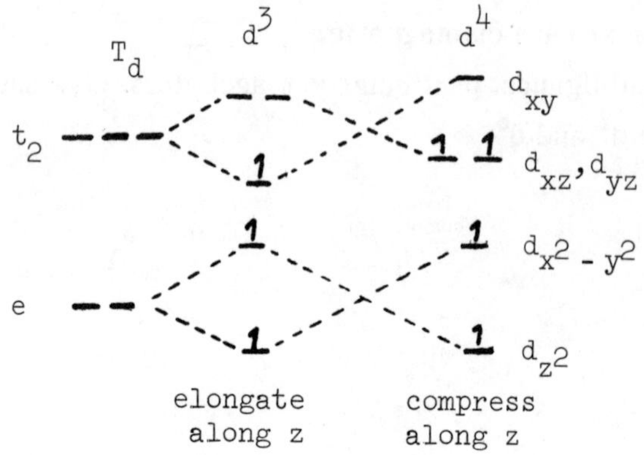

7.7 Problem: (a) Why are low-spin complexes usually not encountered for tetrahedral coordination? (b) Octahedral splitting is expressed as 10 Dq. What should be the splitting for ML_8 with cubic coordination? Assume the same ligands at the same distance as for the octahedral and tetrahedral cases.

7.7 Solution: (a) The LFSE is small for the tetrahedral case since there is small splitting. There are four ligands instead of six for the octahedron and the ligands do not approach along the directions of the d orbital lobes.
(b) Tetrahedral splitting (4L) is $(4/9)(10\ Dq)_{oct}$. Cubic splitting (8L) is twice as great or $(8/9)(10\ Dq)_{oct}$. The cube consists of two interpenetrating tetrahedra.

7.8 Problem: Negative ions might be expected to create stronger ligand fields than neutral molecules. Explain why OH^- is a weaker-field ligand than H_2O. Why is CO such a strong-field ligand?

7.8 Solution: OH^- is a stronger π donor than H_2O and π donation decreases 10 Dq because the metal electrons must be in the antibonding π^* orbitals (see Figure). CO is a π acceptor, the metal t_{2g} orbitals are lowered in energy since they are bonding and electrons are delocalized onto the ligands, increasing 10 Dq.

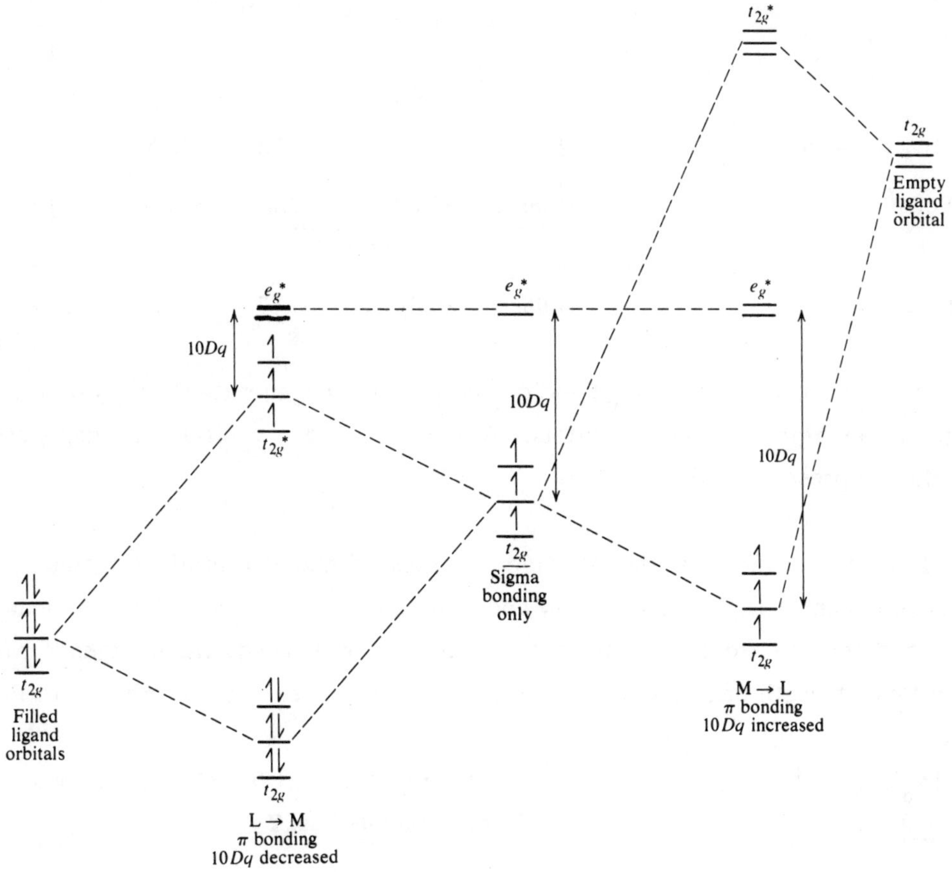

Comparison of the effects of π bonding using *(a)* filled low-energy π ligand orbitals for L → M donation and *(b)* empty ligand orbitals of π symmetry for M → L donation.

7.9 Problem: Calculate the relative energies of the d orbitals for an ML_6 complex with trigonal prismatic coordination (D_{3h}), assuming that the ligands are at the same angle relative to the xy plane as for a regular tetrahedron. (Hint: Start with the tetrahedral case, but allow for three ligands up and down, instead of two. The degeneracy of the d orbitals is the same as for other D_{3h} complexes, such as trigonal, ML_3, and trigonal bipyramidal, ML_5, cases. (See DMA p. 275 or R. Krishnamurthy and W.B. Schaap, J. Chem. Educ. 1969, 46, 799.)

7.9 Solution: We start with the tetrahedral case, 2 ligands above the xy plane and 2 below and multiply by $\frac{3}{2}$ ($\frac{6}{4}$) to give the effect of 3 ligands above the 3 below the xy plane. The relative energies of the d orbitals are given by:

	d_{z^2}	$d_{x^2-y^2}$	d_{xy}	d_{xz}	d_{yz}
ML_4 tetrahedron	-2.67	-2.67	1.78	1.78	1.78 Dq
For 6L $(x\,\frac{3}{2})$	-4.00	-4.00	2.67	2.67	2.67 Dq
		average (of -4.00 and 2.67)			
ML_5 D_{3h}	-4.00	-0.67	-0.67	2.67	2.67 Dq

Since for D_{3h} symmetry the $d_{x^2-y^2}$ and d_{xy} orbitals are degenerate (E' representation) we average the energies for these orbitals. The d_{xz} and d_{yz} orbitals are degenerate as required (E" representation).

7.10 Problem: Calculate the relative energies of the d orbitals for the following complexes, assuming Dq(X) = 1.40 Dq(Y) and that where X and Y are along the axis (trans to one another), the field strength is the same as for two equivalent ligands with the average field strength of X and Y. Use z as the unique axis. (See references in Problem 7.9).

(a) MX_5Y
(b) trans-$[MX_4Y_2]$
(c) cis-$[MX_4Y_2]$ (both Y's in x,y plane)
(d) facial-$[MX_3Y_3]$

7.10 Solution: In each case we can start with an octahedral MX_6 complex, subtract the effects of the missing ligands and add the effects of the Y ligands, or just add the components of the particular case individually.

Dq(X) = 1.40 Dq(Y) or Dq(Y) = 0.714 Dq(X)

(a) MX_5Y Oct. $MX_6 - (X)_z + (Y)_z$

	d_{z^2}	$d_{x^2-y^2}$	d_{xy}	d_{xz}	d_{yz}
Oct. MX_6	6.00	6.00	-4.00	-4.00	-4.00 Dq
$-(X)_z$	-5.14	3.14	3.14	-0.57	-0.57 Dq
$+(Y)_z$	3.67	-2.24	-2.24	0.41	0.41 Dq
MX_5Y	4.53	6.90	-3.10	-4.16	-4.16 Dq(X)

(b) trans-$[MX_4Y_2]$ Oct. $MX_6 - 2(X)_z + 2(Y)_z$ or Square planar $MX_4 + 2(Y)_z$

	d_{z^2}	$d_{x^2-y^2}$	d_{xy}	d_{xz}	d_{yz}
Sq. MX_4, $2xML_2$	-4.28	12.28	2.28	-5.14	-5.14
$+ 2(Y)_z$	7.35	-4.49	-4.49	0.82	0.82
trans-$[MX_4Y_2]$	3.07	7.79	-2.21	-4.32	-4.32 Dq(X)

(c) cis-$[MX_4Y_2]$ Oct. $MX_6 -2(X)_{xy} + 2(Y)_{xy}$ or Linear $MX_2 + 2\ (X)_{xy} + 2(Y)_{xy}$

	d_{z^2}	$d_{x^2-y^2}$	d_{xy}	d_{xz}	d_{yz}
Linear MX_2	10.28	-6.28	-6.28	1.14	1.14
MX_2	-2.14	6.14	1.14	-2.57	-2.57
MY_2	-1.53	4.38	0.81	-1.83	-1.83
cis-$[MX_4Y_2]$	6.61	4.24	-4.33	-3.26	-3.26 Dq(X)

(d) fac-$[MX_3Y_3]$ $(X)_z + (Y)_z + 2(X)_{xy} + 2(Y)_{xy}$

		d_{z^2}	$d_{x^2-y^2}$	d_{xy}	d_{xz}	d_{yz}
MX(z)		5.14	-3.14	-3.14	0.57	0.57
MY (z)	0.714 ×	(5.14	-3.14	-3.14	0.57	0.57)
MX_2 (xy)		-2.14	6.14	1.14	-2.57	-2.57
MY_2(xy)	0.714 ×	(-2.14	6.14	1.14	-2.57	-2.57)
fac-$[MX_3Y_3]$		5.14	5.14	-3.43	-3.43	-3.43 Dq(X)

The X-Y ligands are trans along each axis and the effective field is the average $\frac{Dq(X) + Dq(Y)}{2}$ = 0.86 Dq(X), so the splitting pattern is the same as for the O_h case.

7.11 Problem: Compare the observed energy splittings in the spectra of some Cr(III) or Co(III) complexes of the types given in problem 7.10 with the calculated splittings. Some pertinent spectra are given on page 287 of DMA.

7.11 Solution: In problem 7.10 you calculated energies for d orbitals in some geometries of symmetry lower than O_h. The splitting of the degeneracy of the e_g ($d_{x^2-y^2}$ and d_{z^2}) orbitals in lower symmetries can lead to an increase in the number of d→d bands as compared to the octahedral case. Replacement of one, two or three ligands in MX_6 lowers the symmetry from O_h (cubic). The effective symmetry can be regarded as "tetragonal" when the symmetry is lowered so that the average ligand field along one axis (z) differs from that along the x and y axes. Tetragonal complexes include MX_5Y, _trans_-[MX_4Y_2] and _cis_-[MX_4Y_2] (X-X along z and X-Y along x and y). In these terms _facial_-[MX_3Y_3] is "cubic" having X-Y along all three axes.

Energy difference	$d_{x^2-y^2} - d_{z^2}$
(a) MX_5Y	2.37 Dq
(b) _trans_-[MX_4Y_2]	4.72 Dq, approximately 2 times (a)
(c) _cis_-[MX_4Y_2]	-2.37 Dq, approximately - (a) or - 1/2 (b)
(d) _fac_-[MX_3Y_3]	0, no splitting, the "cubic" case

The usual observation is that splitting of the lower energy band is about twice as great for _trans_-[MX_4Y_2] compared to [MX_5Y] or _cis_-[MX_4Y_2], and no splitting is observed for _facial_-[MX_3Y_3].

7.12 Problem: Calculate the relative energies of the d orbitals (see references in Problem 7.9) for ML_5 in the TBP (D_{3h}) and SP (C_{4v}) geometries. For high-spin d^6-d^9 metal ions, for which case is there a significant preference for one of the two geometries?

7.12 Solution: The energies of the d orbitals for the planar trigonal case are 3/2 times the energies for the ML_2 case with two ligands at 90°, but allowing for the degeneracy of the $d_{x^2-y^2}$ and d_{xy} orbitals and the d_{xz} and d_{yz} orbitals (as required for D_{3h} symmetry):

	d_{z^2}	$d_{x^2-y^2}$	d_{xy}	d_{xz}	d_{yz}
ML_3 ($\frac{3}{2}$ x ML_2)	-3.21	9.21	1.71	-3.85	-3.85 Dq
		average		already degenerate	
Planar ML_3, D_{3h}	-3.21	5.46	5.46	-3.85	-3.85 Dq
Linear ML_2	10.28	-6.28	-6.28	1.14	1.14 Dq
TBP, ML_5, D_{3h}	7.07	-0.82	-0.82	-2.71	-2.71 Dq

The square pyramidal case is the sum of the energies for a square planar arrangement (2 x ML_2 at 90°) plus the contribution for one ligand along z:

M-L (z)	5.14	-3.14	-3.14	0.57	0.57 Dq
M-L (planar) (2 x ML_2, 90°)	-4.28	12.28	2.28	-5.14	-5.14 Dq
ML_5 (SP, C_{4v})	0.86	9.14	-0.86	-4.57	-4.57 Dq

	LFSE (TBP)	LFSE (SP)
d^6	-2.70 Dq	-4.57 Dq
d^7	-5.41	-9.14
d^8	-6.23	-10.00
d^9	-7.05	-9.14

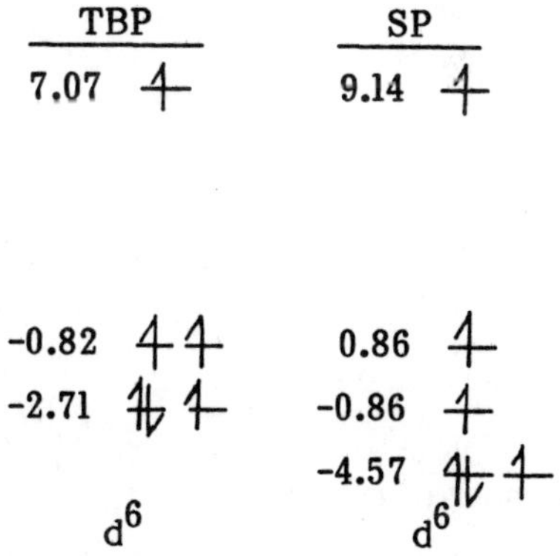

For d^6

TBP -3 (2.71)-2 x 0.82 + 7.07 = -2.70

SP -3 (4.57) - 0.86 + 0.86 + 9.14 = -4.57

For all cases the LFSE is greater for the SP, but particularly for d^7 and d^8. The ligand-ligand repulsion is greater for the SP.

7.13 Problem: Identify the ground state with the spin multiplicity for the following cases in (a) octahedral complexes and (b) tetrahedral complexes:

$$Cu^{2+}, V^{3+}, Cr^{3+}, Mn^{2+}, Fe^{2+}, \text{ and } Ni^{2+}$$

7.13 Solution: We can get the splitting patterns for D and F terms from the Orgel diagrams (here and Problem 7.16) for d^2, d^3, d^8, d^9, and high-spin d^6. For low-spin d^6 (O_h) the six electrons are in the t_{2g} orbitals. The filled t_{2g} orbitals are totally symmetric, $^1A_{1g}$. For low-spin d^5 (O_h) the configuration is t_{2g}^5. There is one vacancy (1 hole) in t_{2g} and the energy state is $^2T_{2g}$ for the ground state. Cu^{2+} (d^9) has the ground state term ^2D, giving ^2E_g for O_h and $^2T_2$ for T_d as ground states. V^{3+} (d^2) has the ground state term ^3F, giving $^3T_{1g}$ for O_h and $^3A_2$ for T_d as ground states.

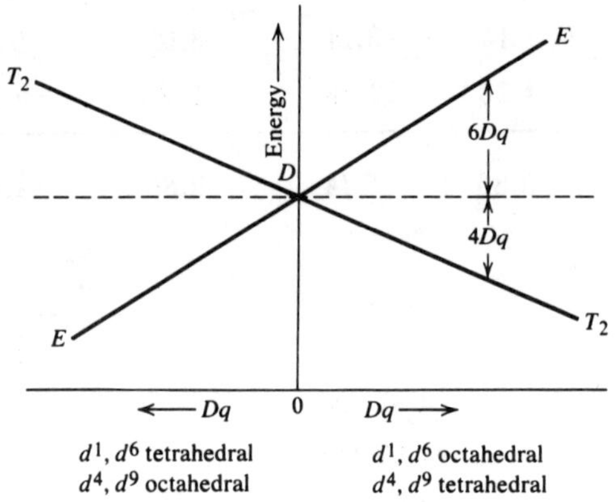

Orgel *D* term splitting diagram.

	Cr^{3+}	Ni^{2+}	Mn^{2+}		Fe^{2+}	
Number of d electrons	3	8	5		6	
Free ion ground state	^4F	^3F	^6S		^5D	
			Hi-spin	Lo-spin	Hi-spin	Lo-spin
O_h	$^4A_{2g}$	$^3A_{2g}$	$^6A_{1g}$	$^2T_{2g}(t_{2g}^5)$	$^5T_{2g}$	$^1A_{1g}(t_{2g}^6)$
T_d	$^4T_1$	$^3T_1$	$^6A_1$		^5E	

Low-spin T_d cases are unlikely.

7.14 Problem: Explain why the ligand field (d-d) bands are shifted only slightly for the $[Co(NH_3)_5X]^{2+}$ ions for different halides, but the charge-transfer bands are shifted greatly.

7.14 Solution: The LF bands shift slightly because the field strengths of the halides differ little. The energy of the charge-transfer band depends on the ease of loss of an electron from X^- (oxidation), and there are great differences in ease of oxidation from F^- to I^-.

7.15 Problem: Common glass used for windows and bottles appears colorless, but when viewed through the edge it appears faintly green. Fe^{3+} causes the color. Why is it so faintly colored? Would you expect one broad, very weak absorption peak or many weak peaks?

7.15 Solution: Fe^{3+} (high-spin d^5) has only spin-forbidden transitions of low intensity. There are many spin-forbidden transitions. There is only one way to place 5 unpaired electrons in the d orbitals, but many ways for spin pairing, giving many energy states.

7.16 Problem: For $Cr(NH_3)_6^{3+}$ there are two absorption bands observed at 21,500 cm^{-1} and 28,500 cm^{-1} and a very weak peak at 15,300 cm^{-1}. Assign the bands and account for any missing spin-allowed bands. Calculate Dq for NH_3 using the Orgel diagram. Account for any discrepancy between the observed position of any of the spin-allowed bands and that expected from the Orgel diagram.

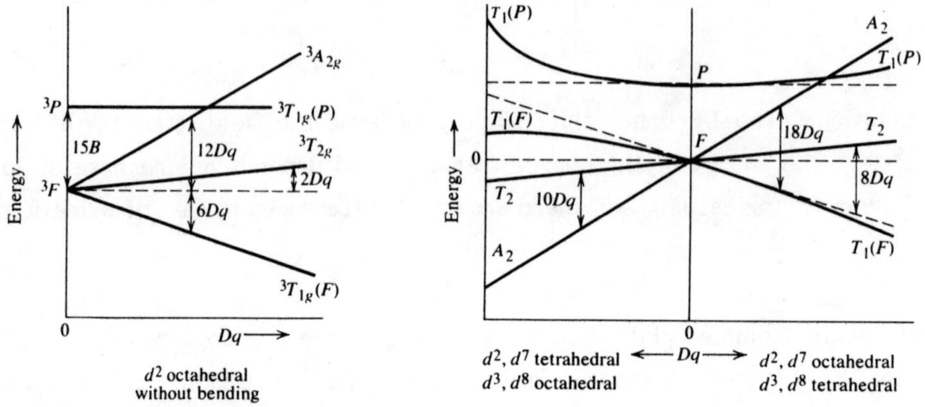

Orgel term splitting diagrams for d^2, d^3, d^7, and d^8 cases in octahedral and tetrahedral fields. (Adapted with permission from L. E. Orgel, *J. Chem. Phys.* 1955, *23*, 1004. Copyright 1955, American Chemical Society.)

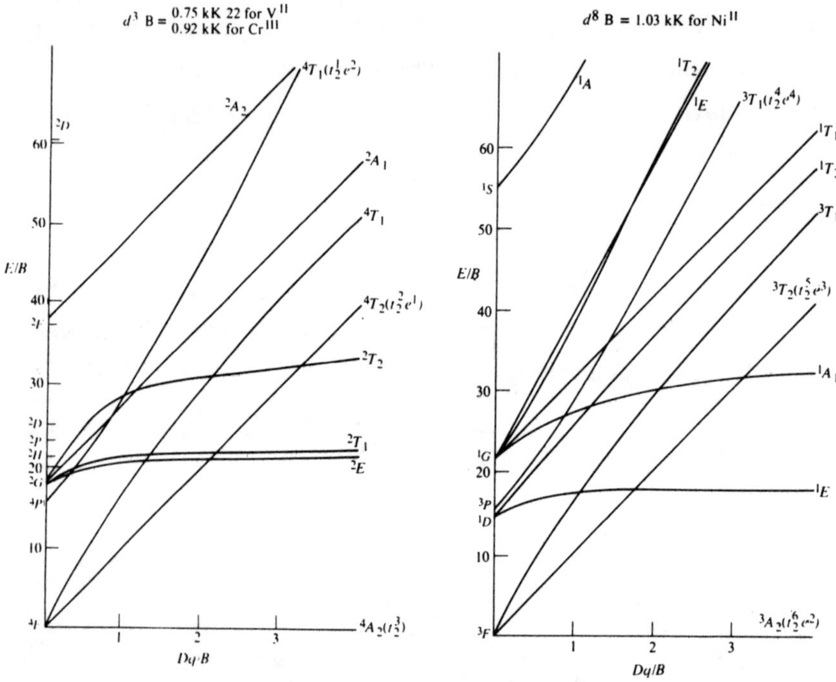

Semiquantitative energy-level diagrams for octahedral symmetry.

(After Y. Tanabe and S. Sugano, *J. Phys. Soc. Japan* 1954, *9*, 753.)

7.16 Solution:

O_h	ν_1	21,500 cm^{-1}	${}^4A_{2g} \rightarrow {}^4T_{2g}$ (lower energy)	Cr^{III} is d^3
	ν_2	28,500 cm^{-1}	${}^4A_{2g} \rightarrow {}^4T_{1g}(F)$	

${}^4A_{2g} \rightarrow {}^4T_{1g}$ (P) can be assumed to be too high in energy to be observed. The weak peak is spin-forbidden. ν_1 is 10 Dq or Dq = 2,150 cm^{-1}. ν_2 is 18 Dq, but lowered because of bending (configuration interaction). 18 Dq = 38,700 cm^{-1}, but 28,500 is observed. The weak low-energy band must be ${}^4A_{2g} \rightarrow {}^2E_g$, ${}^2T_{1g}$ (unresolved). The ${}^4A_{2g} \rightarrow {}^2T_{2g}$ peak would be between the spin-allowed bands and does not appear.

7.17 Problem: The following data are available for $Ni(H_2O)_6^{2+}$ and $Ni(NH_3)_6^{2+}$.

$Ni(H_2O)_6^{2+}$	$Ni(NH_3)_6^{2+}$	
8,600 cm^{-1}	10,700 cm^{-1}	
13,500	17,500	
25,300	28,300	
15,400	15,400	Very weak peaks for
18,400	18,400	both complexes

Assign the bands. Calculate 10 Dq and the expected positions of the spin-allowed bands. Account for any discrepancy between the experimental and calculated energies of the bands. Account for the relative positions of corresponding bands for the two complexes.

7.17 Solution:

d^8	$Ni(H_2O)_6^{2+}$	$Ni(NH_3)_6^{2+}$		
10 Dq = ν_1	8,600	10,700 cm^{-1}	${}^3A_{2g}$	$\rightarrow {}^3T_{2g}$
ν_2	13,500	17,500		$\rightarrow {}^3T_{1g}(F)$
ν_3	25,300	28,300		$\rightarrow {}^3T_{1g}(P)$
	15,400	15,400		$\rightarrow {}^1E_g$
	18,400	18,400		$\rightarrow {}^1A_{1g}$

Dq	860	1,070	
B'	ca. 800	ca. 800	(about 0.8 of the value 1.03 kK from the Tanabe-Sugano diagram for d^8.)
ν_2 (18 Dq)	15,500	19,300	(bending neglected)
ν_3(12 Dq+15B')	22,300	24,800	(bending neglected)

The discrepancies from observed values result from bending (configuration interaction). The bands for NH_3 complex are at higher energy because of the higher field strength. The energies of the spin-forbidden transitions are independent of Dq because these involve pairing of the electrons in the e_g orbitals, not promotion from one set to another.

7.18 Problem: Interpret the following comparisons of intensities of absorption bands for transition metal complexes.

(a) Two isomers of a Co(III) complex believed to be cis- and trans-isomers give the following spectral features:

> Both give two absorption bands in the visible region. Complex A has two symmetrical bands with ε = 60-80. The lower energy band for B is broad with a possible shoulder and has lower intensity. Assign the isomers. Explain.

(b) An octahedral complex of Co(III), with an amine and Cl^- coordinated, gives two bands with ε = 60-80, one very weak peak with ε = 2 and a high energy band with ε = 20,000. What is the presumed nature of these transitions? Explain.

(c) Two complexes of Ni(II) are believed to be octahedral and tetrahedral. Each has three absorption bands, but complex A has $\varepsilon \cong 10$ and B has $\varepsilon \cong 150$. Which probably is the tetrahedral complex? Explain. Measurement of what physical property would exclude the possibility of either complex being square planar?

7.18 Solution: (a) A is the cis isomer because there is smaller splitting (see Problem 7.11) and higher intensity (intensity is usually higher for lower symmetry). B is the trans isomer — larger splitting and lower intensity.

(b) We expect 2 spin-allowed bands with $\varepsilon < 100$. The very weak peak is expected to be spin-forbidden and the very intense band is allowed, presumably a charge-transfer band involving a Cl $\rightarrow$Co transition.

(c) B is expected to be the tetrahedral complex — higher intensity for the non-centrosymmetric complex. Measurement of magnetic susceptibility would distinguish diamagnetic complexes of Ni^{2+} from paramagnetic octahedral or tetrahedral complexes.

7.19 Problem: For a square planar complex with the ligands lying along the x and y axes, indicate all of the metal orbitals that may participate in σ bonds. Sketch one ligand group orbital that could enter into a σ_g molecular orbital and one that could enter into a σ_u molecular orbital. Repeat the above, this time for π bonds.

7.19 Solution: The σ orbitals of a metal in a square planar complex are those directed along the x and y axes, including the spherical s and d_{z^2} with the donut in the xy plane: p_x, p_y, s, $d_{x^2-y^2}$, and d_{z^2}. The ligand group orbitals are those with symmetry corresponding to these orbitals:

σ Group Orbitals

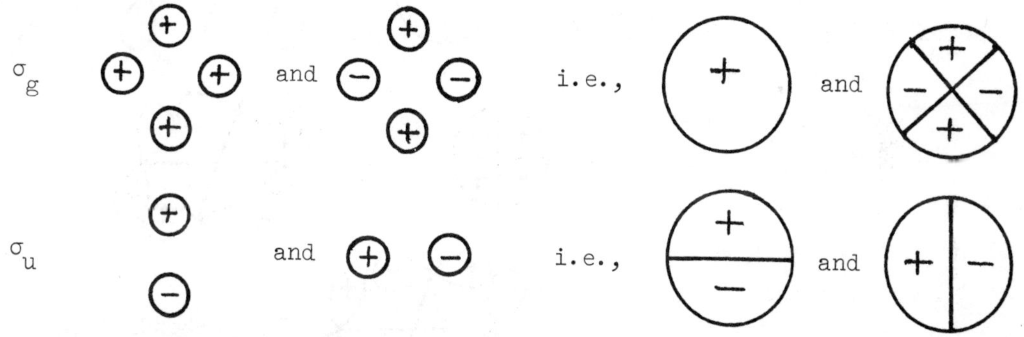

The π orbitals of a metal in a square planar complex are those with opposite signs for the lobes in planes perpendicular to the σ bonds: p_z, d_{xy}, d_{xz}, d_{yz}, p_x, and p_y.

π Group Orbitals

π (vertical)

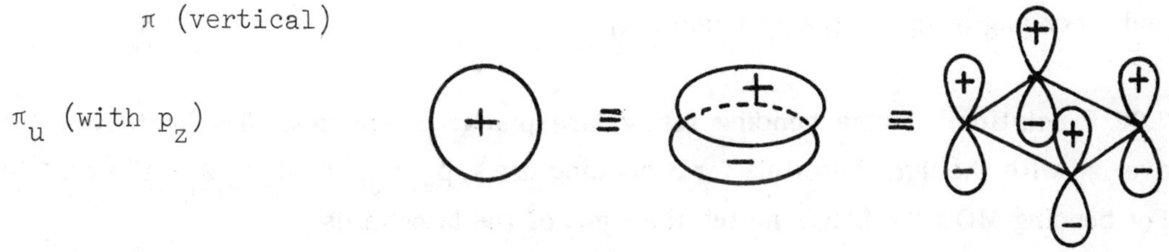

π_g (with d_{xz} and d_{yz})

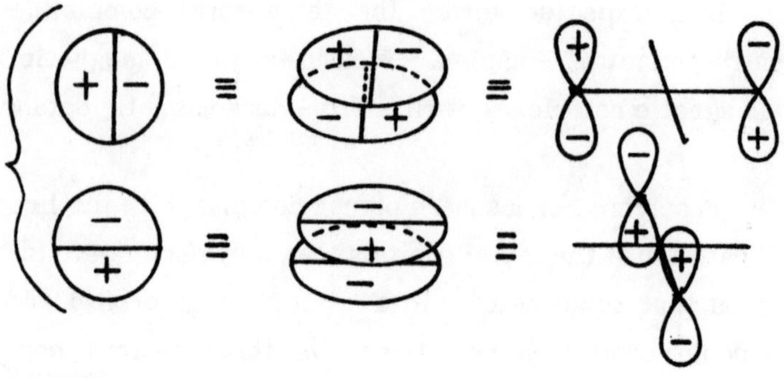

π_u (nonbonding LGO)

$\equiv p_1 - p_2 + p_3 - p_4$ of ligands

π (horizontal)

π_u (with p_x and p_y)

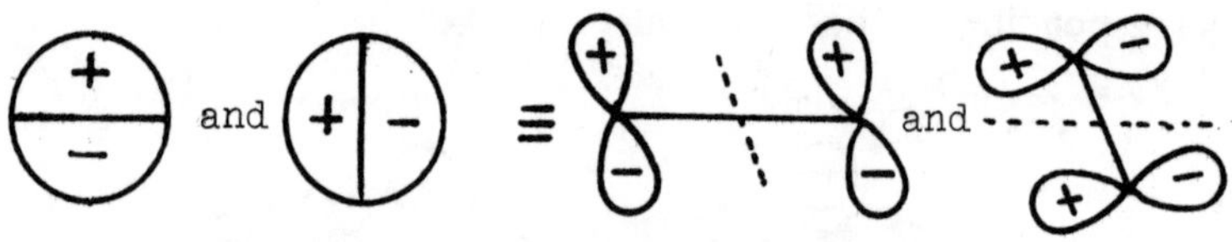

π_g (with d_{xy})

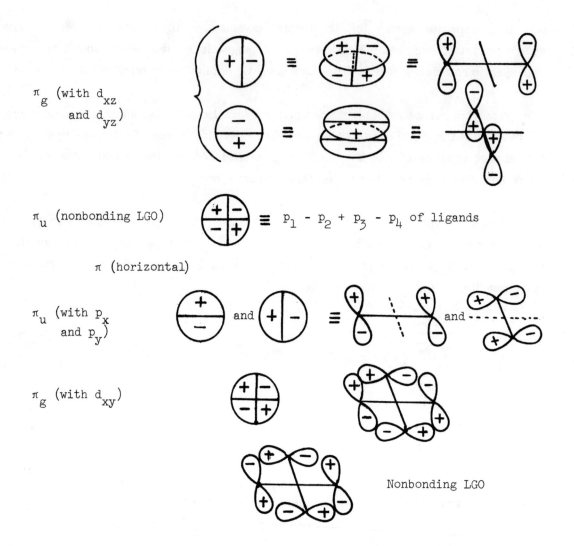

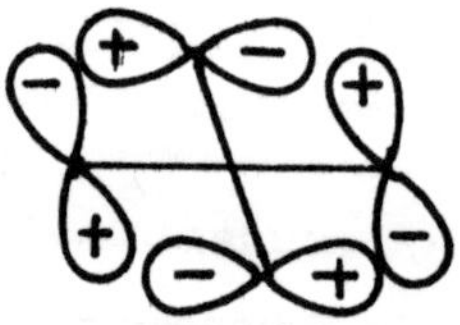

Nonbonding LGO

7.20 Problem: Give a pictorial approach to obtain the bonding molecular orbitals for σ and π bonding in square planar complexes.

7.20 Solution: Sigma bonding for square planar complexes. The metal orbitals that overlap with the ligand orbitals for σ bonding are s, p_x, p_y, and $d_{x^2-y^2}$ — the dsp^2 hybrids. For bonding MO's the LGO's match the signs of the M orbitals.

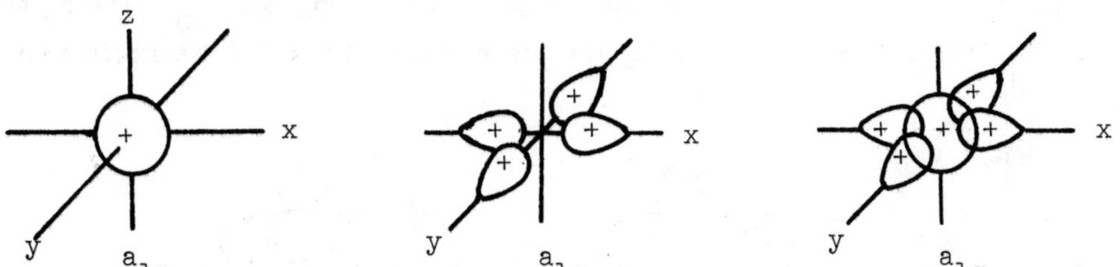

a_{1g} Metal s orbital | a_{1g} LGO | a_{1g} MO

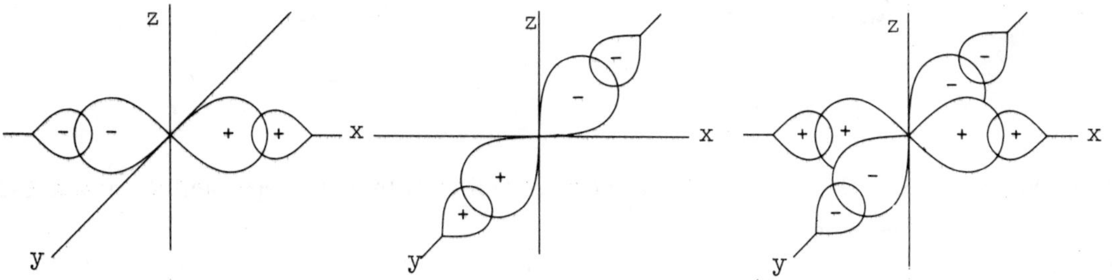

p_x | e_u | p_y | $d_{x^2-y^2}$ b_{1g}

Pi bonding. The M orbitals with π symmetry for overlap with the LGO's parallel to the C_4 axis are p_z and d_{xz}, d_{yz}. Using these orbitals as templates, we can sketch the corresponding LGO's.

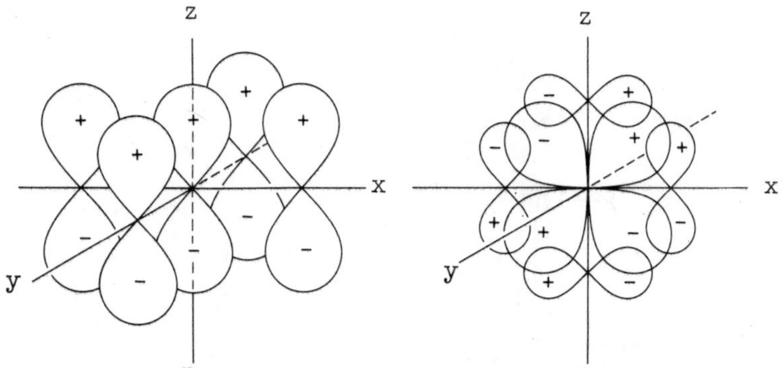

p_z
a_{2u}

d_{xz}
one of the e_g
π orbitals

In the xy plane the M orbitals that can form π bonds are p_x, p_y, and d_{xy}. The p_x and p_y orbitals give better overlap for σ bonding and are expected to be involved primarily in σ bonding.

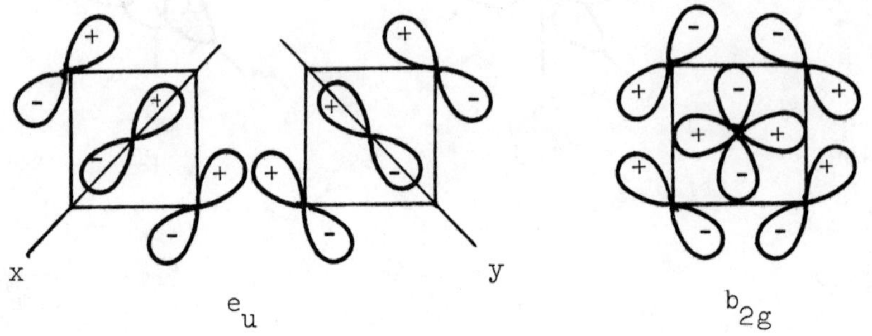

7.21 Problem: Use the group theoretical approach to obtain the representations and LGO for σ bonding in square planar complexes.

7.21 Solution: The vectors left unchanged by the operations of the D_{4h} group are given below. The reducible representation obtained reduces to give $A_{1g} + B_{1g} + E_u$.

D_{4h}	E	C_4	C_2	C_2'	C_2''	i	S_4	σ_h	σ_v	σ_d				
Γ_σ	4	0	0	2	0	0	0	4	2	0	= A_{1g}	+ B_{1g}	+ E_u	
					M orbitals						s	$d_{x^2-y^2}$	p_x, p_y	

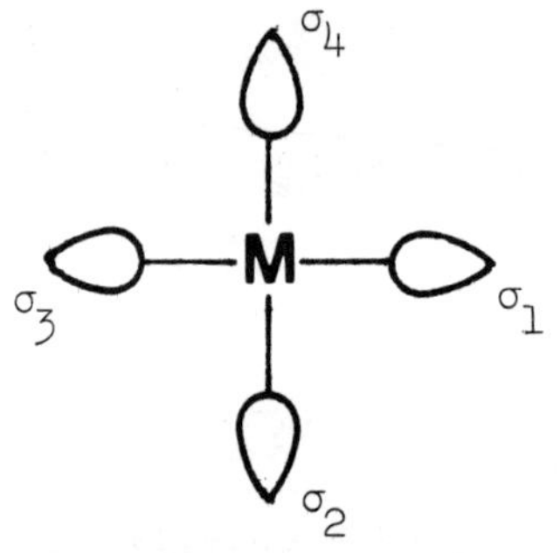

The p_z and other d orbitals are σ nonbonding.

Applying the projection operator:

D_{4h}	E	C_4	C'_4	C_2	$(C'_2)_a$	$(C'_2)_b$	$(C''_2)_a$	$(C''_2)_b$	i	S_4	S'_4	σ_h	σ_v	σ_d	σ'_d
σ_1	σ_1	σ_2	σ_4	σ_3	σ_1	σ_3	σ_2	σ_4	σ_3	σ_2	σ_4	σ_1	σ_3	σ_2	σ_4

Multiplying by the characters for the representations,

a_{1g} $4\sigma_1 + 4\sigma_2 + 4\sigma_3 + 4\sigma_4$ or $a_{1g} = 1/2(\sigma_1 + \sigma_2 + \sigma_3 + \sigma_4)$

b_{1g} $4\sigma_1 - 4\sigma_2 - 4\sigma_4 + 4\sigma_3$ or $b_{1g} = 1/2(\sigma_1 - \sigma_2 + \sigma_3 - \sigma_4)$

e_u $4\sigma_1 - 4\sigma_3$ or $e_u(1) = 1/\sqrt{2}\,(\sigma_1 - \sigma_3)$

rotate $e_u(1)$ by C_4 to get $e_u(2) = 1/\sqrt{2}\,(\sigma_2 - \sigma_4)$

7.22 Problem: The molecular orbitals formed by the six p π orbitals in benzene may be depicted as shown below (top half only shown, the sign of Ψ reverses on going through the plane sheet of paper).

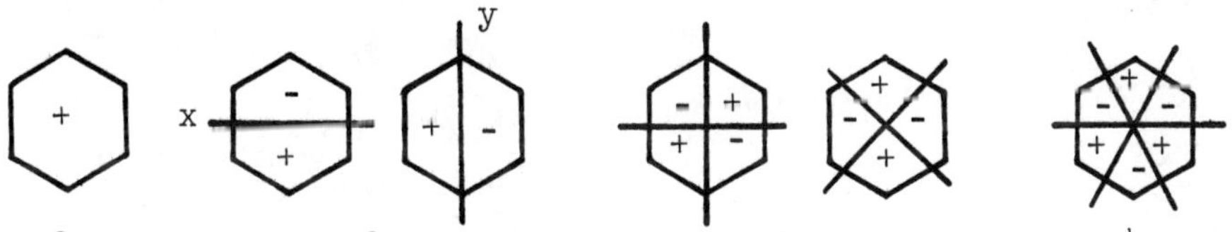

a_{1u} e_{1g} e_{2u} b_{2g}

Consider a molecule of bis(benzene)chromium as having a chromium atom at the origin of a set of Cartesian axes with a benzene ring on each side, centered on, and perpendicular to the z axis. Indicate below each of the above figures the metal orbital(s) that would combine with the group orbitals of the ligands and the type of molecular orbital (σ_g, δ_u, etc.) that would be formed. Draw an energy-level diagram showing ligand orbitals, metal orbitals, and molecular orbitals with appropriate labeling.

7.22 Solution: The sketches are for the π orbitals of benzene. We need to examine the combinations of these in $Cr(C_6H_6)_2$ (D_{6h}) with the Cr at the center of symmetry.

Conserving orbitals, the a_{1u} benzene MO's combine to give a_{1g} and a_{2u} MO's for the complex, as shown.

Metal orbitals	s and d_{z^2}	A_{1g}	Since the s and d_{z^2} orbitals have a_{1g} symmetry, they can combine, separately or together (a hybrid) with the a_{1g} MO sketched. The p_z and a_{2u} MO combine, etc. The e_{2u} group orbital is nonbonding since there is no metal e_{2u} orbital.
(from character	(d_{xz}, d_{yz})	E_{1g}	
table)	$(d_{x^2-y^2}, d_{xy})$	E_{2g}	
	p_z	A_{2u}	
	(p_x, p_y)	E_{1u}	

Combining the π MO's of the rings above and below the Cr, but showing only the signs of the lobes near the Cr, we get:

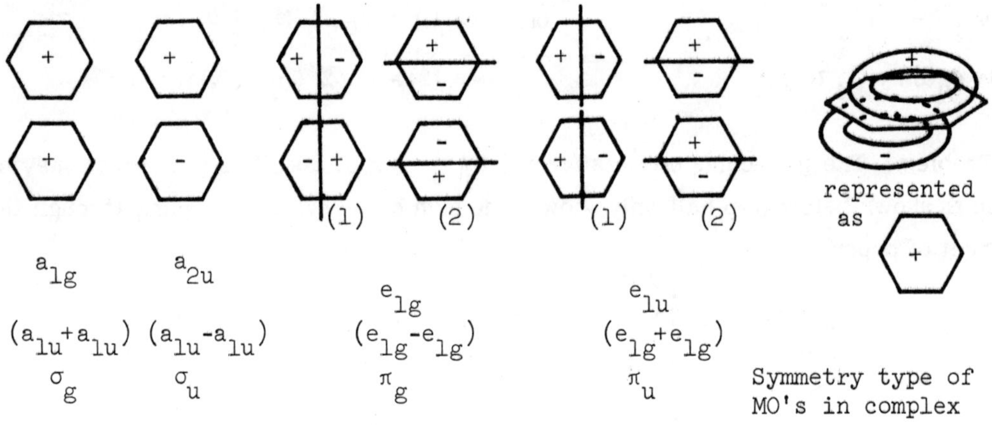

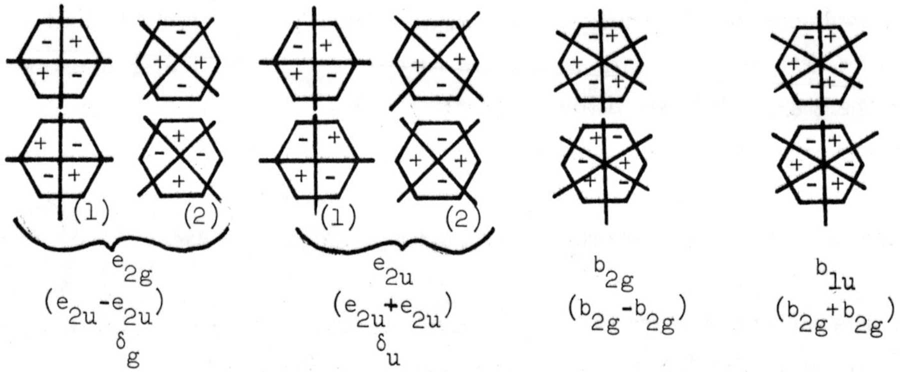

The combinations of the benzene π orbitals (sum and difference) are assumed to have the same energy since the rings are far apart (negligible overlap). The energies of the benzene π orbitals increase with increasing numbers of nodes $a_{1u} < e_{1g} < e_{2u} < b_{2g}$. The antibonding π^* (b^*_{2g}) is not involved in bonding with Cr and is omitted from the diagram.

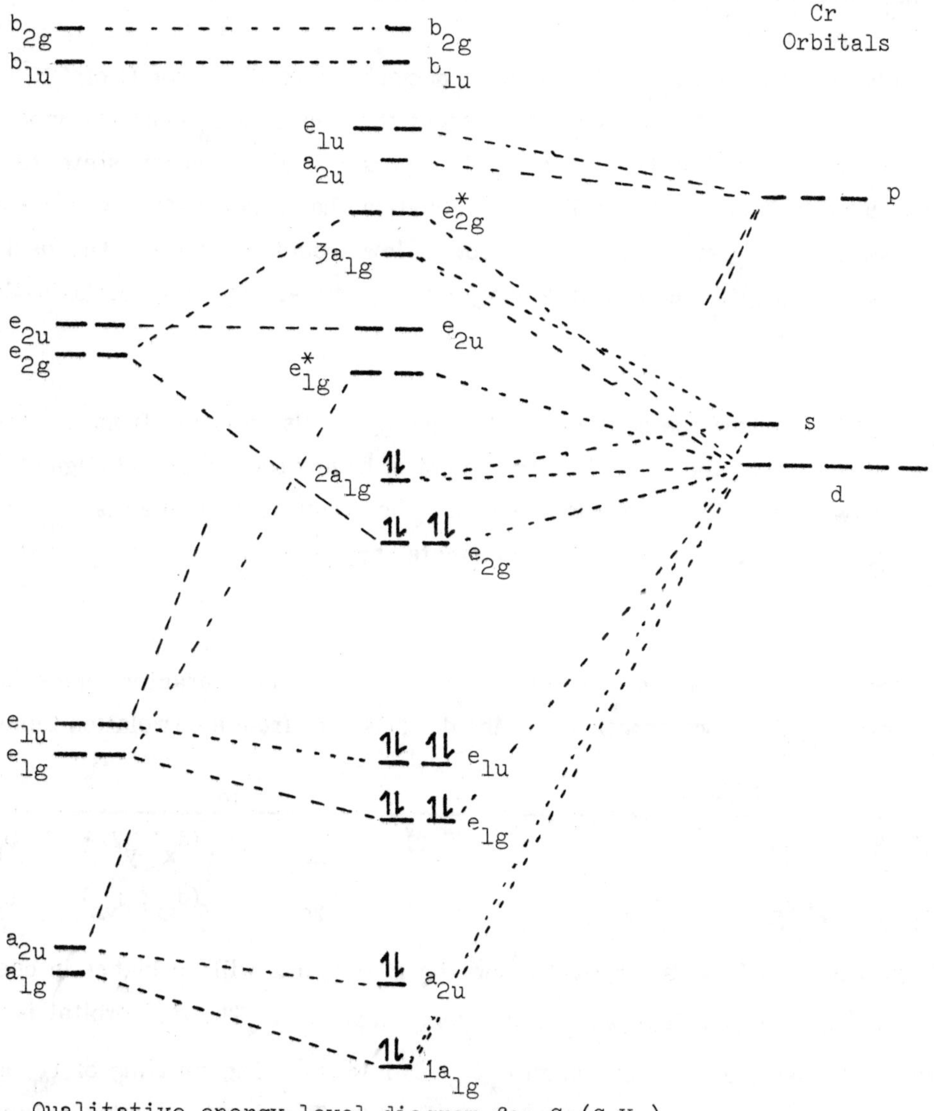

Qualitative energy level diagram for $Cr(C_6H_6)_2$

7.23 Problem: Using Pauling's electroneutrality principle (see Problem 2.14) and considering the degree of covalent character expected from ligand electronegativities, would you expect CN^- as a ligand to favor higher or lower oxidation states in transition metal complexes compared to the ligand NH_3? Does the metal-ligand π bonding of CN^- favor high or low oxidation states? Which oxidation states are favored?

7.23 Solution: Because of the lower electronegativity of the donor C of CN^- compared to N of NH_3, the M-CN bond is more covalent than the M-NH_3 bond and more electron density is transferred to the metal. This favors a high oxidation state to maintain electroneutrality. The metal-to-ligand π donation that greatly stabilizes cyanide complexes uses filled t_{2g} metal orbitals, favoring low oxidation states. The back bonding lowers the accumulation of negative charge on the metal atom and low oxidation states are favored usually.

7.24 Problem: Show the splitting of the energy levels resulting from compressing an octahedral field along the z axis (the equivalent of having four weak field ligands in the xy plane and two strong field ligands along z). The resulting symmetry is D_{4h}. Give the appropriate labels for the d orbitals and indicate how you know which orbitals are lower in energy.

7.24 Solution: The correlations can be obtained from the character tables for O_h and D_{4h} (identifying the representations for the d orbitals) or from a correlation table for O_h:

O_h		D_{4h}			
$(d_{z^2}, d_{x^2-y^2})$	e_g	(d_{z^2})	a_{1g},	$(d_{x^2-y^2})$	b_{1g}
(d_{xy}, d_{xz}, d_{yz})	t_{2g}	(d_{xy})	b_{2g},	(d_{xz}, d_{yz})	e_g

Since the resulting field is stronger along the z axis, d_{z^2} will be higher in energy than $d_{x^2-y^2}$ and (d_{xz}, d_{yz}) will be higher in energy than d_{xy}. The d_{z^2} orbital is raised in energy by the same amount by which $d_{x^2-y^2}$ is lowered. The splitting of t_{2g} is smaller since the orbitals are not directed toward the ligands. The single b_{2g} orbital is lowered by twice as much as the two e_g orbitals are raised (the net stabilization must be zero for three or six t_{2g} electrons).

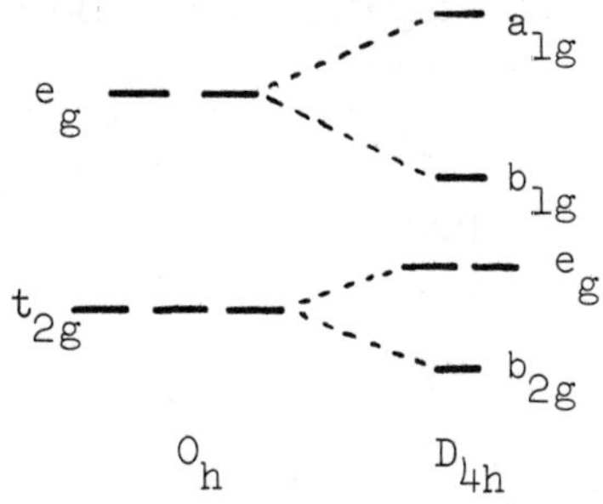

7.25 Problem: Given the relative energies of the d orbitals in Dq units in a ligand field of octahedral symmetry and of square planar symmetry, determine the splitting pattern in a field of pentagonal pyramidal symmetry.

	d_{z^2}	$d_{x^2-y^2}$	d_{xy}	d_{xz}	d_{yz}
Octahedral	6.00	6.00	-4.00	-4.00	-4.00
Square planar	-4.28	12.28	2.28	-5.14	-5.14

At what d-orbital populations would a pentagonal pyramidal geometry give greater crystal field stabilization energy <u>in a weak field</u> than octahedral geometry?

7.25 Solution: Using the procedures of Problems 7.9 and 7.10, we obtain the relative Dq values for a linear MX_2 complex by subtracting the square planar configuration values from those for the octahedral configuration; half of these values gives the values for a linear MX configuration. To the linear MX Dq values we add 5/4 of the values for the square planar configuration (since there are 5 ligands in the base of the pentagonal pyramid), and average the values for orbitals that are degenerate in C_{5v} geometry. The resulting Dq values for the pentagonal pyramidal MX_6 complex are:

d_{z^2}	$d_{x^2-y^2}$	d_{xy}	d_{xz}	d_{yz}
-0.21	5.96	5.96	-5.86	-5.86 Dq

From these values, and those of the octahedral configuration, we may calculate the following ligand field stabilization energies for weak-field complexes:

	LFSE	
	Pentagonal pyramid	Octahedron
d^1, d^6	-5.86	-4.0
d^2, d^7	-11.72	-8.0
d^3, d^8	-11.93	-12.0
d^4, d^9	-5.97	-6.0
d^5, d^{10}	0	0

Thus, we find that the LFSE is greater for the pentagonal pyramid than for the octahedron at d orbital populations of 1, 2, 6 and 7 electrons. The pentagonal pyramid is not favored, however, since ligand-ligand repulsion is much higher than in the octahedral configuration.

7.26 Problem: Using the appropriate Tanabe-Sugano diagram for Cr(III) (See Problem 7.16) in an octahedral complex having a value of Dq/B of approximately 3, compare the following transitions with regard to:

(1) frequency or wavenumber

(2) breadth of band expected

(3) intensity of band expected

(a) $^4A_{2g} \rightarrow {}^2T_{1g}$

(b) $^4A_{2g} \rightarrow {}^4T_{2g}$

Explain why the energy of the $^4T_{2g}$ arising from the ^4F term is linear with ligand field strength, while the $^4T_{1g}$ arising from the ^4F is not.

7.26 Solution: Using a Tanabe-Sugano diagram for a d^3 configuration, we find the following:

Transition (a) is approximately 2/3 the energy of (b) and will thus occur at approximately 2/3 of the frequency. It will be a sharp band since the energy of the transition is relatively independent of field strength and will not be affected by bond stretching. In contrast band (b) should be broad. Transition (a) is both spin and symmetry (parity) forbidden, and so will be very weak; (b) is spin allowed, but symmetry forbidden (g g) so it will be more intense, although still weak.

The $^4T_{1g}$ state from the ^4F term interacts with the $^4T_{1g}$ from the ^4P, so they diverge with increasing Dq. The $^4T_{2g}$ is the only term of this symmetry designation that appears and is thus not influenced by other states.

7.27 Problem: A general knowledge of the features of d–d and charge transfer spectra of octahedral complexes should enable you to match each lettered spectrum with the correct complex.

Complex	Maxima (in kK)			
A	22.4(weak)	25.9(weak)	36.8(strong)	41.7(strong)
B	18.1(weak)	22.2(weak)	30.1(strong)	33.9(strong)
C	23.9(weak)	30.1(weak)		
D	32.7(weak)	39.1(weak)		
E	19.3(weak)	24.3(weak)	39.2(strong)	
F	16.6(weak)	24.9(weak)		

(Weak means $\varepsilon \approx 100$, strong means $\varepsilon \sim 10{,}000$)

Complexes: $[IrBr_6]^{3-}$, $[Co(H_2O)_6]^{3+}$, $[RhBr_6]^{3-}$, $[Rh(NH_3)_6]^{3+}$, $[RhCl_6]^{3-}$, $[Rh(H_2O)_6]^{3+}$

7.27 Solution: All the complexes are expected to be low-spin since most complexes of these metals are low-spin for the d^6 configuration. Low-spin d^6 complexes give two weak (spin-allowed, symmetry-forbidden) bands: $^1A_{1g} \rightarrow {}^1T_{2g}$ and $^1A_{1g} \rightarrow {}^1T_{2g}$. Intense (allowed) charge transfer bands of the $L \rightarrow M$ type might be seen also for the halide complexes. The expected order or increasing energy for the $d \rightarrow d$ bands is:

$Br^- < Cl^- < H_2O < NH_3$ (Spectrochemical series)

Co(III) < Rh(III) < Ir(III)

The complexes with intense bands (A, B, and E) are expected to be the halide complexes. The complexes without charge transfer bands, listed in the order of increasing energy of the first band, are expected to be:

F	C	D
16.6	23.9	32.7 kK
$[Co(H_2O)_6]^{3+}$	$[Rh(H_2O)_6]^{3+}$	$[Rh(NH_3)_6]^{3+}$

Of the remaining complexes $[RhBr_6]^{3-}$ is expected to correspond to the lowest energy first band (B). We expect a small increase in the energy of this band for $[RhCl_6]^{3-}$, so E (19.3) is more likely than A (22.4). Also we expect the Cl→ Rh charge transfer band to be at much higher energy than Br→ Rh, as is the case of E. The spectrum for complex A agrees with that expected for $[IrBr_6]^{3-}$, the d-d bands are at higher energy and the same pattern of charge transfer bands is seen as for $[RhBr_6]^{3-}$.

VIII | Stereochemistry of Coordination Compounds

8.1 Problem: Name the following compounds according to the modified IUPAC rules.

K_2FeO_4

$[Cr(NH_3)_6]Cl_3$

$[Cr(NH_3)_4Cl_2]Cl$

$K[PtCl_3(C_2H_4)]$

$K_3[Al(C_2O_4)_3]$

$K_2[Co(N_3)_4]$

$K[Co(edta)]$

$[Cr(NH_3)_2(H_2O)_3(OH)](NO_3)_2$

$Fe(C_5H_5)_2$

$(CO)_5Mn-Mn(CO)_5$

$$[(NH_3)_4Co\underset{NH_2}{\overset{OH}{\diagup\diagdown}}Co(en)_2]Cl_4$$

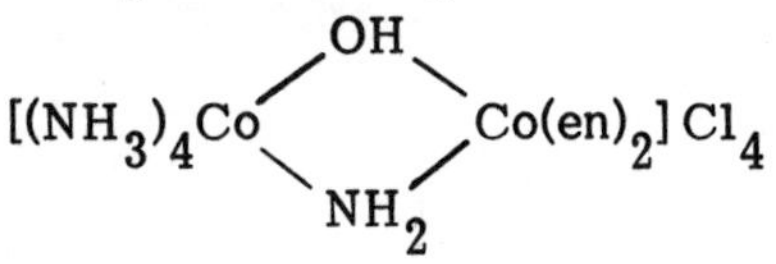

8.1 Solution: (See DMA Appendix B).

Potassium tetraoxoferrate(VI)

Hexaamminechromium(III) chloride

Tetraamminedichlorochromium(III) chloride   (alphabetical order)

Potassium trichloro(ethylene)platinate(II)

Potassium trisoxalatoaluminate

Potassium tetraazidocobaltate(II)   (N_3^- is azide ion)

Potassium ethylenediaminetetraacetatocobaltate(III)   (alphabetical order)

Diamminetriaquahydroxochromium(III) nitrate

Bis(cyclopentadienyl)iron(II)

Bis[pentacarbonylmanganese(0)] or decacarbonyldimanganese(0)

Tetraamminecobalt(III)-μ-amido-μ-hydroxo-bis(ethylenediamine)cobalt(III)chloride oblem: Sketch all of the possible geometrical isomers for the following complexes and indicate which of these would exhibit optical activity.

(a) $[Co(en)(NH_3)_2BrCl]^+$

(b) $[Co(NH_2CH_2CO_2)_2NH_3Cl]^+$

(c) $[Pt(NH_3)BrCl(NO_2)]^-$

(d) $[Co(trien)Cl_2]^+$ (consider the different ways of linking trien to Co).

(e) $[(gly)_2\ Co\overset{\displaystyle OH}{\underset{\displaystyle OH}{<>}}Co(gly)_2]$.

8.2   Solution:

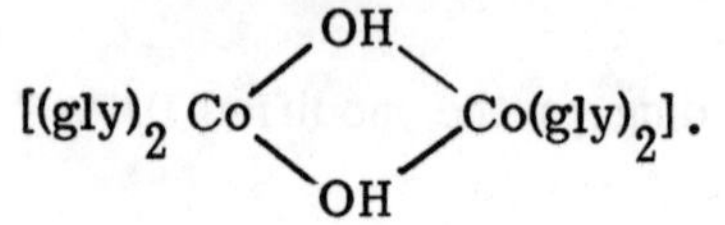

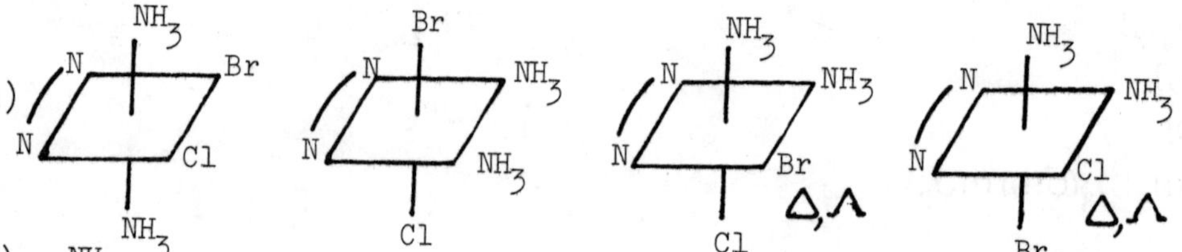

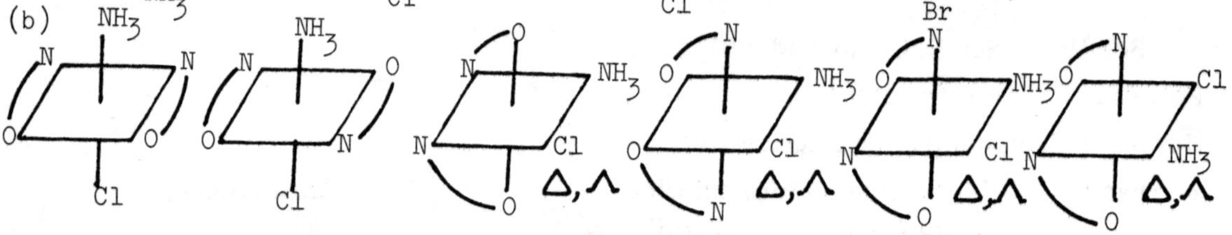

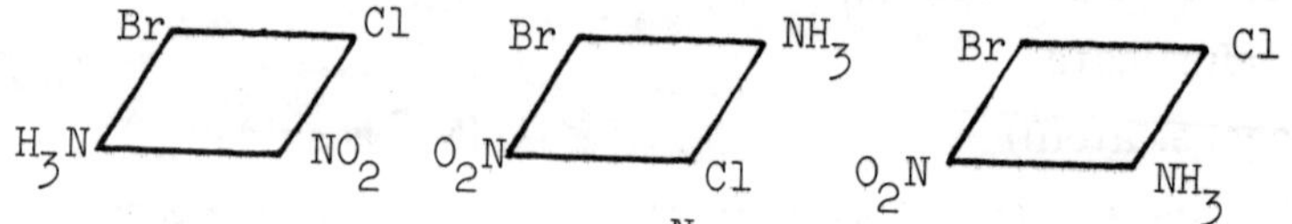

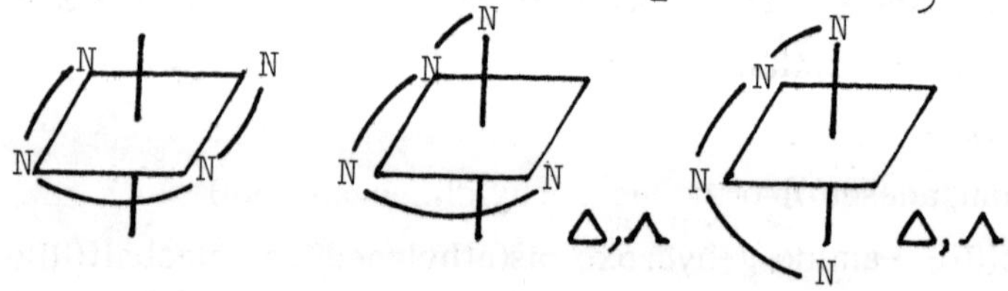

(e)

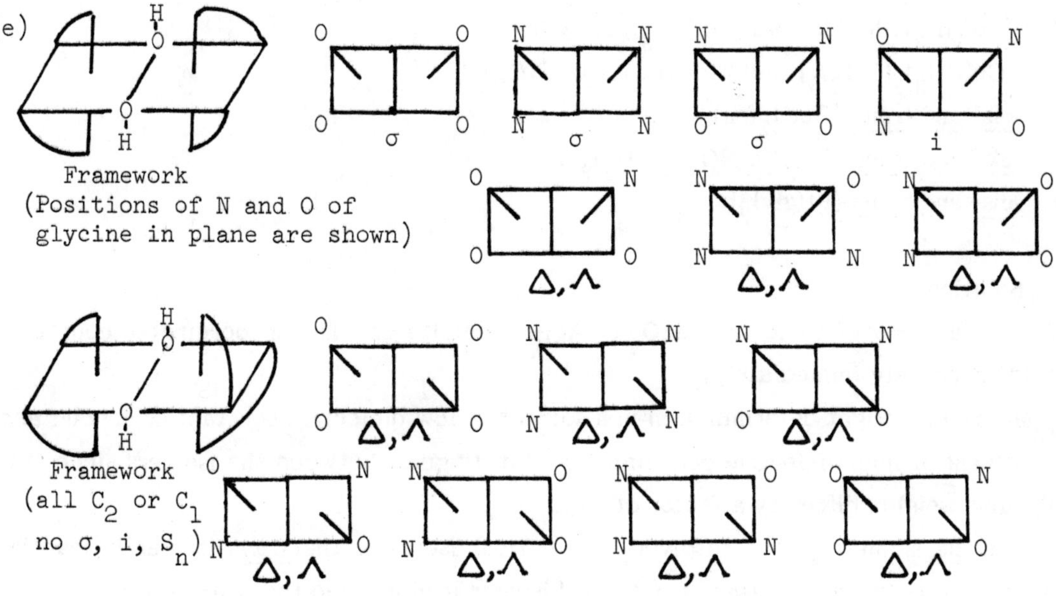

8.3 Problem: Draw all possible isomers for Ma_2bcd assuming the complex forms a square pyramid.

8.3 Solution:

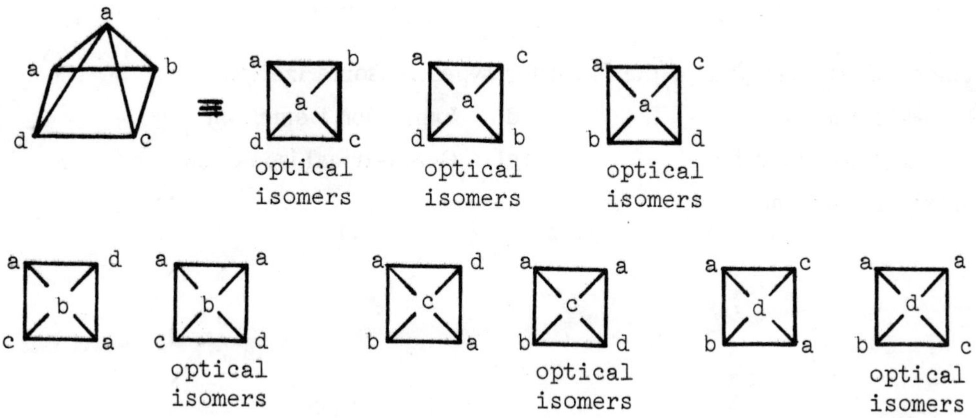

8.4 Problem: How might one distinguish between the following isomers?

(a) $[Co(NH_3)_5Br]SO_4$ and $[Co(NH_3)_5SO_4]Br$.

(b) $[Co(NH_3)_3(NO_2)_3]$ and $[Co(NH_3)_6][Co(NO_2)_6]$.

(c) cis- and trans-$[CoCl_2(en)_2]Cl$.

(d) cis- and trans-$NH_4[Co(NO_2)_4(NH_3)_2]$.

(e) cis- and trans-$[Pt(gly)_2]$.

8.4 Solution:

(a) Add Ba^{2+} or Ag^{+} to detect $BaSO_4$ or AgBr precipitates. The uncoordinated counterion should precipitate immediately.

(b) Molecular weight determination in a solvent of low dielectric constant or conductance in a solvent of high dielectric constant should distinguish between the isomers since their molecular weights differ by a factor of two.

(c) The cis isomer can be resolved into optical isomers, the trans isomer is inactive. Absorption spectra can be used also, the cis isomer is violet and the trans isomer is green.

(d) One possibility is the reaction with $C_2O_4^{2-}$. The trans isomer should give only one product, trans-$[Co(C_2O_4)_2(NH_3)_2]^-$. The cis isomer should give several products, including optical isomers. There would be some differences in the absorption spectra of the original nitro isomers also.

(e) The isomers should differ in absorption spectra. If a suitable solvent can be found, the trans isomer would be found to have zero dipole moment.

8.5 Problem: Give examples of the following types of isomerization:

(a) Hydrate isomerism

(b) Coordination isomerism

(c) Linkage isomerism

(d) Ionization isomerism

(e) Geometrical isomerism

8.5 Solution:

(a) Hydrate isomerism

$[Cr(H_2O)_6]Cl_3$ and $[Cr(H_2O)_5Cl]Cl_2 \cdot H_2O$

(b) Coordination isomerism

$[Co(NH_3)_6][Cr(CN)_6]$ and $[Cr(NH_3)_6][Co(CN)_6]$

$[Pt(NH_3)_4][PdCl_4]$ and $[Pd(NH_3)_4][PtCl_4]$

(c) Linkage isomerism

$[Co(NH_3)_5(NO_2)]Cl_2$ N-bonded and $[Co(NH_3)_5(ONO)]Cl_2$ O-bonded

$[Ir(NH_3)_5NCS]^{2+}$ and $[Ir(NH_3)_5SCN]^{2+}$

(d) Ionization isomerism

$[Pt(NH_3)_4Cl_2]Br_2$ and $[Pt(NH_3)_4Br_2]Cl_2$

(e) Geometrical isomerism

cis-$[CrCl_2(NH_3)_4]^+$ and trans-$[CrCl_2(NH_3)_4]^+$

cis-$[PtCl_2(NH_3)_2]$ and trans-$[PtCl_2(NH_3)_2]$

8.6 Problem: Draw all the isomers possible for octahedral M(abcdef). (Hint: Calculate the number of stereoisomers for [M(abcdef)]. Write the unique pairings trans to a and calculate the number of possible isomers for each of these. Then consider unique trans pairings for the remaining groups.

8.6 Solution: For n different ligands the total number number of stereoisomers possible is $n!/\sigma$, where σ is the order of the rotational group except for linear molecules. For M(abcdef) the rotational group is O, of order 24. The number of stereoisomers is $6!/24 = 30$. For each trans-M(xy) isomer (C_4 is the rotational group of order 4 for the framework) there are $4!/4 = 6$ stereoisomers. The geometrical isomers can be drawn by examining all trans pairings:

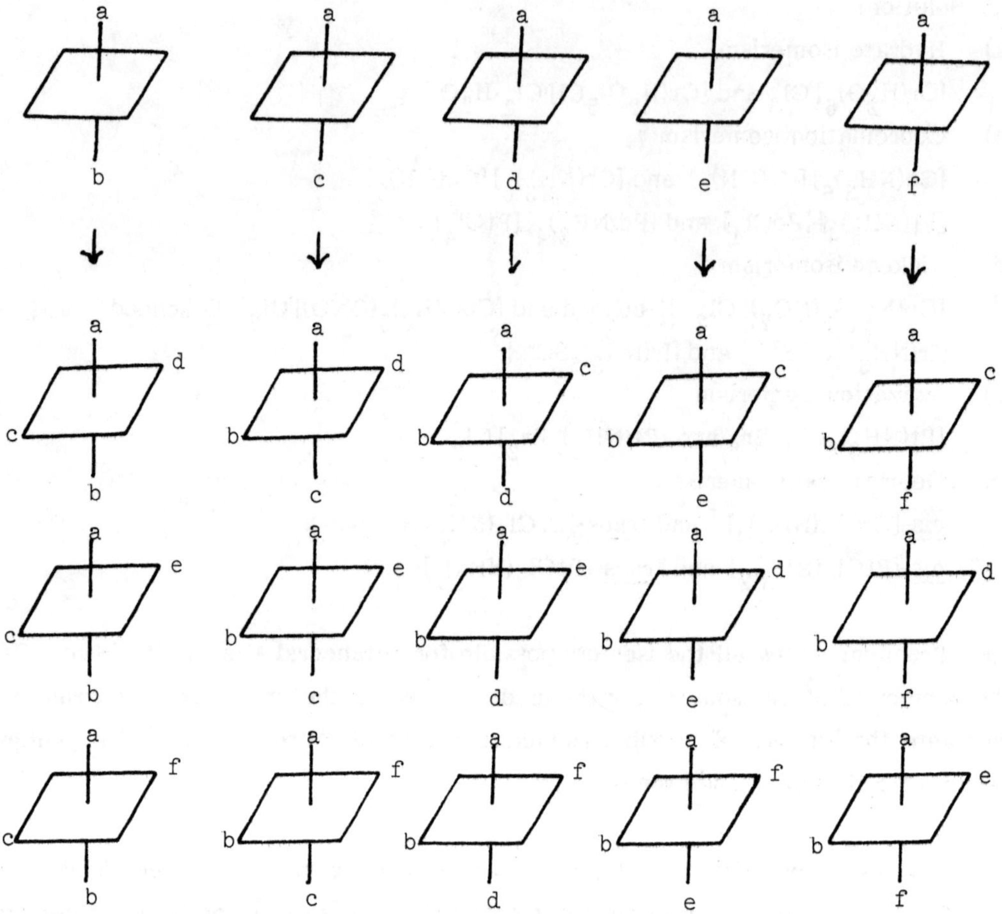

Each of the 15 geometrical isomers is one of a pair of enantiomers. The other enantiomer can be sketched as the mirror image of the first or by interchanging the positions of the last pair of ligands.

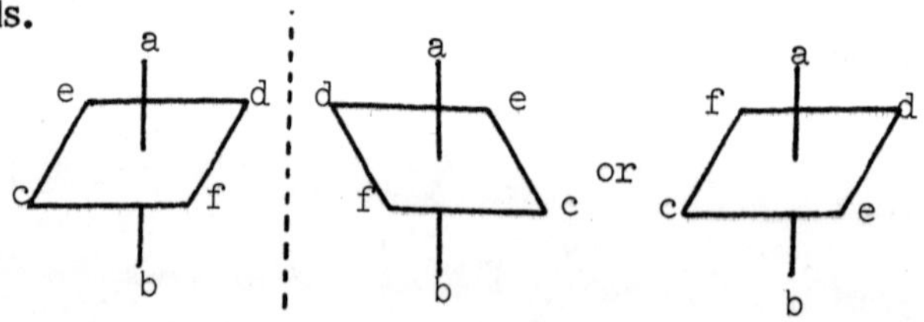

8.7 Problem: Draw all possible isomers for:

(a) Mabcd$\widehat{ef}$ ($\widehat{ef}$ is bidentate).

(b) $M\widehat{aa}\widehat{bc}d_2$ (Hint: leave $\widehat{bc}$ in a fixed position in all drawings except for enantiomers. $\widehat{aa}$ is a symmetrical bidentate ligand and $\widehat{bc}$ is an unsymmetrical bidentate ligand.

8.7 Solution: (a) The framework belongs to the rotational group C_1 or order 1. The number of isomers is 4! /1 = 24.

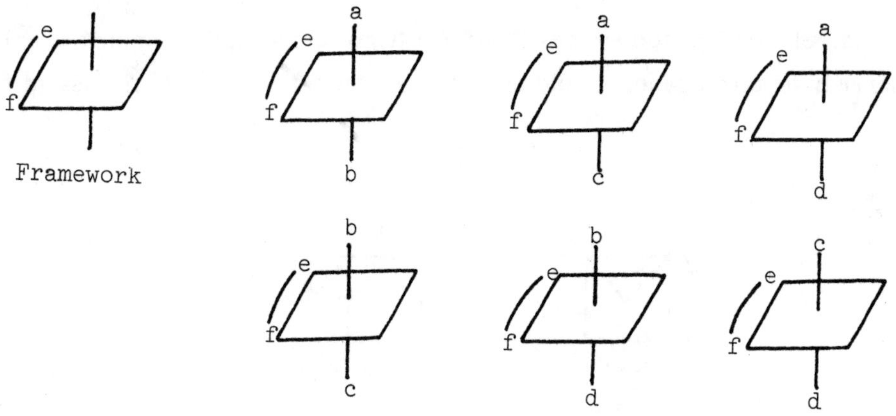

For each of these there are 2 geometrical isomers, for example:

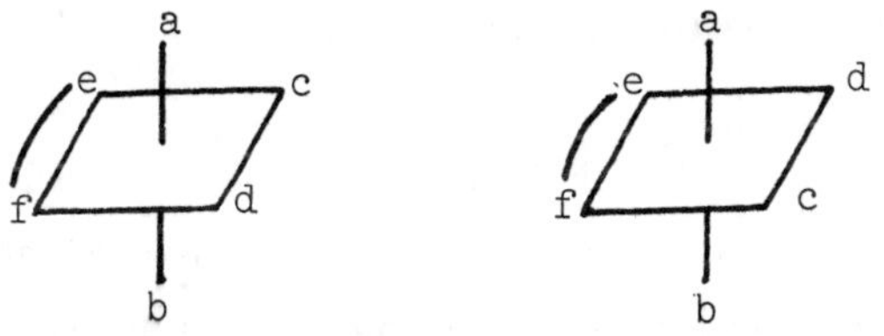

and each is optically active, giving 24 stereoisomers. We can get the mirror image either by reflecting through a mirror plane or by interchanging the axial pair.

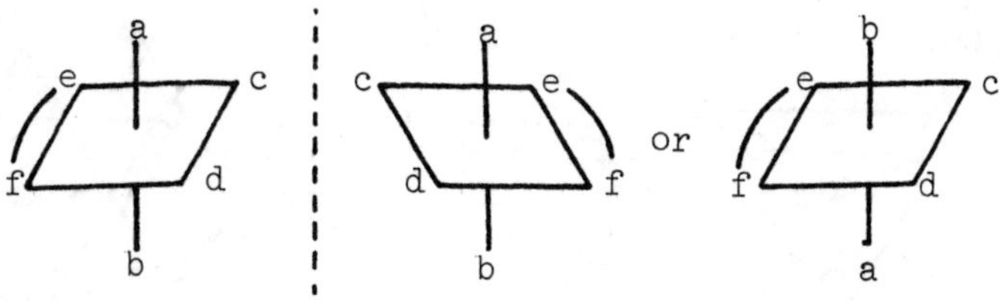

(b) The bidentate ligand $\widehat{aa}$ can be either in the same plane as $\widehat{cb}$, or not. There are 3 geometrical isomers and 5 stereoisomers.

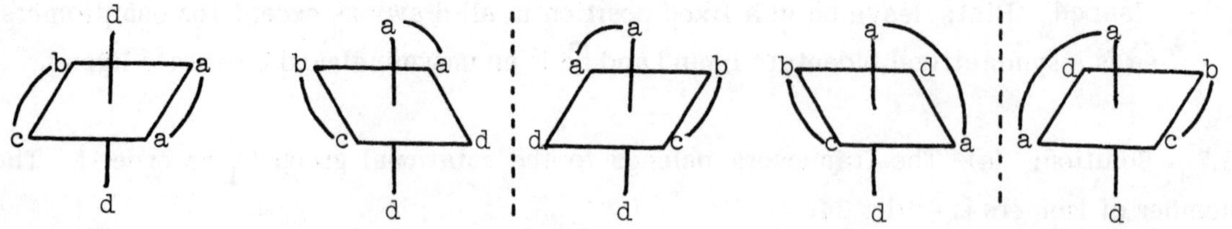

8.8 Problem: Sketch the isomers possible for a trigonal prismatic complex M($\widehat{aa}$)$_3$, where $\widehat{aa}$ is a planar bidentate ligand. Could any of these be optically active? Assign the point groups.

8.8 Solution:

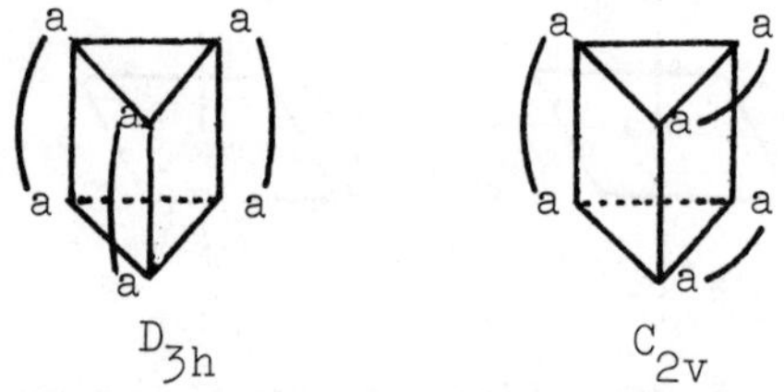

Neither is optically active since both have planes of symmetry. Bidentate ligands cannot span the diagonal positions of the tetragonal faces.

8.9 Problem: Sketch octahedral and trigonal prismatic [M(NO_3)$_3$] complexes to show that the N atoms are in a trigonal planar arrangement in each case. (Hint: Sketch the complexes looking down the threefold axes.) Assign the point groups.

8.9 Solution:

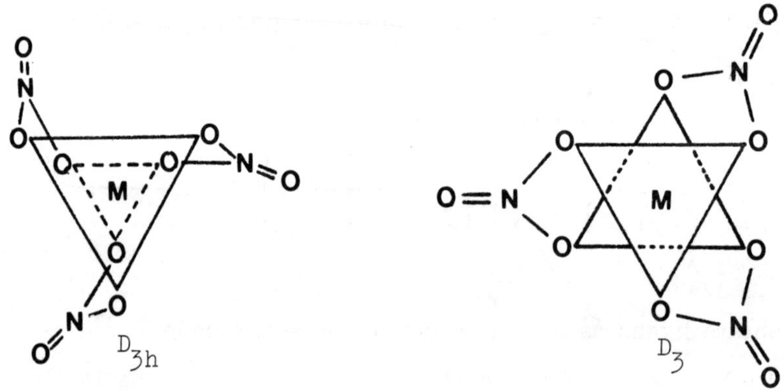

The NO_3^- ligands are in planes slicing through the C_3 axis for the trigonal prism and the NO_3^- ligands are skewed relative to the C_3 axis for the octahedron.

8.10 Problem: Determine the number of Δ and Λ pairs of chelate rings for the isomer of $[Co(edta)]^-$ shown. The pairs of rings attached at a common point and those spanning parallel edges are omitted.

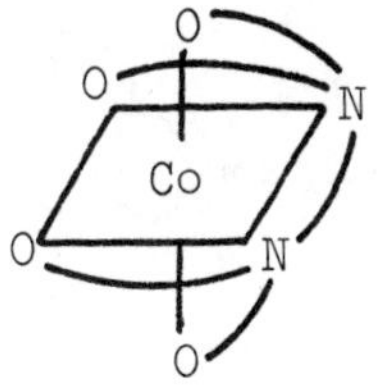

8.10 Solution: The IUPAC rules for designating chirality of octahedral complexes (See DMA p. 310) are based on the helices defined by non-parallel and non-intersecting octahedral edges spanned by chelate rings. A pair of edges viewed as shown (the solid edge B-B is closer to the viewer) defines a right helix (Δ) if B-B tips down to the right and a left helix (Λ) if B-B tips down to the left. The same system applies to the conformation of chelate rings. With the coordinated N atoms of $NH_2C_2H_4NH_2$ spanning an octahedral edge oriented horizontally (A—A), the conformation is δ (right-handed helix) if the C-C bond (replacing B-B) tips down to the right and λ (left handed) if it tips to the left. The drawings of $[Co(edta)]^-$ shown below are all oriented in the same way. They need to be reoriented with the edge farther away from you held horizontally. A model of an octahedron (see Problem 3.7) is helpful.

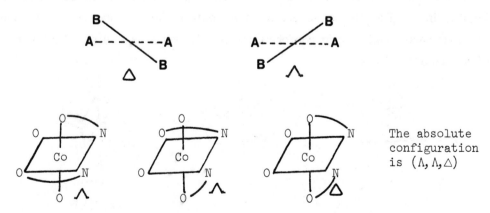

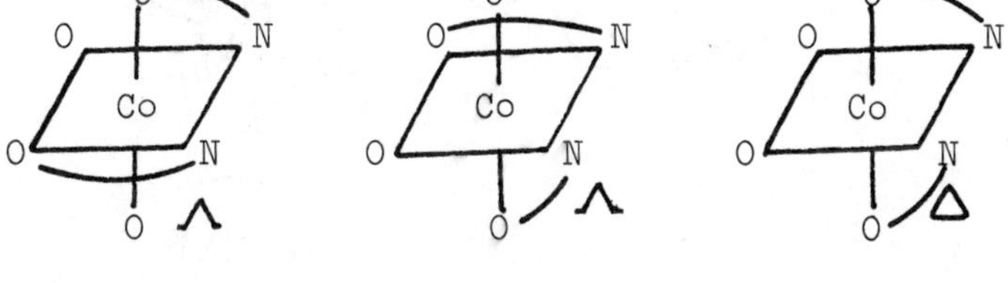

The absolute configuration is $(\Lambda, \Lambda, \Delta)$

8.11 Problem: Which chelate ring conformation is favored for coordinated S-(+)-propylenediamine, and why?

8.11 Solution: The δ conformation is favored in metal complexes of pn because of crowding of the axial methyl group by other ligands if the pn has the λ conformation.

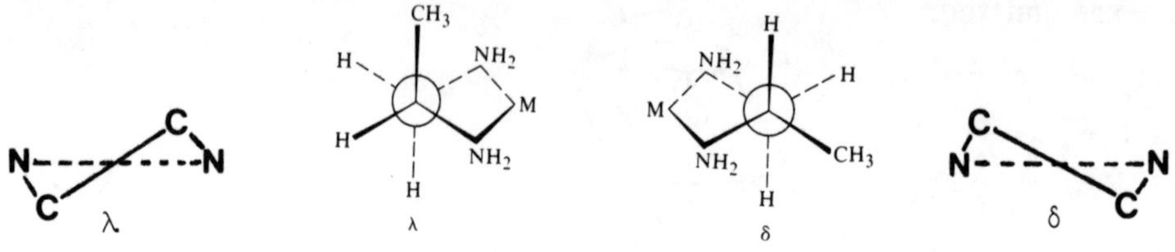

Enantiomeric conformations of a *gauche* chelate ring of *S*-(+)-propylenediamine placing the methyl group in axial (λ) and equatorial (δ) positions. Viewed along the C—C bond.

8.12 Problem: For trans-$[Co(trien)Cl_2]^+$ (trien = $NH_2C_2H_4NHC_2H_4NHC_2H_4NH_2$) there are three isomers possible, depending on the chelate ring conformations. Two are optically active and one is meso with the central diamine ring in an eclipsed "envelope" conformation. Sketch the two optically active trans isomers (all rings gauche) and assign the absolute configurations (δ or λ) to the chelate rings. (See D.A. Buckingham, P.A. Marzilli, and A.M. Sargeson, Inorg. Chem. 1967, 6, 1032.)

8.12 Solution: The three isomers of trans-$[Co(trien)X_2]^{n+}$ differ in the conformations of the chelate rings. For the meso isomer the central chelate ring is achiral and the terminal rings have enantiomeric conformations. See Problems 8.10 and 8.11 for descriptions of ring conformations.

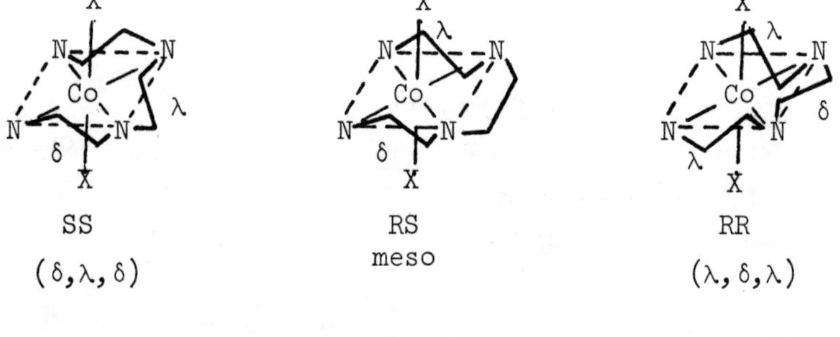

8.13 Problem: What is (are) the number(s) of d electrons for metals usually encountered with the following stereochemistries?

(a) Linear
(b) Square planar
(c) Trigonal prismatic
(d) Dodecahedral

8.13 Solution: (a) Linear complexes are encountered for Ag^+, Cu^+, and Hg^{2+}, all d^{10} ions. An $sp_zd_{z^2}$ hybrid provides strong bonding and a nonbonding orbital for the electron pair from d_{z^2}.

(b) Square planar complexes are found for the following cases: d^7, Co^{2+}; d^8, Ni^{2+}, Pd^{2+}, Pt^{2+}, Ag^{3+}, Rh^+; d^9, Cu^{2+}, Ag^{2+}. Square planar coordination involves one very high energy orbital ($d_{x^2-y^2}$). The most favorable situation is with this orbital empty (d^7 or d^8) or half filled (d^9).

(c) For C.N. 6 the expected geometries are octahedral and trigonal prismatic. The trigonal prism is not expected for d configurations providing high octahedral stabilization energy (d^3, d^6 (low-spin), and d^8). For D_{3h} symmetry, one d orbital (d_{z^2}) is lowest in energy. Trigonal prismatic coordination is found for d^0, Mo^{VI}; d^5, Mn^{II}; d^{10}, Zn^{II}; d^1, Re^{VI}; d^7 (high-spin), Co^{II}.

(d) For the dodecahedron there is one very low energy d orbital, so maximum stabilization is achieved by occupancy of this orbital only: d^1, $[Mo(CN)_8]^{3-}$ and d^2, $[Mo(CN)_8]^{4-}$

8.14 Problem: Selection rules for electric dipole transitions (those usually observed for transition metals) require that the symmetry of the transition (the direct product of the representations for the ground and excited states) must be the same as the representation for one of the electric dipole moment operators, and these correspond to the representations for x, y, and z. Selection rules for magnetic dipole transitions (observed for lanthanide metals) require that the symmetry of the transition be the same as the representation of one of the magnetic dipole moment operators, and these correspond to rotations about x, y, and z (R_x, R_y, and R_z in the character tables). For CD and ORD, the transitions must obey the selection rules for electric and magnetic dipole transitions. For $[Cr(NH_3)_6]^{3+}$ (O_h) the d-d transitions are ${}^4A_{2g} \rightarrow {}^4T_{2g}$ and ${}^4A_{2g} \rightarrow {}^4T_{1g}$. For $[Cr(en)_3]^{3+}$ (D_3) the transitions are ${}^4A_2 \rightarrow {}^4E$ and ${}^4A_2 \rightarrow {}^4A_1$ (both derived from ${}^4A_{2g} \rightarrow {}^4T_{2g}$ by lowering symmetry), and $^4A_2 \rightarrow {}^4E$ and $^4A_2 \rightarrow {}^4A_2$ (both derived from $^4A_{2g} \rightarrow {}^4T_{1g}$ by lowering symmetry). Which of these are electric-dipole-allowed? Which are both electric- and magnetic-dipole-allowed?

8.14 Solution: $\int \psi_{ground}(\text{operator})\psi_{excited}d\tau$

For the integral describing an electronic transition to be nonzero (an allowed transition) the symmetry of the transition (product of the representations for the wave functions for the ground and excited states) must be the same as that of the operator. For electric dipole transitions the operator belongs to the same representations as x, y, and z. For magnetic dipole transitions the operator belongs to the same representations as R_x, R_y, and R_z.

For $[Cr(NH_3)_6]^{3+}$ (O_h)

Transition	Symmetry of transition	Symmetry of operators	
$A_{2g} \rightarrow T_{2g}$	$A_{2g} \times T_{2g} = T_{1g}$	x,y,z	T_{1u}
$A_{2g} \rightarrow T_{1g}$	$A_{2g} \times T_{1g} = T_{2g}$	R_x, R_y, R_z	T_{1g}

Both transitions are electric dipole forbidden (neither is T_{1u}). They appear as low intensity absorption bands because of vibronic coupling. The T_{1g} $(A_{2g} \rightarrow T_{2g})$ transition is magnetic dipole allowed, but $A_{2g} \rightarrow T_{1g}$ is magnetic dipole forbidden. There is no ORD or CD (circular dichroism) for $[Cr(NH_3)_6]^{3+}$ since optical activity requires that both integrals be finite.

For $[Cr(en)_3]^{3+}$ (D_3)

Transition	Symmetry of transition	Symmetry of operators	
$A_2 \rightarrow E$	$A_2 \times E = E$	z, A_2	(x,y), E
$A_2 \rightarrow A_1$	$A_2 \times A_1 = A_2$	R_z, A_2	(R_x,R_y), E
$A_2 \rightarrow A_2$	$A_2 \times A_2 = A_1$		

The transitions of E and A_2 symmetry obey both selection rules and should appear in the CD (or ORD) spectrum. The A_1 $(A_2 \rightarrow A_2)$ transition is electric and magnetic dipole forbidden (See DMA, pp. 284f, 313f).

8.15 Problem: Draw all of the isomers expected for (a) $[B(gly)_2]^+$, (b) $Pt(gly)_2$. What factors account for the difference in your expectations?

8.15 Solution: The boron complex is expected to be tetrahedral since it would have no LFSE and tetrahedral geometry minimizes ligand-ligand repulsion. Alternatively, tetrahedral geometry is expected from sp^3 hybridization. The platinum complex would be expected to be square planar since the LFSE would be high for a d^8 configuration of a heavy metal; or, the geometry would be expected from dsp^2 hybridization. The boron complex should thus give optical isomers whereas the platinum complex should give geometric isomers, as shown below.

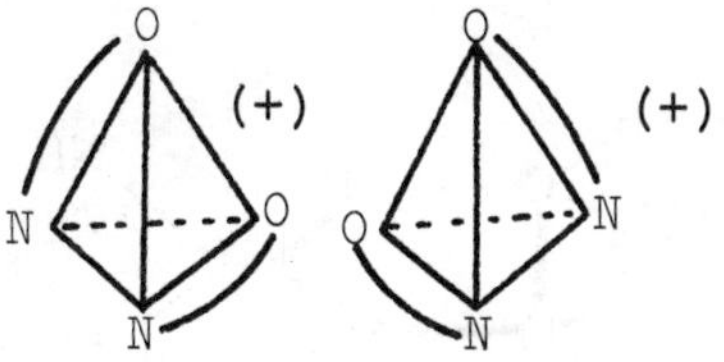

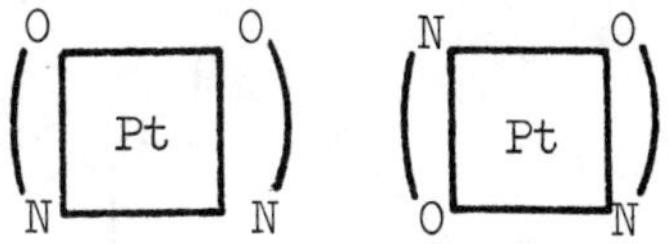

8.16 Problem: Sketch all the possible isomers of [Pt(pn*)BrCl] where pn* is an optically active ligand propylenediamine H - *C(Me)CH$_2$. (with NH_2 on each of the two carbons: NH_2 NH_2)

8.16 Solution: For each geometric isomer a pair of optical isomers exists having the R or S configuration at C*.

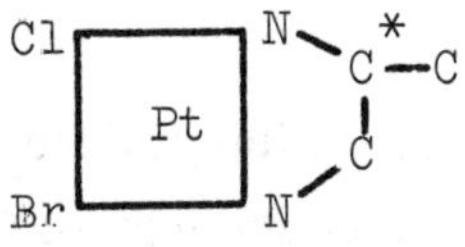

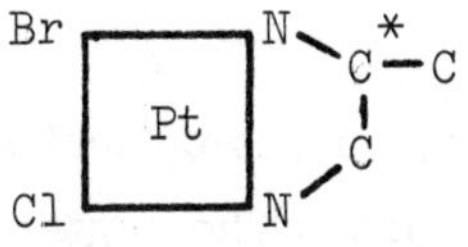

8.17 Problem: Since a square planar complex with four different ligands, Mabcd, is optically inactive, sketch a square planar complex which is optically active even though the free ligands are not.

8.17 Solution: See the solution to Problem 8.18 for the general conditions for optical activity. If we design a complex lacking a center or plane of symmetry it is likely to be optically active. The presence of an S_n ($n > 2$) axis without σ or i is unusual. In order to eliminate the planes of symmetry usually found for a square planar complex, we can use two diamines substituted differently. The use of two different diamines eliminates i and σ between the ligands. In the example shown the diamine on the right has two identical substituents on one carbon only, eliminating the plane of symmetry cutting through both ligands and perpendicular to the PtN_4 plane. Having the same substituent on each carbon of the other diamine eliminates σ in the PtN_4 plane. The ligand on the left has a plane of symmetry itself, so it is an optically inactive meso form. This complex would have a plane of symmetry (and be optically inactive) if it were tetrahedral.

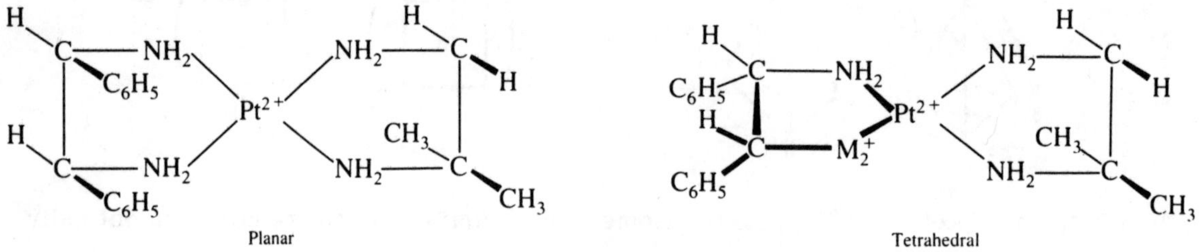

Possible configurations of a Pt(II) complex.

8.18 Problem: The complex $Zr(OC_6H_4CH{=}NC_2H_5)_4$ has S_4 symmetry. It cannot be optically active even though it lacks a center of symmetry and a plane of symmetry. Explain.

8.18 Solution: A compound is optically active (its mirror images are nonsuperimposable) only if it lacks any S_n axis. The center of symmetry is equivalent to S_2 and the plane of symmetry is equivalent to S_1. (See DMA p. 324 or D.C. Bradley, M.B. Hursthouse, and L.F. Randall, Chem. Commun. 1970, 368 for a drawing of the complex.)

IX Reaction Mechanisms for Coordination Compounds

9.1 Problem: In 1.0 M OH^- solution, at least 95% of $[Co(NH_3)_5Cl]^{2+}$ that undergoes base hydrolysis must be present in the pentaammine form. If 5% or more were present as the conjugate base $[Co(NH_3)_4(NH_2)Cl]^+$, it could be detected by departure from second order kinetics which is <u>not</u> observed to happen.

(a) For production of the conjugate base

$$[Co(NH_3)_5Cl]^{2+} + OH^- \rightleftharpoons [Co(NH_2)(NH_3)_4Cl]^+ + H_2O$$

$$K_h = \frac{[Co(NH_2)(NH_3)_4Cl^+]}{[Co(NH_3)_5Cl^{2+}]\,[OH^-]} = \frac{K_a}{K_w}$$

Using this information and the observation described above, show that for $[Co(NH_3)_5Cl]^{2+}$, $K_a \leqslant 5 \times 10^{-16}$

(b) Show the form of the rate law which would result if the conjugate base were present in appreciable amount. (Hint: The concentration of $[Co(NH_3)_5Cl]^{2+}$ used to make up the solution will be partitioned between the complex and its conjugate base).

9.1 Solution:

(a) We must have for the Cl^- substitution:

$$[Co(NH_3)_5Cl]^{2+} + OH^- \rightleftharpoons [Co(NH_3)_4(NH_2)Cl]^+ + H_2O$$

$$K_h = \frac{[Co(NH_3)_4(NH_2)Cl^+]}{[Co(NH_3)_5Cl^{2+}]\,[OH^-]} = \frac{K_a}{K_w} \leqslant \frac{0.05}{(0.95)(1.0)}$$

$$K_a \leqslant K_w \frac{0.05}{(0.95)(1.0)} = 1.0 \times 10^{-14} \times \frac{5 \times 10^{-2}}{0.95} = 5 \times 10^{-16}$$

(b) Rate = $k_{obs}[Co(NH_3)_5Cl^{2+}]_{total}[OH^-]$ since the total concentration of complex is known from preparing the solution.

But $[Co(NH_3)_5Cl^{2+}]_{total} = [Co(NH_3)_5Cl^{2+}] + [Co(NH_3)_4(NH_2)Cl^+]$

From (a): $[Co(NH_3)_4(NH_2)Cl^+] = K_h[Co(NH_3)_5Cl^{2+}][OH^-]$

If $[Co(NH_3)_4(NH_2)Cl^+]$ is the reactive species,

$$\text{rate} = k[Co(NH_3)_4(NH_2)Cl^+]$$

$$= kK_h[Co(NH_3)_5Cl^{2+}][OH^-]$$

$$= k_{obs}[Co(NH_3)_5Cl^{2+}]_{total}[OH^-]$$

$$= k_{obs}\left\{[Co(NH_3)_5Cl^{2+}] + K_h[OH^-][Co(NH_3)_5Cl^{2+}]\right\}[OH^-]$$

$$\therefore k_{obs} = \frac{kK_h}{1+K_h[OH^-]}$$

and the deviations will appear when $K_h[OH^-] \approx 1$.

Another way of viewing this problem is as a special case of the mechanism

$$A + B \underset{k_{-1}}{\overset{k_1}{\rightleftharpoons}} C + D$$

$$C \xrightarrow{k_2} E + F$$

Applying an "improved steady-state approximation" to [C], we get (See D.H. McDaniel and C.R. Smoot, J. Phys. Chem. 1956, 60, 966.):

$$\frac{d[E]}{dt} = \frac{k_1k_2[A][B]}{k_1[B] + k_{-1}[D] + k_2}$$

In our case, A = $[Co(NH_3)_5Cl]^{2+}$, B = OH^-, C = $[Co(NH_2)(NH_3)_4Cl]^+$, D = H_2O, E = $[Co(NH_2)(NH_3)_4]^{2+}$ and F = Cl^-. The reaction of E with water is fast and so does not appear in the rate law. When the first step is an equilibrium established prior to the second step, $k_1, k_{-1} >> k_2$. Since D = H_2O, $k_{-1}[D] = k'_{-1}$, and $k_1/k'_{-1} = K_h$. So,

$$\frac{d[E]}{dt} = \frac{k_1k_2[Co(NH_3)_5Cl^+][OH^-]}{k_1[OH^-] + k'_{-1}}$$

Dividing by k'_{-1}, we get

$$\frac{d[E]}{dt} = \frac{k_2K_h}{1 + K_h[OH^-]}[Co(NH_3)_5Cl^{2+}][OH^-]$$

9.2 Problem: The following are approximate rates of exchange of solvent water with bound water for some metal aqua complexes:

Metal	Substitution rate (sec^{-1})
Li^+, Na^+, K^+, Rb^+, Cs^+	$10^8 - 10^{10}$
Ca^{2+}, Sr^{2+}, Ba^{2+}	$10^8 - 10^9$
Mg^{2+}	10^5
Hg^{2+}	10^9
Cd^{2+}	10^8
Zn^{2+}	10^7
In^{3+}	10^2
Al^{3+}	1
Cu^{2+}, Cr^{2+}	10^9
Ni^{2+}	10^4
Ru^{2+}	1

Explain how these data are compatible with a dissociative (d) mechanism for water exchange.

9.2 Solution: A d mechanism is one in which the breaking of the bond between a metal and the leaving ligand exerts a dominant influence on the rate. The rates given tend to be slower for smaller, highly charged cations which should form strongest M-O bonds, thus supporting a $\underline{d}$ mechanism. This trend is somewhat complicated by effects of d electron configuration in the transition series.

9.3 Problem: Account for the difference in rate constants for the following two reactions:

$$[Fe(H_2O)_6]^{2+} + Cl^- \rightarrow [Fe(H_2O)_5Cl]^+ + H_2O \quad k(M^{-1}\,sec^{-1}) = 10^6$$

$$[Ru(H_2O)_6]^{2+} + Cl^- \rightarrow [Ru(H_2O)_5Cl]^+ + H_2O \quad k(M^{-1}\,sec^{-1}) = 10^{-2.0}$$

9.3 Solution: The Fe^{2+} complex involving a first-row transition metal is high-spin d^6 and labile. Ru^{2+}, like other second- and third-row transition metal ions, forms low-spin complexes. Low-spin d^6 species are kinetically inert.

9.4 Problem: Distinguish between the intimate and stoichiometric mechanism of a reaction.

9.4 Solution: The intimate mechanism deals with the factors which contribute primarily to the activation energy - whether bond breaking by the leaving group in substitution reactions or bond making by the entering group. The stoichiometric mechanism is the sequence of detectable elementary steps and describes the occurrence or non-occurrence of reaction intermediates. (We use lower case letters in referring to intimate mechanisms and upper case for stoichiometric mechanisms.)

9.5 Problem: What is the significance of the following facts for the mechanism of substitution at Co(III) in aqueous solution?

(a) The rates of aquation are always given the expression: rate = $k_{aq}[Co(NH_3)_5X^{2+}]$.

(b) No direct replacement of X^- by Y^- is ever observed. Instead, water enters first and is subsequently replaced by Y^-.

9.5 Solution:

(a) Since water is the solvent, its concentration cannot be varied. Hence, it never appears as an observable in the rate law.

(b) The observation implies that energetic significance of bond making to the entering group is small. Hence, water is most often the entering ligand for statistical reasons. Ultimately, the thermodynamically stable product is formed with Y^- replacing water.

9.6 Problem: The following data have been obtained at 50°C for aquation of $[Cr(NH_3)_5X]^{2+}$ (k_{aq}) and anation by X^- of $[Cr(NH_3)_5(H_2O)]^{3+}$ (k_{an}):

X^-	k_{aq} (sec^{-1})	k_{an} ($M^{-1}sec^{-1}$)
NCS^-	0.11×10^{-4}	4.16×10^{-4}
$CCl_3CO_2^-$	0.37×10^{-4}	11.81×10^{-4}
$CF_3CO_2^-$	0.50×10^{-4}	1.37×10^{-4}
Cl^-	1.75×10^{-4}	0.69×10^{-4}
Br^-	12.5×10^{-4}	——
I^-	102×10^{-4}	——
$HC_2O_4^-$	——	6.45×10^{-4}
H_2O	13.7×10^{-4} (exchange)	

What can you say about the intimate mechanism of these reactions? (See T. Ramasani and A.G. Sykes, Chem. Comm. 1976, 378.)

9.6 Solution: The nature of the entering group makes very little difference to the rate of replacement of H_2O by the anion (the anation rate). (See Problem 9.5.) By changing the leaving group, the aquation rate varies over a range of 10^3. Hence, the intimate mechanism is probably d.

9.7 Problem: The following data have been reported for aquation at 298.1K of complexes trans-$[Co(N_4)LCl]^{n+}$ where N_4 is a quadridentate chelate. What do these data suggest about the intimate mechanism for aquation? (See W.-K. Chau, W.-K. Lee and C.-K. Poon, J. Chem. Soc. Dalton Trans., 1974, 2419.)

L	k(sec^{-1}) (N$_4$) = cyclam	(N$_4$) = tet-b
Cl^-	1.1×10^{-6}	9.3×10^{-4}
NCS^-	1.1×10^{-9}	7.0×10^{-7}
NO_2^-	4.3×10^{-5}	4.1×10^{-2}
N_3^-	3.6×10^{-6}	2.1×10^{-2}
CN^-	4.8×10^{-7}	3.4×10^{-4}
NH_3	7.3×10^{-8}	---

9.7 Solution: The rather wide variation in rate as a function of leaving group suggests a d mechanism as in Problem 9.6.

9.8 Problem: The reactions $[Cr(NCS)_6]^{3-} + solv \rightarrow [Cr(NCS)_5(solv)]^{2-} + NCS^-$ have been investigated and found to have the following constants near 70°C.

solvent	k(sec^{-1})
dimethylacetamide	9.5×10^{-5}
dimethylformamide	12.4×10^{-5}
dimethylsulfoxide	6.2×10^{-5}

What do these values suggest about the intimate mechanism of these reactions? (See S.T.D. Lo and D.W. Watts, Aust. J. Chem. 1975, 28, 1907.)

9.8 Solution: The different entering groups all have about the same rate. Hence a d mechanism is indicated.

9.9 Problem: The figure shows plots of k_{obs} vs. $[X^-]$ for the anation reactions:

$$[Co(en)_2(NO_2)(dmso)]^{2+} + X^- \rightarrow [Co(en)_2(NO_2)X]^+ + dmso$$

(dmso = dimethylsulfoxide)

All three reactions are presumed to have the same mechanism.

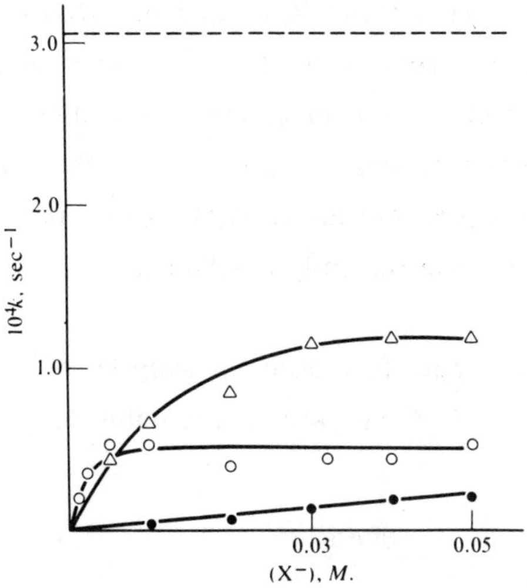

Rate of anation of *cis*-$[Co(en)_2$-$NO_2dmso]^{2+}$ as a function of the concentration of the entering anion X^-: △,X^- = NO_2^-; ○,X^- = Cl^-; ●,X^- = SCN^-. The broken line shows the rate of the dmso-exchange reaction. (Reprinted with permission from W. R. Muir and C. H. Langford, *Inorg. Chem.* 1968, 7, 1032. Copyright 1968, American Chemical Society.)

(a) What is the significance of the shapes of the curves?

(b) What is the significance of the fact that the first-order limiting rate constants are smaller than that for dmso exchange?

(c) If the mechanism were D, to what would the limiting rate constants correspond?

(d) If the mechanism were I_d, to what would the limiting rate constants correspond?

(e) The limiting rate constants are 0.5×10^{-4} sec^{-1} and 1.2×10^{-4} sec^{-1} for Cl^- and NO_2^-, respectively. For NCS^-, the limiting rate constant can be estimated as 1×10^{-4} sec^{-1}. Do these values constitute evidence for a D or an I_d mechanism? (See W.R. Muir and C.H. Langford, Inorg. Chem. 1968, 7, 1032.)

9.9 Solution:

(a) From the lack of linearity, we conclude that for X^- = NO_2^- and Cl^-, the data show mixed first- and second-order kinetics. Presumably NCS^- would also behave similarly at higher concentrations. This makes the reactions candidates for I_a, D or I_d mechanisms.

(b) Unless some other entering group could be found which entered faster than dmso, the anations should be regarded as having a d intimate mechanism.

(c) k_1 (the rate constant for Co–O bond breaking). (See Table 9.7 in DMA.)

(d) kK (the product of the outer sphere complexation constant K and the rate constant for ligand interchange). (See Table 9.7 in DMA.)

(e) No appreciable discrimination among entering groups is observed. Unless some other group could be found which gives a very different rate, the mechanism should be regarded as $\underline{I_d}$ since the five-coordinate intermediate of a $\underline{D}$ mechanism should be able to discriminate among entering ligands. The observed behavior indicates that the breaking of the Co–dmso bond is rate-determining and the entering group is whatever happens to be in the second coordination sphere - the defining criteria for an $\underline{I_d}$ mechanism.

9.10 Problem: The following data were obtained for base hydrolysis of $(+)_{589}$-$[Co(en)_2(NH_3)X]^{n+}$ in the presence of added N_3^- (1.0 M) and are reproducible to $\pm$ 0.50%. (R = configuration retention, I = inversion.)

Complex	Products					
	$[Co(en)_2(NH_3)OH]^{2+}$, %			$[Co(en)_2(NH_3)N_3]^{2+}$, %		
	trans	cisR	cisI	trans	cisR	cisI
$[Co(en)_2(NH_3)Cl]^{2+}$	17	48	11	7	13	4
$[Co(en)_2(NH_3)Br]^{2+}$	17	48	12	7	12	4
$[Co(en)_2(NH_3)(NO_3)]^{2+}$	17	46	11	8	13	5
$[Co(en)_2(NH_3)(dmso)]^{3+}$ [a]	16	45	9	10	17	4
$[Co(en)_2(NH_3)(tmp)]^{3+}$ [b]	16	46	8	10	18	3

[a] dmso= dimethylsulfoxide [b] tmp = trimethylphosphate

What do these data reveal about the stoichiometric mechanism of these reactions? (See D.A. Buckingham, I.I. Olsen and A.M. Sargeson, J. Am. Chem. Soc. 1968, 90, 6654; D.A. Buckingham, C.R Clark and T.W. Lewis, Inorg. Chem. 1979, 18, 1985.)

9.10 Solution: The chloro, bromo and nitrato compounds must proceed through the same intermediate, which constitutes evidence in favor of a $\underline{D}$ mechanism. The same argument can be made for the dmso and tmp complexes, but the structure of the common intermediate must be different from that in the other three cases. Perhaps ion pairing occurs in the first three cases.

9.11 Problem: The complexes trans-$[Rh(en)_2LX]^+$ react with various Y^-, giving $[Rh(en)_2LY]^+$. The rate law for the appearance of the product is:

$$\text{rate} = (k_1 + k_2[Y^-])[Rh(en)_2LX]^+.$$

Do these data constitute evidence for an a mechanism? Why or why not? (See A.J. Poë and C.P.J. Vuik, Inorg. Chem. 1980, 19, 1771.)

9.11 Solution: Not necessarily. The form of the rate law is compatible with two parallel paths, one of which involves ion pairing. k_2 might be equal to some kK.

9.12 Problem: The anation of trans-$[Rh(en)_2(H_2O)_2]^{3+}$ with chloride has recently been studied. Two plots of k_{obs} vs. concentration of species in solution are reproduced below:

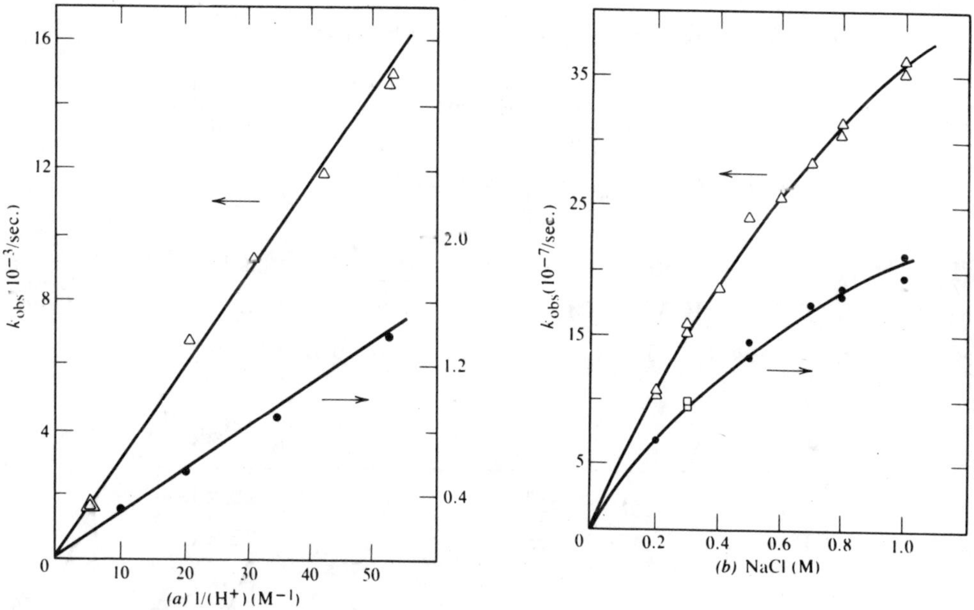

(a) Dependence of k_{obs} on reciprocal concentration of H^+, with I = 1.00 *M*, (complex) = 0.001 *M*: △,65°,(NaCl) = 0.801 *M*; ○,50°,(NaCl) = 0.641 *M*. Arrows indicate the corresponding ordinate. (*b*) Dependence of k_{obs} on concentration of chloride with T = 40°, *I* = 1.00 *M*, and (complex) = 0.001 *M*: △,(H^+) = 0.0190 *M*; ○,(H^+) = 0.199 *M*. Arrows indicate the corresponding ordinate. (Reprinted with permission from M. J. Pavelich, *Inorg. Chem.* 1975, *14*, 982. Copyright 1975, American Chemical Society.)

(a) Account for the linearity of the plots in Figure a.

(b) What does the non-zero intercept for each of the two lines in Figure a indicate about the reaction mechanism?

(c) Account for the shape of the two curves in Figure b. Do these curves alone allow you to distinguish what the stoichometric mechanism is?

9.12 Solution:

(a) The reaction rate increases linearly as $[H^+]$ decreases. This implies that the reactive species is a deprotonated one.

(b) Because the intercept is non-zero, a second pathway involving slower reaction of the parent (protonated) species is also operative. From this information the rate law is:

$$\text{rate} = \left(a + \frac{b}{[H^+]}\right)[Rh(en)_2(H_2O)_2{}^{3+}]$$

(c) These curved plots indicate the transition from first-order to mixed first-and zero-order behavior in $[Cl^-]$. Thus, the complex rate law has the form:

$$\text{rate} = \left(a + \frac{b}{[H^+]}\right)[Rh(en)_2(H_2O)_2{}^{3+}]\frac{[Cl^-]}{1+d[Cl^-]}$$

The rate law is compatible with D, I_d or I_a mechanisms.

9.13 Problem: Rate constants for the reaction trans-$[Pt(py)_2Cl_2]$ + $XC_6H_4SC_6H_4Y \rightarrow$ trans-$[Pt(py)_2Cl(XC_6H_4SC_6H_4Y)]$ + Cl^- were measured in methanol at 30°C. From these data calculate $n_{Pt} = \log k_2(L) - \log k_2(MeOH)$ for the following ligands:

X	Y	$k_2(M^{-1}sec^{-1})$
NH_2	NH_2	7.25×10^{-3}
OH	OH	2.72×10^{-3}
CH_3	CH_3	1.84×10^{-3}
NH_2	NO_2	0.78×10^{-3}
NO_2	NO_2	0.096×10^{-3}

(See J.R. Gaylor and C.V. Senoff, Can. J. Chem. 1971, 49, 2390.)

The value of k_2 for L = MeOH is $2.7 \times 10^{-7} M^{-1} sec^{-1}$. (See Table 9.21 in DMA.)

9.13 Solution:

X	Y	$n_{Pt} = \log k_2(L) - \log k_2(MeOH)$
NH_2	NH_2	$\log(7.25 \times 10^{-3}) - \log(2.7 \times 10^{-7}) = 4.43$
OH	OH	$\log(2.72 \times 10^{-3}) - \log(2.7 \times 10^{-7}) = 4.00$
CH_3	CH_3	$\log(1.84 \times 10^{-3} - \log(2.7 \times 10^{-7}) = 3.83$
NH_2	NO_2	$\log(0.78 \times 10^{-3}) - \log(2.7 \times 10^{-7}) = 3.46$
NO_2	NO_2	$\log(0.096 \times 10^{-3}) - \log(2.7 \times 10^{-7}) = 2.55$

9.14 Problem: A plot of k_{obs} vs. $[X^-]$ is shown for the reactions $[Pd(Et_4dien)Br]^+ + X^- \rightarrow [Pd(Et_4dien)X]^+ + Br^-$. Account for the shape of the plot. In particular, what mechanism can you propose to account for the zero slope when $X^- = N_3^-, I^-, NO_2^-, SCN^-$?

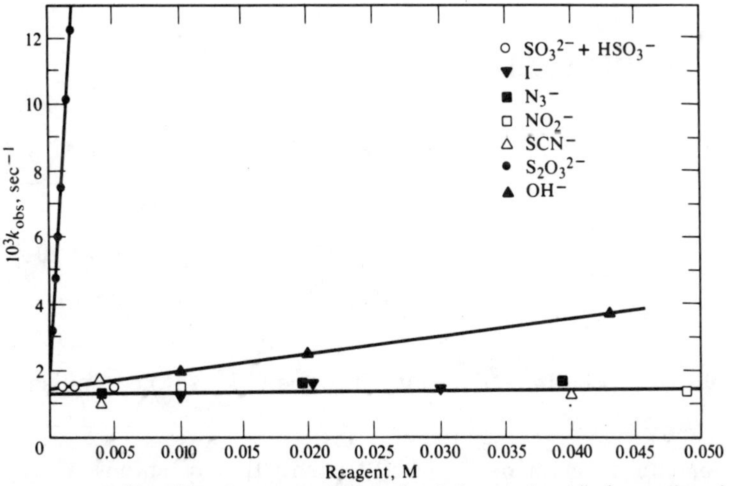

Plot of k_{obs} vs. concentration of entering nucleophile for anation of $[Pd(Et_4dien)Br]^+$ in water at 25°C. (Reprinted with permission from J. B. Goddard and F. Basolo, *Inorg. Chem.* 1968 7, 936. Copyright 1968, American Chemical Society.)

9.14 Solution: The linear dependence on $[S_2O_3^{2-}]$ and $[OH^-]$ indicates direct attack by these species. The intercepts of all lines are the same and indicate a substitution path independent of the nature of the incoming ligand. This is presumably not water attack followed by water displacement by X^-. If it were, we would have to postulate that SO_3^{2-}, I^-, N_3^-, NO_2^-, and SCN^- are all much less nucleophilic toward the Pd compound than water since the rate never depends on their concentrations. This would be in marked contrast to their behavior toward Pt complexes. Probably the sterically crowded Et_4dien complex undergoes rate-determining dissociation of Br^- followed by rapid ligand attack for SO_3^{2-}, I^-, N_3^-, NO_2^-, and SCN^- which are less nucleophilic than $S_2O_3^{2-}$ and OH^-. For $S_2O_3^=$ and OH^- parallel a and d paths operate.

9.15 Problem: Base hydrolysis is a well known reaction of octahedral complexes and is generally about 10^6 times faster than acid hydrolysis. Given the following facts, what can you say about the existence and importance, if any, of a special mechanism for base hydrolysis of square planar complexes?

(a) Au(III) complexes which are also d^8 typically react about 10^4 times as fast as Pt(II) complexes. The following data are representative:

$$[M(dien)Cl]^{n+} + Br^- \xrightarrow{H_2O} [M(dien)Br]^{n+} + Cl^-$$

M	n	$k_1(sec^{-1})$	$k_2(M^{-1}sec^{-1})$
Pt	1	8.0×10^{-5}	3.3×10^{-3}
Au	2	0.5	154

(b) The reactivity of $[Au(dien)Cl]^{2+}$ and $[Pt(dien)Cl]^+$ toward anions increases in the order $OH^- << Br^- < SCN^- < I^-$.

(c) The replacement of chloride in $[Au(Et_4dien)Cl]^+$ by anions Y^- is independent of $[Y^-]$. The rate constant (25°C) increases with pH from $k = 1.9 \times 10^{-6}$ to a maximum of $1.3 \times 10^{-4} sec^{-1}$.

9.15 Solution: (a) The data imply the usual a mechanism for square planar substitution since rates increase with electrophilicity of the metal complex in the order Pt(II) < Au(III).

(b) We can conclude that OH^- is a poor entering group toward square planar complexes. Its mechanistic role, if any, would be confined to formation of a conjugate base. But this does not happen with the dien complexes since OH^- is the least reactive.
(c) The fact that the substitution rate of the Et_4dien complex is orders of magnitude smaller than for dien complexes and independent of $[Y^-]$ shows that rate-determining very slow dissociation of Cl^- must be occurring in this sterically crowded complex. The rate increase with pH implies that the conjugate base is more effective at dissociating Cl^- than the protonated parent complex.

9.16 Problem: The following activation parameters have been measured for the reactions:

$$[PtL_2ClX] + py \xrightarrow[\text{methanol}]{30^\circ C} [PtL_2Cl(py)]^+ + X^-$$

Complex	$k_2(M^{-1}sec^{-1})$	$\Delta H_2^\ddagger$ (kJ/mole)	$\Delta S_2^\ddagger$ (J/Kmole)	$\Delta V^\ddagger (cm^3/mole)$
trans-$[Pt(py)_2Cl(NO_2)]$	7.35×10^{-3}	49.3	-94	-8.8
cis-$[Pt(py)_2Cl(NO_2)]$	0.150×10^{-3}	55.2	-110	-19.8
trans-$[Pt(PEt_3)_2Cl_2]$	0.53×10^{-3}	53.9	-100	-13.6

What mechanistic information can be extracted from these values? (See M. Kotowski, D.A. Palmer, and H. Kelm, Inorg. Chem. 1979, 18, 2555.)

9.16 Solution: A negative $\Delta S_2^\ddagger$ is consistent with an associative (a) mechanism since we expect degrees of motional freedom to be restricted by incipient formation of a new metal-ligand bond before the old one is broken. Likewise, the negative $\Delta V_2^\ddagger$ is compatible with the formation of a closely associated species occupying less volume than the separate individual components. The relatively small dependence of $\Delta H_2^\ddagger$ on the nature of the leaving group also points to an a mechanism.

9.17 Problem: Show that for a redox reaction of the type:

$$A^+ + B \overset{K}{\rightleftharpoons} \text{Intermediate} \xrightarrow{k} A + B^+$$

where B is present in excess, the rate law will be of the form:

$$\text{rate} = \frac{a[A^+]_o[B]}{1 + b[B]}$$

whether the intermediate is a precursor or successor complex. $[A]_o$ is the total initial concentration of A. (Hint: A will be partitioned between free A and the Intermediate.)

9.17 Solution: $[A^+]_o = [A^+] + [\text{Intermediate}]$

If B is in excess, the concentration will not be appreciably diminished by being tied up in the intermediate.

$$[\text{Intermediate}] = K[A^+][B]$$

$$\text{rate} = k_{obs}[A^+]_o[B] = k[\text{Intermediate}]$$

$$= k_{obs}([A^+] + K[A^+][B])[B] = kK[A^+][B]$$

$$k_{obs} = \frac{kK}{1 + K[B]} \quad \text{and rate} = \frac{kK}{1 + K[B]} \quad [A^+]_o[B]$$

where $kK = a$ and $K = b$.

Again, this is a special case of the mechanism discussed in Problem 9.5.

9.18 Problem: From the Marcus Equation, $k_{12} = \sqrt{k_{11}k_{22}K_{12}f}$ where:

$$\log f = \frac{(\log K_{12})^2}{4 \log \left(\frac{k_{11}k_{22}}{Z^2}\right)}$$

and Z is the number of collisions per second at the temperature of interest (10^{11} at 25^oC), calculate k_{12} for the following outer sphere redox reactions. Measured values of k_{12} are given for comparison.

Reaction	$k_{11}(M^{-1}sec^{-1})$	$k_{22}(M^{-1}sec^{-1})$	E^o(volts)	k_{12}^{meas} $(M^{-1}sec^{-1})$
$Cr^{2+} + Fe^{3+}$	$\leqslant 2 \times 10^{-5}$	4.0	1.18	2.3×10^3
$[W(CN)_8]^{4-} + Ce(IV)$	$> 4 \times 10^4$	4.4	0.90	$> 10^8$
$[Fe(CN)_6]^{4-} + MnO_4^-$	7.4×10^2	3×10^3	0.20	1.7×10^5
$[Fe(phen)_3]^{2+} + Ce(IV)$	$> 3 \times 10^7$	4.4	0.36	1.4×10^5

9.18 Solution: K_{eq} is obtained from E^o through the Nernst equation $\Delta G^o = -nFE^o = -RT\ell nK_{12}$

or $\log K_{12} = \frac{E^o}{0.059}$ at 25°. Substitutions in the Marcus equation then gives:

	f	k_{12}(calc)
$Cr^{2+} + Fe^{3+}$	1.5×10^{-4}	$\leqslant 1 \times 10^6$
$[W(CN)_8]^{4-}$ + Ce(IV)	3.4×10^{-4}	$> 3 \times 10^8$
$[Fe(CN)_6]^{4-} + MnO_4^-$	6.6×10^{-1}	6×10^4
$[Fe(phen)_3]^{2+}$ + Ce(IV)	2.1×10^{-1}	$> 6 \times 10^6$

9.19 Problem: Assign an outer- or inner-sphere mechanism for each of the following:

(a) The main product of the reaction between $[Cr(NCS)F]^+$ and Cr^{2+} is CrF^{2+}. (See F.N. Welch and D.E. Pennington, Inorg. Chem. 1976, 15, 1515.)

(b) When $[VO(edta)]^{2-}$ reacts with $[V(edta)]^{2-}$ a transient red color is observed. (See F.J. Kristine, D.R. Gard and R.E. Shepherd, Chem. Comm. 1976, 994.)

(c) The rates of reduction of $[Co(NH_3)_5(py)]^{3+}$ by $[Fe(CN)_6]^{4-}$ are insensitive to substitution on py. (See A.J. Miralles, R.E. Armstrong and A. Haim, J. Am. Chem. Soc., 1976, 99, 1416.)

(d) The rate of reduction of $[Co(NH_3)_5(NCS)]^{2+}$ by Ti^{3+} is 36,000 times smaller than the rate of $[Co(NH_3)_5(N_3)]^{2+}$ reduction. (See J.P. Birk, Inorg. Chem. 1975, 14, 1724.)

(e) Activation parameters for some reductions by V^{2+} are:

Complex	$\Delta H^\ddagger$(kJ/mole)	$\Delta S^\ddagger$(J/mole K)
$[Co(NH_3)_5F]^{2+}$	46.4	-77.4
$[Co(NH_3)_5Cl]^{2+}$	31.4	-120
$[Co(NH_3)_5Br]^{2+}$	30.1	-115
$[Co(NH_3)_5I]^{2+}$	30.5	-103
$[Co(NH_3)_5N_3]^{2+}$	48.9	-58.5
$[Co(NH_3)_5SO_4]^+$	48.5	-54.8

(See M.R. Hyde, R.S. Taylor and A.G. Sykes, J. Chem. Soc. Dalton Trans. 1973, 2730.)

(f) A series of Co(III) carboxylato complexes is known to be reduced by Cr^{2+} in an inner-sphere mechanism. The rate constants for Eu^{2+} give the following log-log plot vs. $k_{Cr^{2+}}$. (See F.-R. Fan and E.S. Gould, Inorg. Chem. 1974, 13, 2639.)

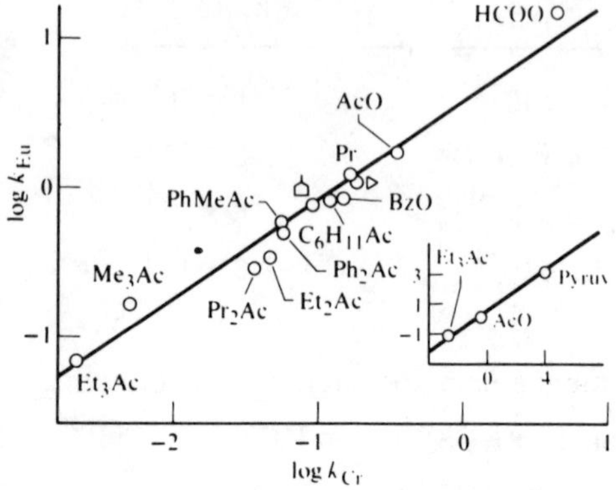

Log-log plot comparing specific rates of reductions of carboxylatopentaamminecobalt(III) complexes $R(NH_3)_5Co^{2+}$ by Eu^{2+} and Cr^{2+} at 25° (Reprinted with permission from F.-R. Fan and E. S. Gould, *Inorg. Chem.* 1974, *13*, 2639. Copyright 1974, American Chemical Society.)

(g) Reduction of b by Cr^{2+} is much faster than reduction of a.

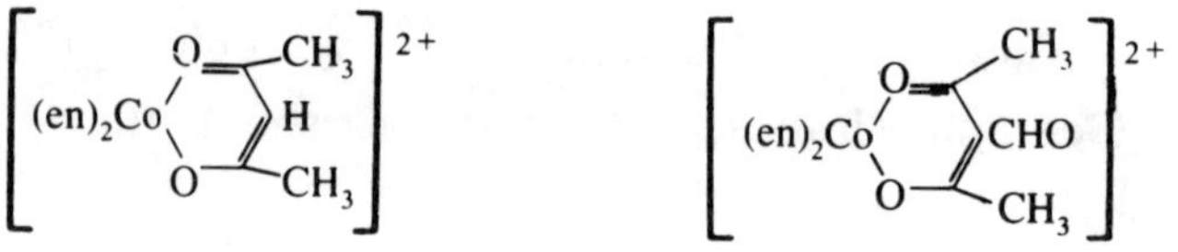

(a) (b)

(See R.J. Balahura and N.A. Lewis, Can. J. Chem. 1975, 53, 1154.)

(h) The reduction rates of $[(NH_3)_5Co\text{-}O\text{-}C(O)R]^{2+}$ (R = Me, Et) by Eu^{2+}, V^{2+} and Cr^{2+} decrease as the pH decreases. (See J.C. Thomas, J.W. Reed and E.S. Gould, Inorg. Chem. 1975, 14, 1696.)

(i) On mixing 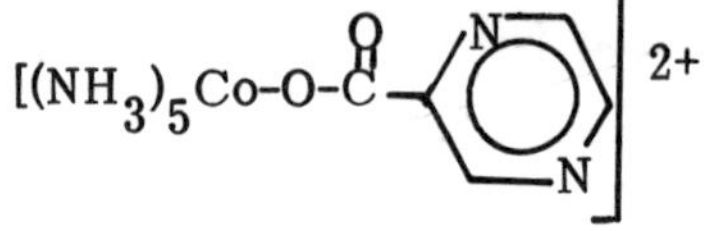 and Cr^{2+} a transient ESR signal could be observed. The g-value indicated that the odd electron resided mainly in the aromatic ring. (See H. Spiecker and K. Wieghardt, Inorg. Chem. 1977, 16, 1290.)

9.19 Solution: See DMA Table 9.30.

(a) Inner sphere. As expected for an inner-sphere mechanism, the bridging F ligand is transferred to the product, $[Cr(H_2O)_5F]^{2+}$.

(b) Inner sphere. The transient red color could be that of the precursor or successor complex which disappears on account of electron transfer or breakup of the complex to products.

(c) Outer sphere. We might expect that substituents on a py ligand would affect the ability of py to form bridges either for steric or electronic reasons. Since all py complexes react at a similar rate, it is likely that none forms the bridged complex required for an inner-sphere mechanism.

(d) Inner sphere. The difference in the ability of S and N to form a bridge with Ti^{3+} could account for the observed difference in rates. This difference is in the direction expected since Ti^{3+} is a hard acid and N a hard base. (See DMA, Chapter 12.)

(e) The data divide into two groups - one with $\Delta S^{\ddagger} \sim -55$ J $mole^{-1}K^{-1}$ and one with more negative $\Delta S^{\ddagger}$. Presuming that these differences indicate a difference in mechanism, we would ascribe an inner-sphere path to all the complexes except the sulfato and azido on account of more negative $\Delta S^{\ddagger}$. See Problem 9.16.

(f) The linear plot indicates that $\log k_{Eu} = 0.67 \log k_{Cr} + 0.54$ or, in other words

$$\log\left(\frac{3k_{Eu}}{2k_{Cr}}\right) = 0.54 \text{ and } k_{Eu} = 2.31\, k_{Cr}$$

A constant ratio exists between the constants for reduction by both Eu^{2+} and Cr^{2+}. Since the Cr^{2+} reductions are known to be inner-sphere, the Eu^{2+} ones are also inner-sphere. (See DMA, Section 9.8.)

(g) The C=Ö: group in b should make the acac ligand a better bridging group because of its available lone pairs. The fact that the rate goes up as the bridging ability of the ligand should increase is evidence that the reduction of b is inner-sphere. (The order of reactivity of a and b was in error in the first printing of DMA. If the inverted order were real, this would suggest an outer-sphere path.)

(h) As pH decreases, $[H^+]$ increases and protons can compete with Eu^{2+}, V^{2+}, or Cr^{2+} for the lone pairs on the carbonyl O. Hence, an inner-sphere mechanism is indicated.

(i) The transfer of an electron to the ring indicates an inner-sphere radical ion mechanism. The ESR signal decays as the electron is transferred to Co^{3+}.

9.20 Problem: The complex $[(bipy)_2ClRu-N$  $N-Ru(bipy)_2Cl]^{5+}$ displays an intervalence transfer absorption band at 7.69×10^3 cm^{-1}. Estimate E_{th}, the energy barrier to thermal electron transfer. If E_{th} can be very approximately equated to $\Delta G^{\ddagger}$ for thermal electron transfer, estimate the rate constant. The thermal rate constant has been reported to be ~3×10^{10} sec^{-1}.

9.20 Solution: $E_{op} = \frac{1}{4} E_{th} = \frac{1}{4}(7.69 \times 10^3 \text{cm}^{-1}) = 1.92 \times 10^3 \text{cm}^{-1}$

For a mole of complex ions:

$$\Delta G^{\ddagger} \sim NE_{op} = 6.022 \times 10^{23} \text{mole}^{-1} (1.92 \times 10^3 \text{cm}^{-1})(6.63 \times 10^{-34} \text{J sec}) \times (3.00 \times 10^{10} \text{cm sec}^{-1})$$

$$= 2.30 \times 10^4 \text{J mole}^{-1}$$

From activated complex theory:

$$k = \frac{kT}{h} e^{-\Delta G^{\ddagger}/RT}$$

$$= \frac{(1.38 \times 10^{-23} \text{ J K}^{-1})(3.00 \times 10^2 \text{K})}{6.63 \times 10^{-34} \text{ J sec}} \exp\left(\frac{-2.30 \times 10^4 \text{ J mol}^{-1}}{8.33 \text{ J mol}^{-1} \text{K}^{-1} \times 300\text{K}}\right)$$

$= 6.24 \times 10^{12} \text{sec}^{-1} \exp(-9.20) = 6.24 \times 10^{12} \text{ sec}^{-1} (1.01 \times 10^{-4})$

$= 6.3 \times 10^8 \text{sec}^{-1}$ not in very good agreement with the quoted estimate.

9.21 Problem: According to the classification suggested by Taube, octahedral complexes of d^0, d^1, and d^2 ions are expected to be labile and those of d^3 and low-spin d^6 are expected to be particularly inert with respect to substitution. What mechanism do these observations suggest? Explain why d^3 and low-spin d^6 would be expected to be inert from ligand field considerations by either a dissociative or associative means.

9.21 Solution: An obvious difference between the d^0, d^1, and d^2 cases and the d^3 and low-spin d^6 cases is the availability of (at least) one additional d orbital for d^{0-2} to accommodate an incoming ligand. This suggests an associative mechanism. However, later evidence has not confirmed this view. The d^3 (t_{2g}^3) and low-spin d^6 (t_{2g}^6) configurations give very high octahedral ligand field stabilization energy (LFSE). A change to C.N. 5 or 7 involves a great loss in LFSE for these configurations. This suggests high activation energy for either mechanism.

9.22 Problem: How might you account for the observation that the rate of hydrolysis of

$[Co(en)_2F_2]NO_3$ is greatly enhanced by either acid or base — it increases linearly with $[H^+]$ below pH = 2 or with $[OH^-]$ above pH = 6?

9.22 Solution: The linear increase with $[OH^-]$ above pH 6 is base hydrolysis. Below pH 2 $[H^+]$ is sufficient to protonate an F ligand assisting its departure as HF. The reaction has a three-term rate law in which different terms become predominant in different pH regions. (See DMA p. 348.)

$$\text{rate} = k_H [Co(en)_2F_2^+][H^+] + k [Co(en)_2F_2^+] + k_{OH} \frac{[Co(en)_2F_2^+]}{[H^+]}$$

9.23 Problem: What mechanistic implications may be drawn from the observation that base hydrolysis of Λ-cis-$[Co(en)_2(NH_3)Cl]^{2+}$ yields 84% cis and 16% trans-$[Co(en)_2(NH_3)(OH)]^{2+}$ whereas acid hydrolysis in the presence of $AgClO_4$ yields 75% cis and 25% trans isomer?

9.23 Solution: The 84%:16% cis:trans ratio is the product distribution which would be expected if the square pyramidal species can undergo substitution before rearrangement and also rearrange to all possible trigonal bipyramids before substitution with equal probability. See the figure below. The 75%:25% cis:trans ratio would result if only a square pyramid and one trigonal bipyramid (the result of L,N_3 collapse) could form prior to substitution.

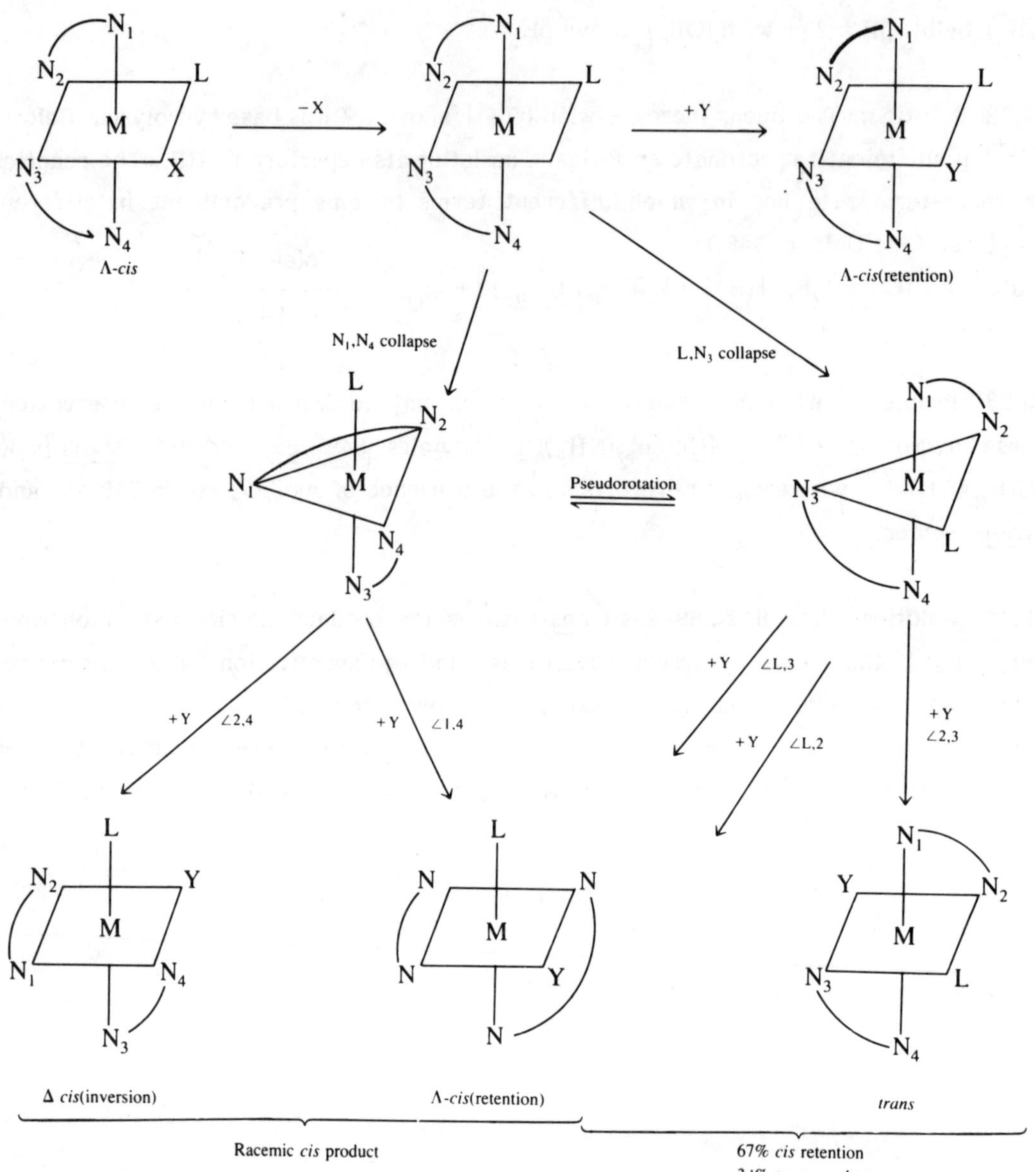

Stereochemical course of substitution in *cis*- and *trans*-$[M(N\text{-}N)_2LX]^{n+}$ complexes (After *Ligand Substitution Processes*, 1966 by C. H. Langford and H. B. Gray with permission of the publisher, Benjamin/Cummings, Inc, Reading, Mass.)

9.24 Problem: Recently the optically active Co complex:

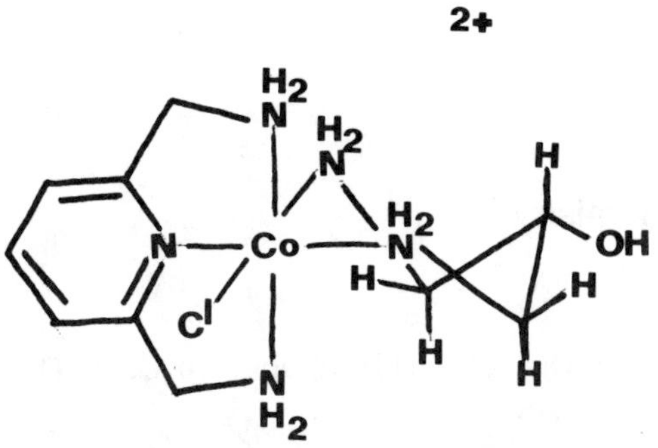

was prepared and found to undergo base hydrolysis with complete retention of stereochemistry. What does this result suggest about the geometry of the five-coordinate intermediate which is usually presumed to be involved in the D-CB mechanism? (See U. Tinner and W. Marty, Inorg. Chem. 1981, 20, 3750.)

9.24 Solution: The π-orbital of the pyridine ring is in an ideal position for overlap with an empty Co orbital vacated by the departure of Cl^-. Hence, it would be expected to stabilize a TBP intermediate. (See DMA Figure 9.4.) Water would have equal probability of entering on either side of the pyridine ring which should lead to complete racemization in the product. Since this is not observed, the geometry of the intermediate cannot be that described above.

X Organometallic Chemistry

10.1 Problem: Show that the EAN rule applies to $Os_3(CO)_{12}$ whose molecular geometry appears below.

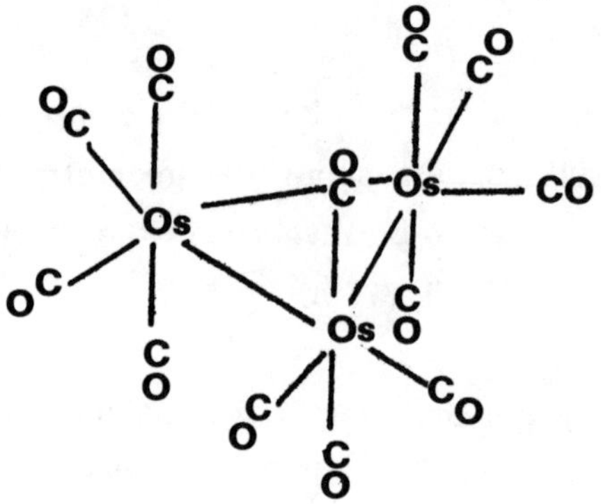

10.1 Solution: The EAN rule requires that each transition metal be surrounded by a total of eighteen electrons. For each Os we have:

Os(0)	+ 4 CO	+ 2 Os-Os bonds	
8e	+ 4 x 2e	+ 2 x 1e	= 18e

10.2 Problem: Give an electron count for $Ir(PPh_3)_2(CO)Br(tcne)$. (tcne=tetracyano-ethylene. See DMA Figure 10.7.) Does it conform to the EAN rule?

10.2 Solution:

Ir(I)	+ 2 PPh_3	+ CO	+ Br^-	+ tcne	
8e	+ 2 x 2e	+ 2e	+ 2e	+ 2e	= 18e

Yes, the EAN rule applies.

10.3 Problem: Show that each of the following complexes conforms to the EAN rule.

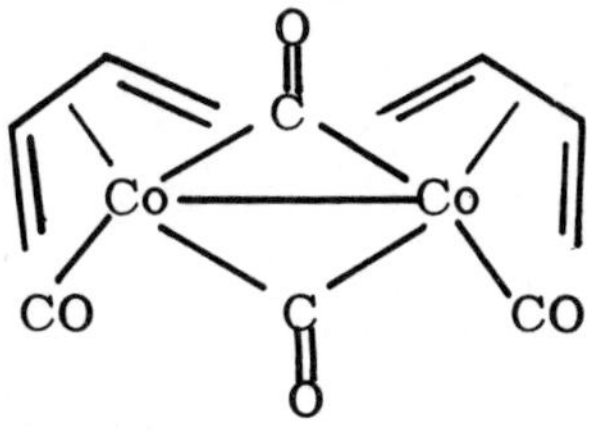

(a)

(b)

10.3 Solution: For each Co in (a):

Co(0)	+ C_4H_6	+ CO	+ 2μ-CO	+ Co-Co bond	
9e +	4e	+ 2e	+ 2 x le	+ le	= 18e

For Fe in (b):

Fe(0)	+ C_4H_6	+ 3 CO	
8e	+ 4e	+ 3 x 2e	= 18e

10.4 Problem: Using the molecular orbital energy diagram given on page 178, write down electron configurations which account for the number of unpaired electrons (given in parentheses) for each of the following (Cp = η^5-C_5H_5): Cp_2V(3), Cp_2V^+(2), Cp_2Cr (2), Cp_2Mn(5), Cp_2Fe(0), Cp_2Fe^+(1), Cp_2Co(1), Cp_2Co^+(0), Cp_2Ni(2), Cp_2Ni^+(1). Keep in mind that the pairing energies may sometimes be comparable in magnitude to orbital separations.

10.4 Solution: All compounds have $(1a_{1g})^2(1a_{2u})^2(1e_{1g})^4(1e_{1u})^4$...

$Cp_2V...(1e_{2g})^2(2a_{1g})^1$

$Cp_2V^+...(1e_{2g})^2$

$Cp_2Cr...(2a_{1g})^2(1e_{2g})^2$

$Cp_2Mn...(1e_{2g})^2(2a_{1g})^1(2e_{1g})^2$

$Cp_2Fe...(1e_{2g})^4(2a_{1g})^2$

$Cp_2Fe^+...(1e_{2g})^4(2a_{1g})^1$

$Cp_2Co...(1e_{2g})^4(2a_{1g})^2(2e_{1g})^1$

$Cp_2Co^+...(1e_{2g})^4(2a_{1g})^2$

$Cp_2Ni...(1e_{2g})^4(2a_{1g})^2(2e_{1g})^2$

$Cp_2Ni^+...(1e_{2g})^4(2a_{1g})^2(2e_{1g})^1$

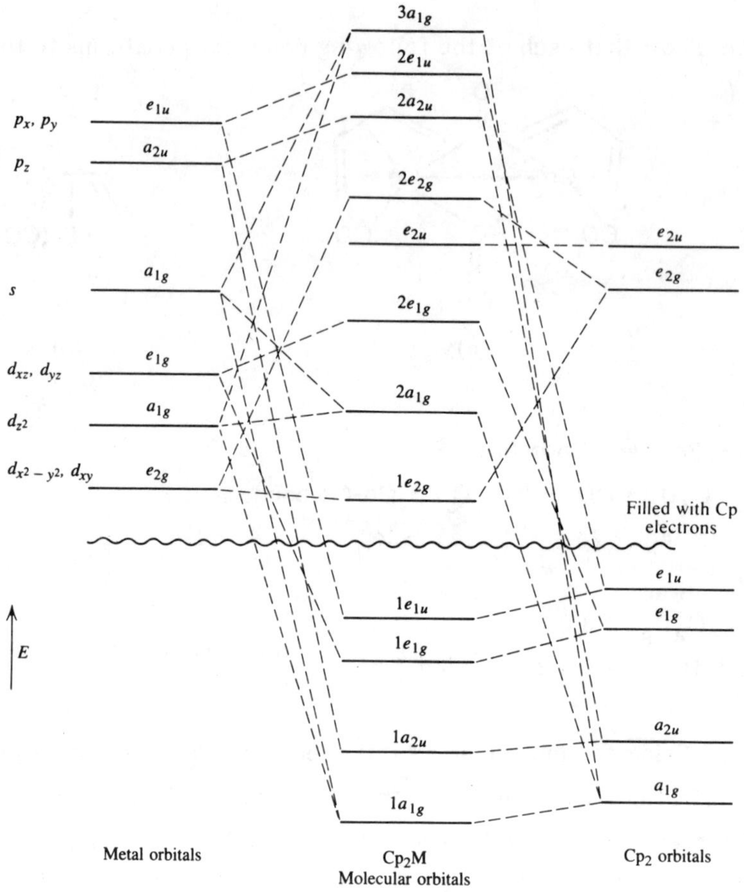

Molecular orbital energy diagram for metallocene complexes. (Reproduced with permission from S. Sohn, D.N. Hendrickson, and H.B. Gray, J. Am. Chem. Soc. 1971, 93, 3603. Copyright 1971, American Chemical Society.)

10.5 Problem: (a) The butene product from the reaction $CpFe(CO)(PPh_3)(n\text{-}C_4H_9) \rightarrow Cp\text{-}Fe(CO)(PPh_3)H$ + butene consists of 1-butene as well as cis- and trans-2-butene. Keeping in mind the reversibility of the β-hydride elimination, write a mechanism to account for this. (b) When the reaction in (a) is run with $CpFe(CO)(PPh_3)(CD_2CH_2Et)$, complete scrambling of the D occurs. Write a mechanism to account for this observation.

10.5 Solution:

(a) $CpFe(CO)(PPh_3)CH_2CH_2Et \rightleftharpoons CpFe(CO)CH_2CH_2Et + PPh_3$

$$CpFe(CO)CH_2CH_2Et \rightleftharpoons CpFe\overset{CO}{\underset{H}{\diagdown}} \longleftarrow \overset{CHEt}{\underset{CH_2}{||}} \overset{PPh_3}{\longrightarrow} CpFe\overset{CO}{\underset{PPh_3}{\diagdown}}-H + \text{1-butene}$$

$$\updownarrow$$

$$CpFe-\underset{CH_3}{\underset{|}{CHEt}} \rightleftharpoons CpFe\overset{CO}{\underset{H}{\diagdown}} \longleftarrow \overset{CHCH_3}{\underset{CHCH_3}{||}} \longrightarrow CpFe\overset{CO}{\underset{PPh_3}{\diagdown}}-H + \text{2-butene}$$

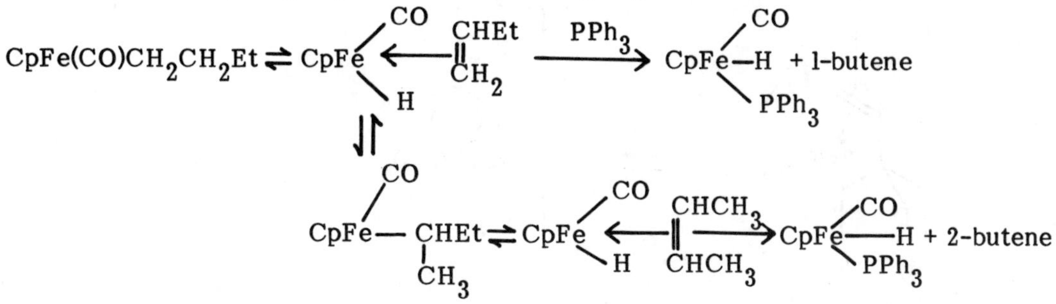

The β-elimination from the <u>sec</u>-C_4H_9 species can produce <u>cis</u>- or <u>trans</u>-2-butene depending on the rotamer from which the elimination occurs.

(b) A reversible mechanism such as that in (a) provides a pathway for transferring D to Fe and transferring it back to different C's.

10.6   Problem: (a) A reaction thought to be important in catalysis is the so-called 1,3-hydride shift. An alkyl C–H bond in a propene π-bonded ligand adds oxidatively to the metal, giving an η^3-allyl hydride complex. Show how the reversibility of this process can afford a mechanism for shifting H from C_1 to C_3. (b) Account for the following observation by a mechanism involving the 1,3-hydride shift: in the

complex [structure] $Fe(CO)_4$, the olefin is isomerized to give [structure with (1/3 D) and (1/3D)(1/3D)] $Fe(CO)_4$

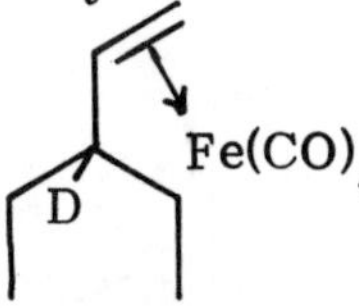

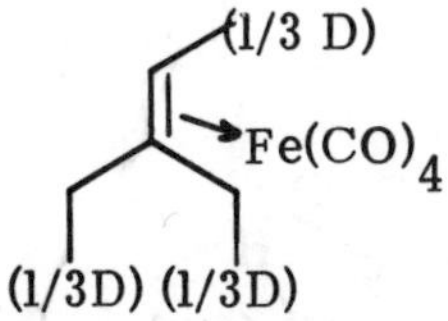

and the D is scrambled to all terminal methyl groups. (C.P. Casey and C.R. Cyr., <u>J</u>. <u>Am</u>. <u>Chem</u>. <u>Soc</u>. 1973, <u>95</u>, 2248).

10.6 Solution:

(a) [reaction scheme structures]

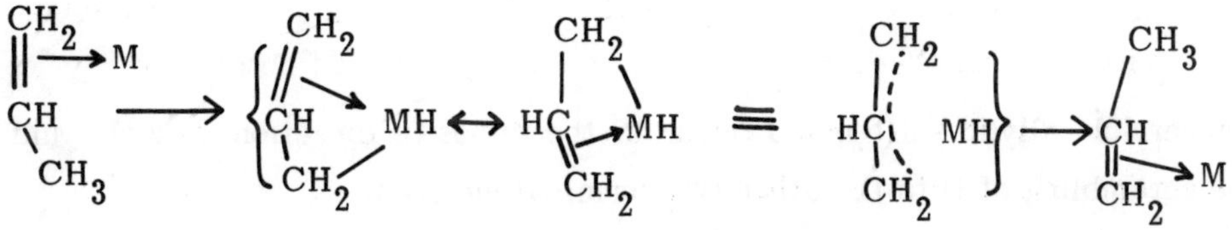

(b)

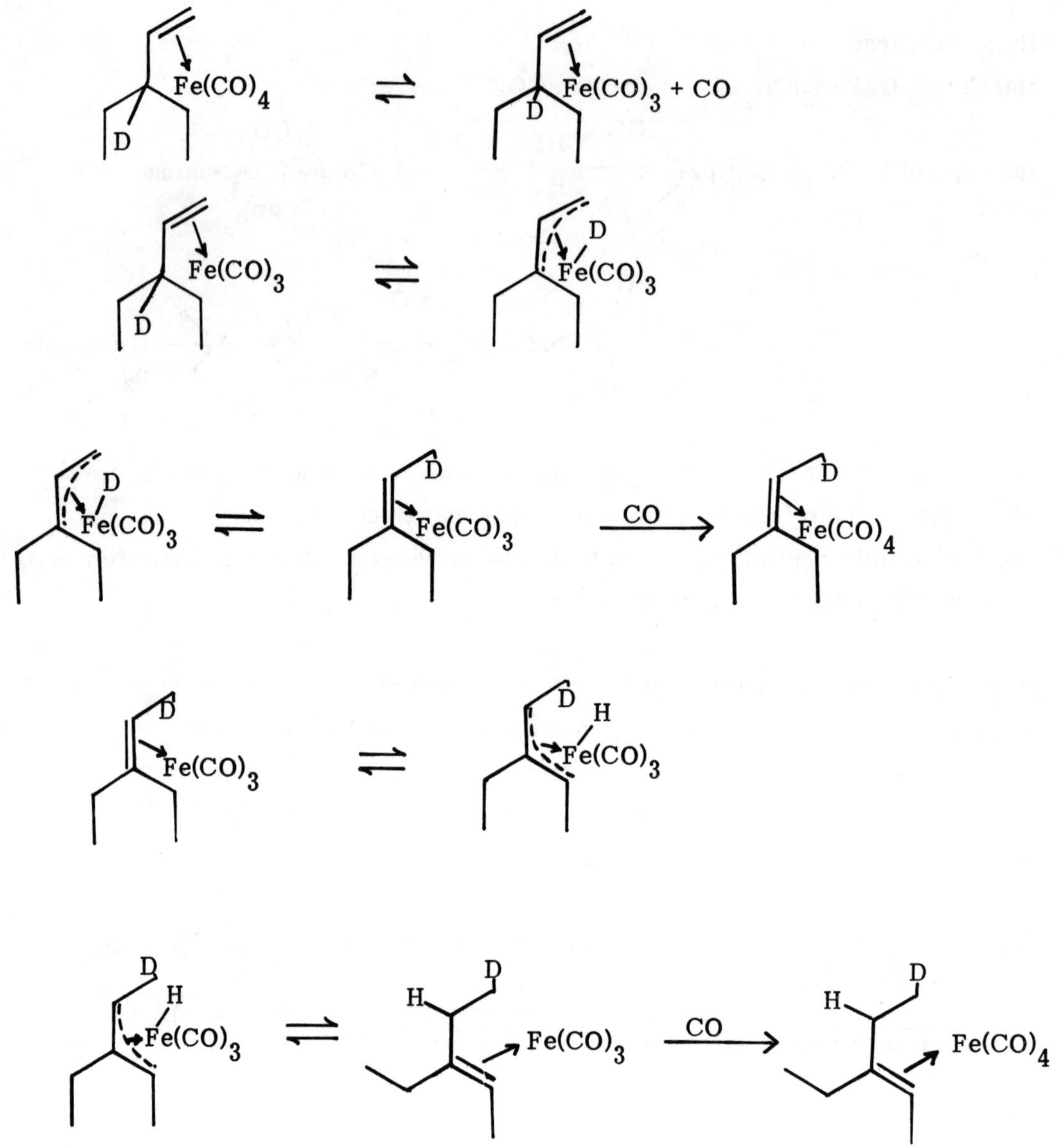

The last reversible allylic shift could occur in the opposite direction as well, and these lead to the scrambling of D to the other two terminal methyls.

10.7 Problem: Write a catalytic cycle for the production of acetone from propylene via the Wacker Process.

10.7 Solution:

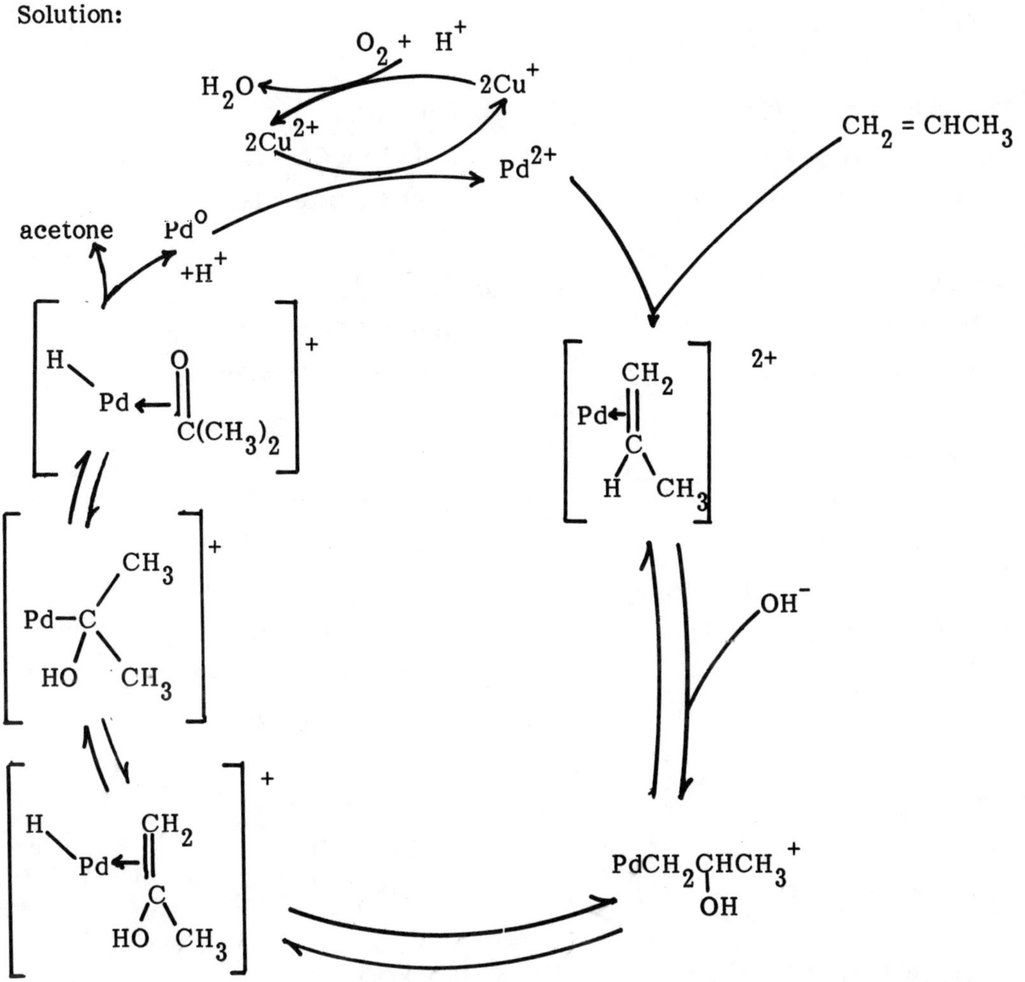

10.8 Problem: Give the valence electron count for the following species. Which ones conform to the EAN rule?

(a) $Co_2(\mu_2\text{-}CO)_2(CO)_6$	(d) $[Fe(CN)_6]^{4-}$	(g) $Fe(CO)_4Br_2$	(j) $HRh(CO)_4$
(b) $W(CO)_6$	(e) $[Cr(CO)_5]^{2-}$	(h) $[Mn(CO)_5]^{2-}$	(k) $Ru_3(CO)_{12}$
(c) $Cr(CNMe)_6$	(f) $Mn(CO)_5(CH_2C_6H_5)$	(i) $Ni(PPh_3)_4$	(l) $[Co(CN)_5]^{2-}$

(m) $H_2Fe(CO)_4$

10.8 Solution: The EAN rule requires 18 valence electrons. All species except $[Co(CN)_5]^{2-}$ conform to the EAN rule.

(a) For each Co in $Co_2(CO)_8$

Co(0) + 2 μ-CO + 3 CO + Co-Co

9e + 2e + 3 x 2e + 1e = 18e

(b) $W(CO)_6$

W(0) + 6 CO

6e + 6 x 2e = 18e

c) $Cr(CNMe)_6$

Cr(0) + 6 CNMe

6e + 6 x 2e = 18e

(d) $[Fe(CN)_6]^{4-}$

Fe(II) + 6 CN^-

6e + 6 x 2e = 18e

(e) $[Cr(CO)_5]^{2-}$

Cr(-II) + 5 CO

8e + 5 x 2e = 18e

(f) $Mn(CO)_5(CH_2C_6H_5)$

Mn(I) + 5 CO + $benzyl^-$

6e + 5 x 2e + 2e = 18e

(g) $Fe(CO)_4Br_2$

Fe(II) + 4CO + 2 Br^-

6e + 4 x 2e + 2 x 2e = 18e

(h) $[Mn(CO)_5]^-$

Mn(-I) + 5 CO

8e + 5 x 2e = 18e

(i) $Ni(PPh_3)_4$

Ni(0) + 4 PPh_3

10e + 4 x 2e = 18e

(j) $HRh(CO)_4$

Rh(I) + H^- + 4CO

8e + 2e + 4 x 2e = 18e

(k) For each Ru in $Ru_3(CO)_{12}$

Ru(0) + 4 CO + 2 Ru-Ru

8e + 4 x 2e + 2 x 1e = 18e

(l) $[Co(CN)_5]^{2-}$

Co(III) + 5 CN^-

6e + 5 x 2e = 16e

(m) $H_2Fe(CO)_4$

Fe(II) + 2 H^- + 4 CO

6e + 2 x 2e + 4 x 2e = 18e

10.9 Problem: Name each of the species in Problem 10.8.

10.9 Solution:

(a) di-μ-carbonylhexacarbonyldicobalt
(b) hexacarbonyltungsten
(c) hexakis(methyl isocyanide)chromium
(d) hexacyanoferrate(4-)
(e) pentacarbonylchromate(2-)
(f) benzylpentacarbonylmanganese
(g) dibromotetracarbonyliron
(h) pentacarbonylmanganate(1-)
(i) tetrakis(triphenylphosphine)nickel
(j) tetracarbonylhydridorhodium
(k) dodecacarbonyltriruthenium
(l) pentacyanocobaltate(2-)
(m) tetracarbonyldihydridoiron

10.10 Problem: Give the valence electron count for the following species. Which ones conform to the EAN rule?

(a) $Cp_2Ru_2(CO)_4$
(b) $[\eta^3\text{-}CH_2C(CH_3)CH_2]_2Ni$
(c) $(\eta^4\text{-cot})Fe(CO)_3$
(d) $CH_3C(O)Re(CO)_4(CNMe)$
(e) $[(\eta^4\text{-cod})Rh(\mu\text{-}Cl)]_2$
(f) $Ru(Me_2PCH_2CH_2PMe_2)_2$
(g) $CpMo(PPh_3)(CO)_2CH_3$
(h) $CpTa(CO)_4$
(i) $Rh(\eta^2\text{-}C_2H_4)(PPh_3)_2Cl$
(j) $(\eta^6\text{-}C_6H_6)_2Mo$
(k) $HIr(CO)_3(PPh_3)$
(l) $Rh_2(CO)_4(\mu\text{-}Cl)_2$

(cot = cyclooctatetraene; cod = cyclooctadiene)

10.10 Solution:

(a) For each Ru in $Cp_2Ru(CO)_4$

Ru(I) + Cp^- + 2 CO + 1 Ru-Ru

7e + 6e + 2 x 2e + 1e = 18e

(b) $[\eta^3\text{-}CH_2C(CH_3)CH_2]_2Ni$

Ni(II) + 2 $CH_2C(CH_3)CH_2^-$

8e + 2 x 4e = 16e

(c) $(\eta^4\text{-cot})Fe(CO)_3$

Fe(0) + 2 C=C + 3 CO

8e + 2 x 2e + 3 x 2e = 18e

(d) $CH_3C(O)Re(CO)_4(CNMe)$

Re(I) + $acetyl^-$ + 4 CO + CNMe

6e + 2e + 4 x 2e + 2e = 18e

(e) For each Rh in $[(\eta^4\text{-cod})Rh(\mu\text{-Cl})]_2$
$Rh(I) + \eta^4\text{-cod} + \mu\text{-Cl} + Cl^-$
8e + 4e + 2e + 2e = 16e

(f) $Ru(Me_2PCH_2CH_2PMe_2)_2$
Ru(0) + 2 diphosphine
8e + 2 x 2 x 2e = 16e

(g) $CpMo(PPh_3)(CO)_2CH_3$
$Mo(II) + Cp^- + PPh_3 + 2\ CO + CH_3^-$
4e + 6e + 2e + 2 x 2e + 2e = 18e

(h) $CpTa(CO)_4$
$Ta(I) + Cp^- + 4\ CO$
4e + 6e + 4 x 2e = 18e

(i) $Rh(C_2H_4)(PPh_3)_2Cl$
$Rh(I) + C_2H_4 + 2\ PPh_3 + Cl^-$
8e + 2e + 2 x 2e + 2e = 16e

(j) $(\eta^6\text{-}C_6H_6)_2Mo$
$Mo(0) + 2\ \eta^6\text{-}C_6H_6$
6e + 2 x 6e = 18e

(k) $HIr(CO)_3(PPh_3)$
$Ir(I) + H^- + 3\ CO + PPh_3$
8e + 2e + 3 x 2e + 2e = 18e

(l) For each Rh in $Rh_2(CO)_4(\mu\text{-Cl})_2$
$Rh(I) + 2\ CO + \mu\text{-Cl} + Cl^-$
8e + 2 x 2e + 2e + 2e = 16e

All except (b), (e), (f), (i) and (l) conform to the EAN rule.

10.11 Problem: Name each of the species in Problem 10.10.

10.11 Solution:

(a) tetracarbonyldicyclopentadienyldiruthenium
(b) bis[η^3-(2-methyl)allyl] nickel
(c) tricarbonyl(η^4-cyclooctatetraene)iron
(d) acetyltetracarbonyl(methyl isocyanide)rhenium
(e) di-μ-chlorobis(η^4-cyclooctadiene)dirhodium
(f) bis[1,2-bis(dimethylphosphino)ethane] ruthenium
(g) dicarbonylcyclopentadienylmethyl(triphenylphosphine)molybdenum
(h) tetracarbonylcyclopentadienyltantalum
(i) chloro(η^2-ethylene)bis(triphenylphosphine)rhodium
(j) bis(η^6-benzene)molybdenum
(k) tricarbonylhydrido(triphenylphosphine)iridium
(l) tetracarbonyldi-μ-chlorodirhodium

10.12 Problem: The following kinetic data have been reported by Covey and Brown for the reactions $[Mo(CO)_5(amine)] + PPh_3 \xrightarrow{35^o} [Mo(CO)_5(PPh_3)] + amine$.

Amine	$k_1(sec^{-1})$	$k_2(M^{-1}sec^{-1})$
Quinuclidine	0.9×10^{-5}	3.0×10^{-5}
Piperidine	1.8×10^{-5}	1.24×10^{-3}
Cyclohexylamine	9.9×10^{-5}	3.9×10^{-3}

Other data for carbonyl substitution are

Complex	Lewis Base	$T(^oC)$	Solvent	k_1x10^4 (sec^{-1})	k_2x10^4 $(M^{-1}sec^{-1})$
$Cr(CO)_6$	PPh_3	130.7	decalin	1.38	0.450
$W(CO)_6$	PPh_3	165.7	decalin	1.15	0.888
$Mo(CO)_6$	PPh_3	112.0	decalin	2.13	1.77
$Mo(CO)_6$	$(\underline{n}\text{-}Bu)_3P$	112.0	decalin	2.13	20.5
$Mo(CO)_6$	$C_6H_5CH_2NH_2$	112.0	decalin	2.44	4.4
$Mo(CO)_6$	Br	55.0	C_6H_5Cl	—	32.4
$Mo(CO)_5py$	$P(OCH_2)_3CCH_3$	47.9	$ClCH_2CH_2Cl$	9.3	~4.8
$[Mo(CO)_5Br]^-$	PPh_3	19.6	diglyme	1.34	4.1
$[Mo(CO)_5I]^-$	PPh_3	29.8	diglyme	1.13	1.8
$[Mo(CO)_5I]^-$	$P(p\text{-}C_6H_4F)_3$	29.8	diglyme	1.13	2.8
$[Mo(CO)_5I]^-$	$P(p\text{-}C_6H_4Cl)_3$	29.8	diglyme	1.13	3.6

Considering all these data, what can you say about the $\underline{a}$ or $\underline{d}$ character of the k_2 path for substitution at Mo? Are the data given here consistent with a $\underline{d}$ path for the k_1 term? (W.D. Covey and T.L. Brown, Inorg. Chem. 1973, 12, 2820.)

10.12 Solution: The rates for the k_1 path depend only on the nature of the leaving group consistent with a $\underline{D}$ mechanism for this pathway. The rates for the k_2 path for the same leaving group and different entering groups do display a dependence on the nature of the entering group as would be required for an a intimate mechanism. Nevertheless, considering the very wide range of nucleophilicity and basicity of the entering groups, the spread in k_2 is quite small. Consistent with this would be an I_d mechanism with some bond making in the transition state.

10.13 Problem: Rationalize the trends in the following sets of IR-active CO stretching frequencies (in cm^{-1}):

(a)	$(\eta^6\text{-}C_6H_6)Cr(CO)_3$	1980, 1908
	$CpMn(CO)_3$	2027, 1942
(b)	$CpV(CO)_4$	2030, 1930
	$CpMn(CO)_3$	2027, 1942
	$[CpFe(CO)_3]^+$	2120, 2070
(c)	$W(CO)_5(P\underline{n}\text{-}Bu_3)$	2068, 1936, 1943
	$W(CO)_5(PPh_3)$	2075, 1944, 1944
	$W(CO)_5(P(OBu)_3)$	2079, 1947, 1957
	$W(CO)_6$	ca. 2000
(d)	$Ni(CO)_4$	2046
	$[Co(CO)_4]^-$	1883
	$[Fe(CO)_4]^{2-}$	1788

10.13 Solution:

(a) Mn(I) is in a higher oxidation state than Cr(0). Hence, it attracts electrons more strongly and less electron density can be back-donated into CO π^* orbitals. Hence, the C–O bond in the Mn compound is weakened to a lesser extent and ν_{C-O} is higher.

(b) Same argument as in (a). The oxidation state of Fe(II) is greater than that in V(I) and Mn(I).

(c) The Lewis basicity of the phosphine ligands lies in the order $P(OBu)_3 < PPh_3 < P(\underline{n}\text{-}Bu)_3$. The electron density supplied to W increases in this order. The metal has increasing electron density to back-donate to CO. Thus, ν_{C-O} decreases in the observed order.

(d) The isoelectronic complexes have increasing negative charges to delocalize in the order $Ni(CO)_4 < [Co(CO)_4]^- < [Fe(CO)_4]^{2-}$. The method of delocalization is back donation into π*CO which causes a shift of ν_{CO} to progressively lower energies in the order given above.

10.14 Problem:

(a) Draw at least two possible structures of $Os_3(CO)_9(PPh_3)_3$. (b) the IR spectrum of this compound in CH_2Cl_2 has CO stretches at 1962 and 1917 cm^{-1}. How does this knowledge help to narrow the possible structures?

10.14 Solution:

(a)

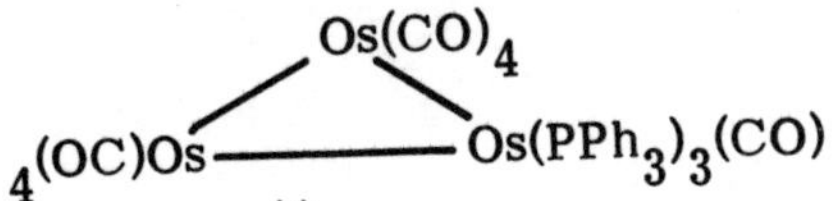

(I)

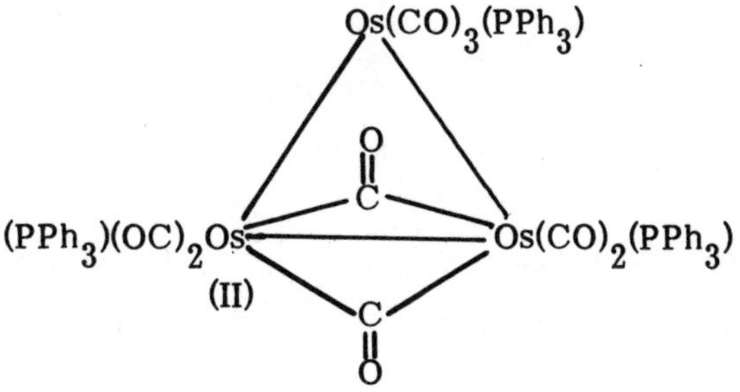

(II)

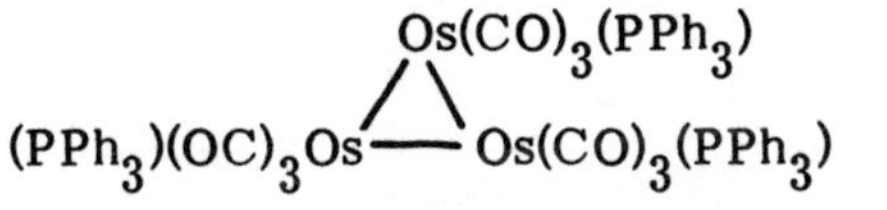

(III)

are some possibilities

(b) Structure (II) is not possible because no stretching frequencies are seen in the μ-CO region. Structure (I) would display a very low frequency terminal CO stretch due to the unique CO. Structure (III) is consistent with the two frequencies observed.

10.15 Problem: The IR spectrum of $Rh_2I_2(CO)_2(PPh_3)_2$ has CO stretches as 2061 and 2005 cm^{-1}. Suggest a structure consistent with this.

10.15 Solution:

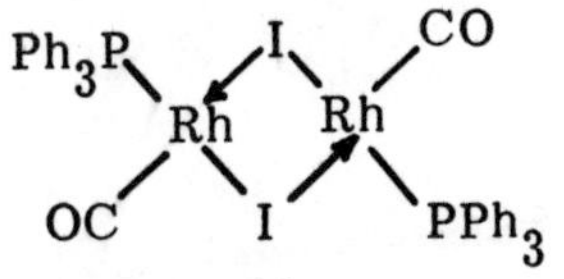

(I)

(I) Symmetric and antisymmetric CO stretches could account for the two observed bands.

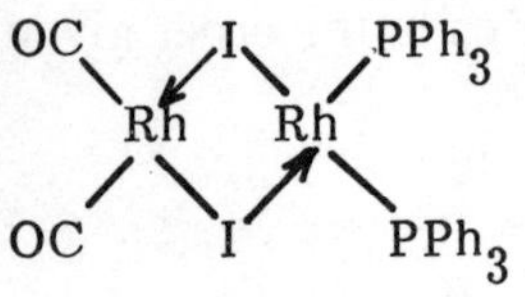

(II)

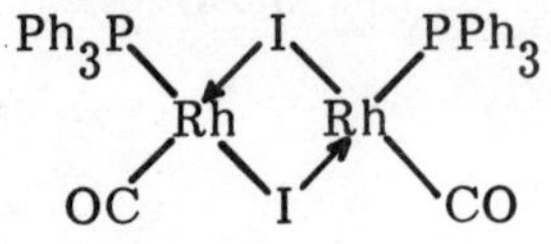

(III)

(II) is also a possibility from the IR data. The two stretches could be assigned as symmetric and antisymmetric. However, it is less likely to exist than (I) because of less favorable entropy.

(III) is also a possibility, but unlikely for same reasons as (II).

Structures with one or two μ-CO are ruled out because all stretches observed are in the terminal CO stretching region.

10.16 Problem: Explain in your own words how the VB and MO descriptions of the electronic structure of $C_3H_5^-$ are equivalent.

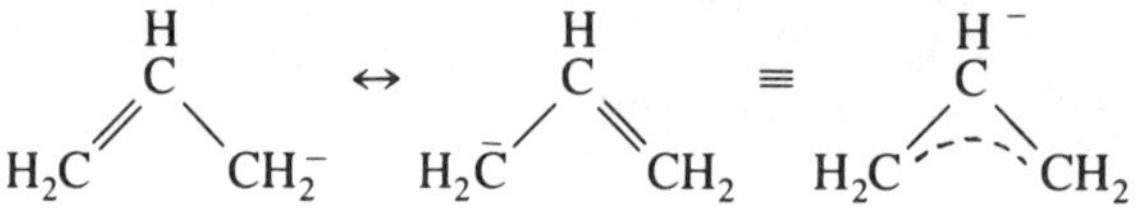

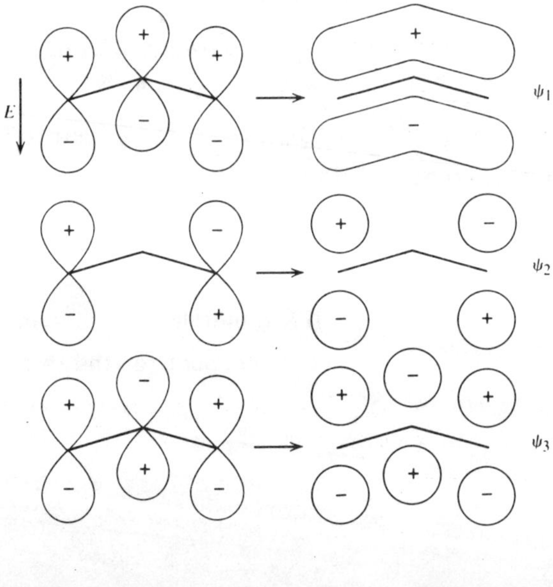

MO's for $C_3H_5^-$

10.16 Solution: The two VB resonance structures indicate the equivalence of the two end C's and of both C-C bonds. In the MO description, occupation of ψ_1 and ψ_2 leads to equivalence of C-C bonds. Implicit in the construction of the MO's is the equivalence of end C's which is a consequence of the C_{2v} symmetry.

10.17 Problem: Coordination of ligands to transition metals often makes them subject to nucleophilic attack. Make a list of reactions which involve nucleophilic attack on coordinated ligands.

10.17 Solution: Some examples (from DMA Chapter 10) are:

$13\ Mn_2(CO)_{10} + 40\ OH^- \rightarrow 24[Mn(CO)_5]^- + 2\ Mn^{2+} + 10\ CO_3^{2-} + 20\ H_2O$

$Fe_2(CO)_9 + 4\ OH^- \rightarrow [Fe_2(CO)_8]^{2-} + 2\ H_2O + CO_3^{2-}$

$[Mn(CO)_6]^+ + LiCH_3 \rightarrow CH_3C(O)Mn(CO)_5 + Li^+$

$[Mn(PPh_3)(CO)_5]^+ + 2\ CH_3NH_2 \xrightarrow[THF]{} \underline{cis}\text{-}CH_3NHC(O)Mn(PPh_3)(CO)_4 + NH_3CH_3^+$

$[Mn(CO)_5(CH_3CN)]^+ + CH_3NO_2 + 3\ py \rightarrow \underline{fac}\text{-}[Mn(CO)_3py_3]^+ + CH_3CN + CO + CO_2 + CH_2NOH$

$\underline{cis}\text{-}Pt(PPh_3)Cl_2(\eta^2\text{-}C_2H_4) + Et_2NH \rightarrow \underline{cis}\text{-}Pt(PPh_3)Cl_2(CH_2CH_2NHEt_2)$

$[CpFe(CO)_2(\eta^2\text{-}C_2H_4)]^+ + CN^- \rightarrow CpFe(CO)_2CH_2CH_2CN$

$[CpW(CO)_3(\eta^2\text{-}C_2H_4)]^+ + CN^- \rightarrow CpW(CO)_3CH_2CH_2CN$

$Cp_2Rh + NaBH_4 \rightarrow CpRh(\eta^4\text{-}C_5H_6)$

$(\eta^6\text{-}C_6H_5Cl)Cr(CO)_3 + OMe^- \rightarrow (\eta^6\text{-}C_6H_5OMe)Cr(CO)_3 + Cl^-$

$CpFe(\eta^6\text{-}C_6H_6)^+ + H^- \rightarrow CpFe(\eta^5\text{-}C_6H_7)$

$$W(CO)_6 \xrightarrow[\substack{3\ H^+ \\ 4\ CH_2N_2}]{\substack{1\ PhLi \\ 2\ Me_4N^+Cl^-}} (CO)_5W{=}CPh$$

$$(CO)_5W{=}C(OMe)Ph \xrightarrow[-78^\circ]{PhLi} [(CO)_5W\text{-}CPh_2]^-$$

Insertions of olefins or acetylenes into M-H bonds can be regarded as internal nucleophilic attack by coordinated hydride on coordinated olefins e.g., $Pt(H)(PEt_3)_2Cl + C_2H_4 \rightarrow Pt\text{-}(C_2H_5)(PEt_3)_2Cl$.

10.18 Problem: Give the formula of the most stable compound of the type $M(olefin)(CO)_x$ to be expected for each of the following metals with each olefin listed.

Metals	Olefins
Cr, Mn, Fe	$C_3H_5^-$, Cp^-, C_6H_6, $C_7H_7^+$

10.18 Solution: The most stable compounds are expected to be those which conform to the EAN rule. Formulas which satisfy this criterion are:

$[(C_3H_5)Cr(CO)_4]_2$
$[CpCr(CO)_3]_2$
$(C_6H_6)Cr(CO)_3$
$[(C_7H_7)Cr(CO)_3]^+$
$(C_3H_5)Mn(CO)_4$
$CpMn(CO)_3$
$[(C_6H_6)Mn(CO)_3]^+$
$[(C_7H_7)Mn(CO)_3]^{2+}$
$[(C_3H_5)Fe(CO)_3]_2$
$[CpFe(CO)_2]_2$
$(C_6H_6)Fe(CO)_2$
$[(C_7H_7)Fe(CO)_2]^+$

10.19 Problem: Identify the following reactions by type and predict the products:

(a) $Re_2(CO)_{10}$ + Na/Hg
(b) $W(CO)_6$ + ($\underline{n}$-Bu_4N)I
(c) $CH_3Mn(CO)_5 + C_6H_5NH_2$
(d) $CpCo(CO)_2 + PPh_3$
(e) $[Rh(CO)_2Cl_2]^- + CH_3I$
(f) $CH_3Mn(CO)_5 + SO_2$
(g) 4EtMgI + $SnCl_4$
(h) $Me_2Hg + Cp_2TiCl_2$
(i) MeLi(excess) + $CuCl_2$
(j) $[CpFe(CO)_2]^- + C_6H_5Cl$
(k) $(\eta^1\text{-}C_3H_5)Co(CO)_4 \xrightarrow{\Delta}$
(l) Cp_2Fe + $\underline{n}$-BuLi
(m) $Mo(CO)_5py + PPh_3$
(n) $Rh(PPh_3)_3Br + Cl_2$
(o) $Pt(PPh_3)_3 + CF_3I$
(p) $RuH_2(PPh_3)_4 + C_2H_4$
(q) $HIrCl_2(PPh_3)_2 \xrightarrow{\Delta}$
(r) $Ge(OMe)_2 + CH_3I$
(s) $RuH(CO)_2(PPh_3)_2C(O)R \xrightarrow{\Delta}$
(t) $Ru(PPh_3)_2(CO)_3 + H_2$
(u) $PtMe_2Cl(PMe_2Ph)_2C(O)CH_3 \xrightarrow{\Delta}$
(v) $AlMe_3 + CrCl_3 \xrightarrow{THF}$
(w) LiMe + WCl_6
(x) $CpMo(CO)_3Me + CN^-$
(y) $CpFe(CO)_2CH_2CH_2CH{=}CH_2 \xrightarrow{h\nu}$

10.19 Solution:

(a) Reduction; $Na^+[Re(CO)_5]^-$

(b) Substitution; $[W(CO)_5I]^-$ + CO

(c) CO insertion; $CH_3C(O)Mn(CO)_4(NH_2C_6H_5)$

(d) CO substitution; $CpCo(CO)(PPh_3)$ + CO

(e) Oxidative addition; $[CH_3Rh(CO)_2Cl_2I]^-$

(f) SO_2 insertion; $CH_3S(O)_2Mn(CO)_5$

(g) Alkylation; Me_4Sn + 4MgClI

(h) Alkylation; $Cp_2TiMe_2 + HgCl_2$

(i) Alkylation and reduction; $Li[Me_2Cu]$

(j) Halide displacement; $CpFe(CO)_2C_6H_5 + Cl^-$

(k) Decarbonylation; $(\eta^3\text{-}C_3H_5)Co(CO)_3$ + CO

(l) Metallation; $(LiC_5H_4)Fe(Cp) + C_4H_{10}$

(m) Ligand displacement; $Mo(CO)_5(PPh_3)$ + py

(n) Oxidative addition; $Rh(PPh_3)_3Cl_2Br$

(o) Oxidative addition with ligand dissociation; $Pt(PPh_3)_2(CF_3)I + PPh_3$

(p) Ligand displacement followed by insertion and reductive elimination with ligand uptake; $H_2Ru(PPh_3)_3(\eta^2\text{-}C_2H_4) \rightarrow HRuEt(PPh_3)_3 \rightarrow HRuEt(PPh_3)_4 \rightarrow C_2H_6$ + [Ru-$(PPh_3)_4]$ $\xrightarrow{C_2H_4}$ $(C_2H_4)Ru(PPh_3)_4$

(q) Reductive elimination; HCl + $1/2[(PPh_3)_2Ir(\mu\text{-}Cl)]_2$

(r) Oxidative addition; $CH_3Ge(OMe)_2I$

(s) Ligand dissociation followed by decarbonylation, reductive elimination and ligand uptake; $HRu(CO)_2(PPh_3)C(O)R + PPh_3 \rightarrow HRu(CO)_3R(PPh_3) \rightarrow RH + [Ru(CO)_3(PPh_3)]$ $\xrightarrow{+PPh_3}$ $Ru(CO)_3(PPh_3)_2$

(t) Ligand dissociation followed by oxidative addition; $Ru(PPh_3)(CO)_3 + PPh_3 \xrightarrow{H_2}$ $(H)_2Ru(PPh_3)(CO)_3$

(u) Ligand dissociation, decarbonylation and reductive elimination; $PtMe_2Cl$-$(PMe_2Ph)C(O)CH_3 + PMe_2Ph \longrightarrow PtMe_2Cl(PMe_2Ph)(CO)Me \longrightarrow C_2H_6$ + $PtMeCl(PPh_2Me)(CO)$

(v) Alkylation; $CH_3CrCl_2\cdot THF + Me_2AlCl$

(w) Alkylation; WMe_6 + 6 LiCl

(x) Carbonylation; $[CpMo(CO)_2(CN)\overset{O}{\overset{\|}{C}}Me]^-$

(y) Decarbonylation; $CpFe(CO)(\eta^3\text{-}C_3H_5)$ + CO

10.20 Problem: $CpMoH_2$ reacts with diphenylacetylene giving $CpMo(PhC_2Ph)$ and cis-PhCH=CHPh. Propose a mechanism for this reaction. (See A. Nakamura and S. Otsuka, J. Am. Chem. Soc. 1972, 94, 1886.)

10.20 Solution: The mechanism might involve coordination of diphenylacetylene followed by insertion into one of the M-H bonds giving an alkenyl hydride complex. This could undergo reductive elimination to afford the olefin and CpMo which could then coordinate a molecule of diphenylacetylene.

$$CpMo(H)_2 + PhC{\equiv}CPh \longrightarrow CpMo(H)_2(\eta^2\text{-}PhC{\equiv}CPh)$$
$$CpMo(H)_2(\eta^2\text{-}PhC{\equiv}CPh) \longrightarrow CpMo(H)(CPh{=}CHPh)$$
$$CpMo(H)(CPh{=}CHPh) \longrightarrow CpMo(\eta^2\text{-}CHPh{=}CHPh)$$
$$CpMo(\eta^2\text{-}CHPh{=}CHPh) \longrightarrow [CpMo] + CHPh{=}CHPh$$
$$[CpMo] + PhC{\equiv}CPh \longrightarrow CpMo(PhC{\equiv}CPh)$$

10.21 Problem: The rate of reaction of O_2 with trans-$IrX(CO)(PPh_3)_2$ in benzene decreases in the order X = $NO_2 > I > ONO_2 > Br > Cl > N_3 > F$. Explain this observation. (L. Vaska and C.V. Senoff, Science 1971, 174, 587.)

10.21 Solution: The rates decrease as the electron-withdrawing ability of X increases. This reduces the electron density on Ir and increases the enthalpy of activation since the transition state must involve transfer of electron density from Ir to O_2.

10.22 Problem: Propose a mechanism for the stoichiometric decarbonylation of $C_6H_5CH_2C(O)Cl$ by $Rh(PPh_3)_3Cl$ giving benzyl chloride. Keep in mind the 16- and 18-electron rule.

10.22 Solution:

$$Rh(PPh_3)_3Cl + C_6H_5CH_2C(O)Cl \longrightarrow C_6H_5CH_2C(O)Rh(PPh_3)_3Cl_2 \quad \text{oxidative addition}$$
$$C_6H_5CH_2C(O)Rh(PPh_3)_3Cl_2 \rightleftharpoons C_6H_5CH_2C(O)Rh(PPh_3)_2Cl_2 + PPh_3 \quad \text{ligand dissociation}$$

$C_6H_5CH_2C(O)Rh(PPh_3)_2Cl_2 \longrightarrow C_6H_5CH_2Rh(CO)(PPh_3)_2Cl_2$ alkyl migration

$C_6H_5CH_2Rh(CO)(PPh_3)_2Cl_2 \longrightarrow C_6H_5CH_2Cl + Rh(CO)(PPh_3)_2Cl$ reductive elimination

All species proposed in the mechanism are either 16- or 18 -e species.

10.23 Problem: The ease of oxidative addition and the tendency toward five coordination for d^8 metals increase as shown:

⟵ Tendency to five coordination

Fe(0)	Co(I)	Ni(II)	↓
Ru(0)	Rh(I)	Pd(II)	Ease of oxidative addition
Os(0)	Ir(I)	Pt(II)	↓

Explain these trends.

10.23 Solution: Because the additions result in formal increase in oxidation state of the metal their ease should parallel ease of removal of two metal electrons. Ionization energies decrease on descending each group. Earlier members of Group VIII have lower effective nuclear charges. Hence, their electrons are more available for back-donation to π-acid ligands, thereby strengthening the fifth metal-ligand bond.

10.24 Problem: Write a catalytic cycle for the production of ethyl acetate via the Monsanto acetic acid synthesis.

10.24 Solution: The usual cycle for the carbonylation of methanol could be used to produce acetic acid which could be esterified separately. If a mixed H_2O/EtOH/MeOH solvent were used in the catalytic reactions, CH_3CO_2Me, $CH_3CH_2CO_2Me$, $CH_3CH_2CO_2Et$ as well as the corresponding acids are possible in the reaction mixture besides CH_3CO_2Et.

10.25 Problem: $[HNi(P(OEt)_3)_4]^+$ is known to be a catalyst for olefin isomerization. Write a catalytic cycle for isomerization of 1-butene catalyzed by this species. Keep in mind the 16- and 18-electron rule. How would you formulate the electronic structure of the Ni cation in the most reasonable way?

10.25 Solution:

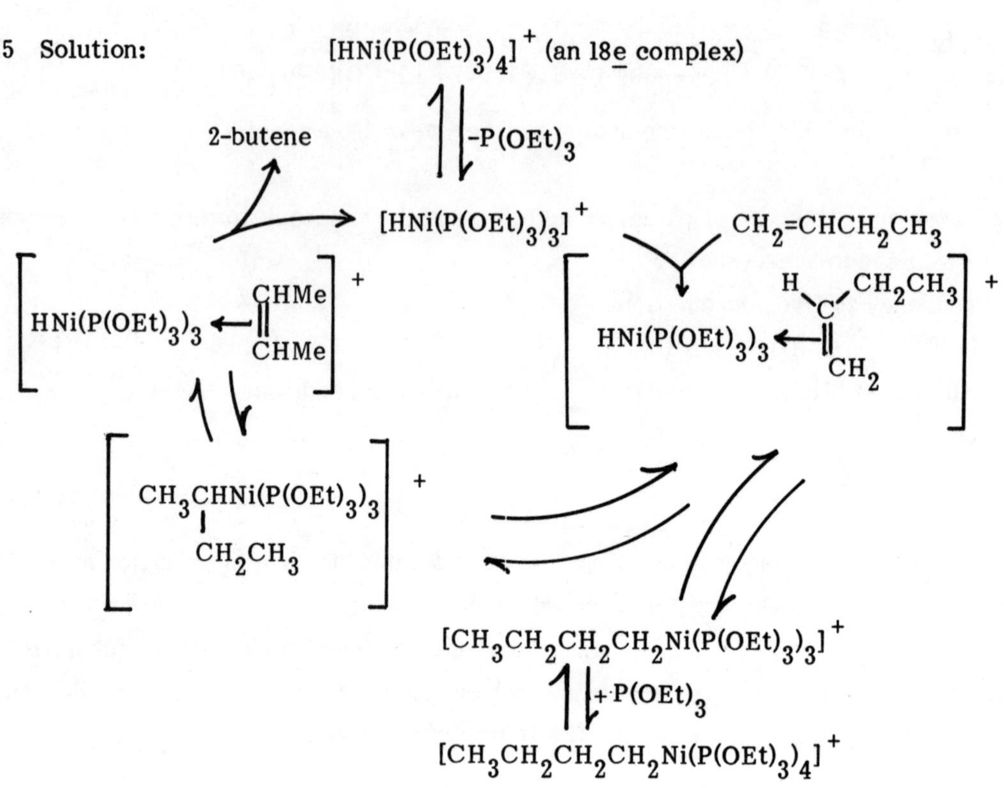

The cation $[HNi(P(OEt_3)_4]^+$ could be regarded as containing protonated Ni(0) or as a hydride complex of d^8 Ni(II). The latter is probably the better formulation since the insertion of olefin into the Ni-H bond postulated in the cycle is known with other d^8 hydrides.

10.26 Problem: Give electron counts for all the species postulated to be involved in the catalytic cycle for the hydroformylation reaction shown below.

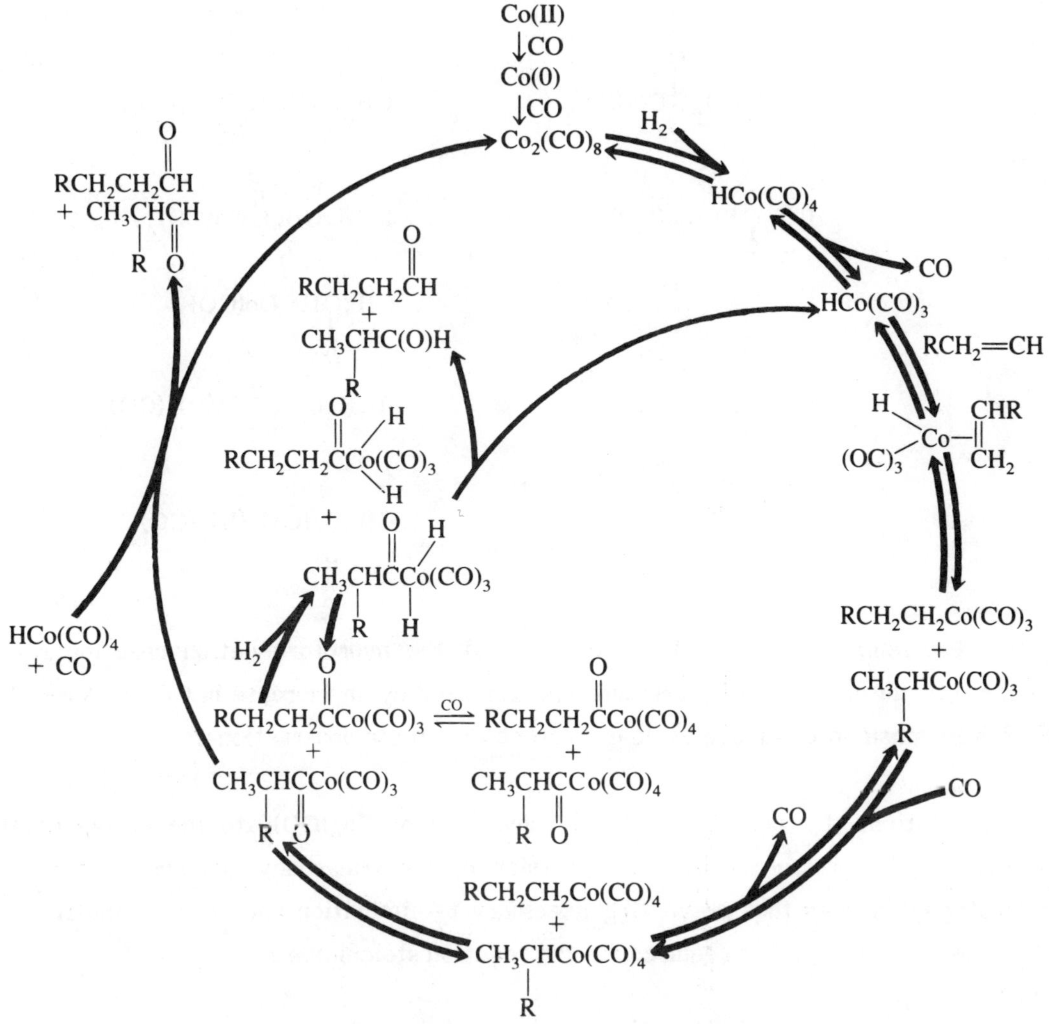

Catalytic cycle for hydroformylation of olefins.

10.26 Solution:

Sixteen Electrons	Eighteen Electrons
$HCo(CO)_3$	$Co_2(CO)_8$
$RCH_2CH_2Co(CO)_3$	$HCo(CO)_4$
$CH_3\underset{R}{C}HCo(CO)_3$	$HCo(CO)_3(\eta^2\text{-}CH_2{=}CHR)$

$$RCH_2CH_2\overset{\overset{O}{\|}}{C}Co(CO)_3 \qquad\qquad CH_3\underset{\underset{R}{|}}{C}HCo(CO)_4$$

$$CH_3\underset{\underset{R}{|}}{C}H\overset{\overset{O}{\|}}{C}Co(CO)_3 \qquad\qquad RCH_2CH_2\overset{\overset{O}{\|}}{C}Co(CO)_4$$

$$CH_3\underset{\underset{R}{|}}{C}H\overset{\overset{O}{\|}}{C}Co(CO)_4$$

$$RCH_2CH_2\overset{\overset{O}{\|}}{C}Co(H)_2(CO)_3$$

$$CH_3\underset{\underset{R}{|}}{C}H\overset{\overset{O}{\|}}{C}Co(H)_2(CO)_3$$

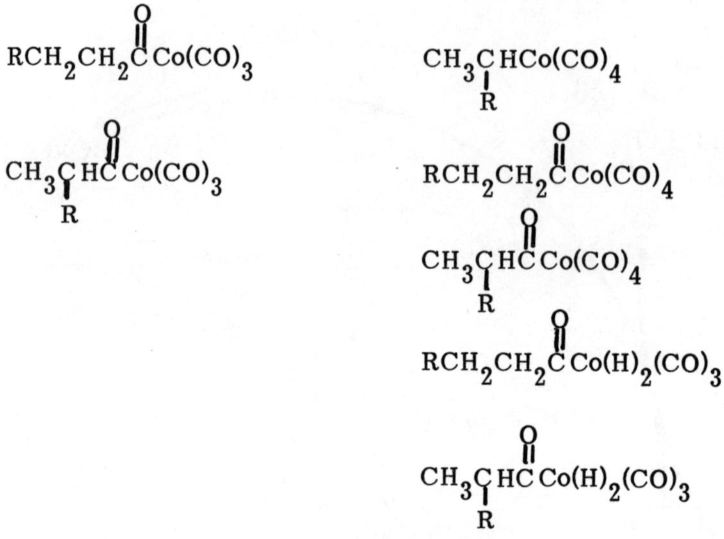

10.27 Problem: Kinetic studies indicate that the hydroformylation reaction rate is enhanced by an increase in H_2 pressure and inhibited by an increase in CO pressure. How is the mechanism in the above cycle consistent with these observations?

10.27 Solution: H_2 pressure leads to conversion of $Co_2(CO)_8$ to the active catalyst $HCo(CO)_4$. CO pressure leads to conversion of 16e tricarbonyl species to stable 18e tetracarbonyl species thus preventing necessary β-elimination and/or H_2 oxidative addition steps. However, CO is required for the reaction stoichiometry.

10.28 Problem: Propose a mechanism for the following reaction:

$$IrCl_3(PEt_3)_3 + C_2H_5O^- \xrightarrow[KOH]{EtOH} HIrCl_2(PEt_3)_3 + CH_3CHO$$

(J. Chatt and B.L. Shaw, Chem. Ind. (London) 1960, 931.)

10.28 Solution:

$$C_2H_5OH + KOH \rightleftharpoons K^+(OC_2H_5)^- + H_2O \qquad\qquad \text{deprotonation}$$

$$IrCl_3(PEt_3)_3 + K^+(OC_2H_5)^- \longrightarrow Ir(OC_2H_5)Cl_2(PEt_3)_3 + KCl \quad \text{ligand displacement}$$

$$Ir(OC_2H_5)Cl_2(PEt_3)_3 \rightleftharpoons Ir(OC_2H_5)Cl_2(PEt_3)_2 + PEt_3 \qquad \text{dissociation}$$

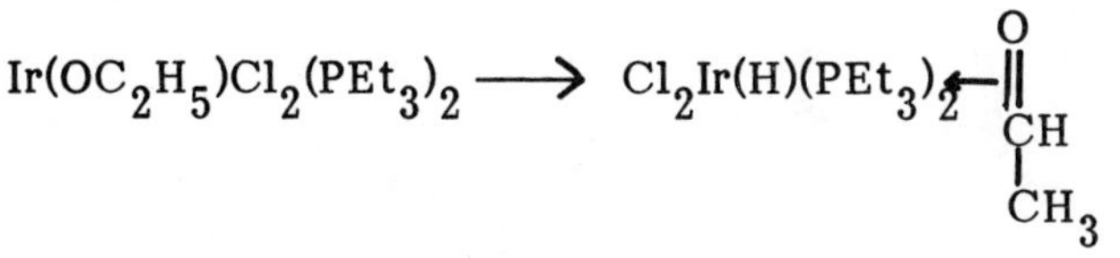

β-elimination

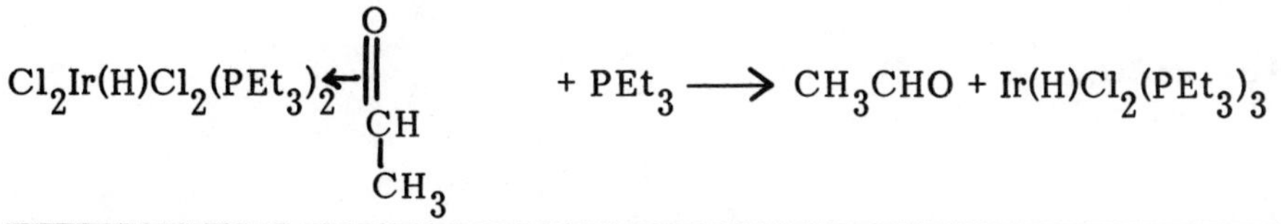

substitution

$$IrCl_3(PEt_3)_3 + C_2H_5OH + KOH \longrightarrow Ir(H)Cl_2(PEt_3)_3 + CH_3CHO + KCl + H_2O \text{ (in sum)}$$

10.29 Problem: What products would you expect from the hydroformylation of $C_3H_7CD(CH_3)CH{=}CH_2$? Show how each is obtained.

10.29 Solution: Of importance here is the reversability of the β-elimination.

$$HCo(CO)_4 \rightleftharpoons HCo(CO)_3 + CO$$

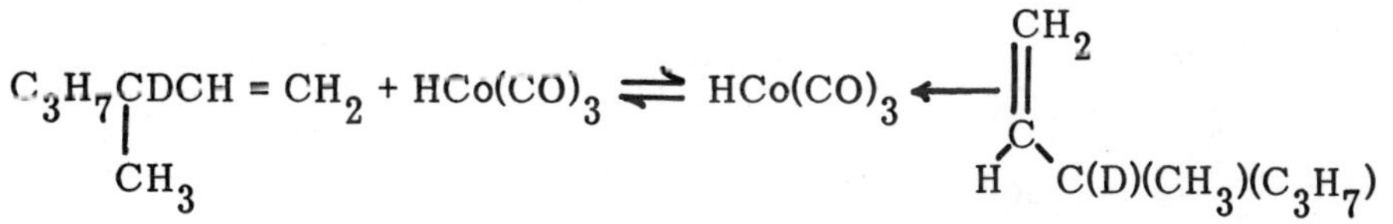

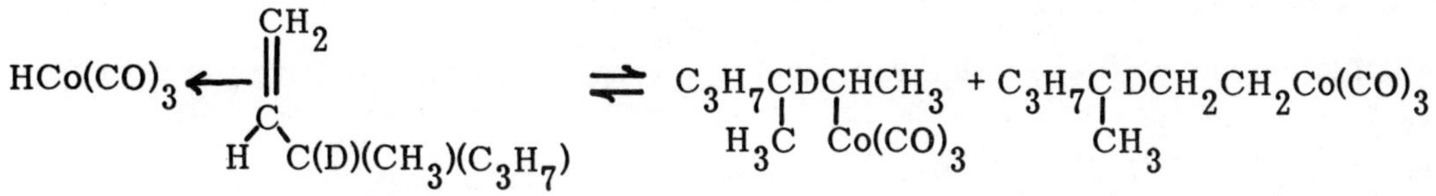

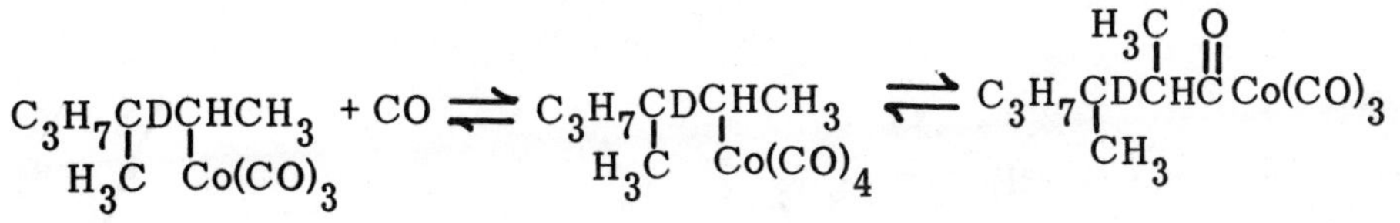

$$C_3H_7\underset{\underset{CH_3}{|}}{C}DCH_2CH_2Co(CO)_3 + CO \rightleftharpoons C_3H_7\underset{\underset{CH_3}{|}}{C}DCH_2CH_2Co(CO)_4 \rightleftharpoons$$

$$C_3H_7\underset{\underset{CH_3}{|}}{C}DCH_2CH_2\overset{\overset{O}{\parallel}}{C}Co(CO)_3$$

$$C_3H_7\underset{\underset{CH_3}{|}}{C}DCH\underset{\underset{O}{\parallel}}{C}Co(CO)_3 + H_2 \rightleftharpoons C_3H_7\underset{\underset{CH_3}{|}}{C}DCH-\overset{\overset{H_3C}{|}}{\underset{\underset{H}{|}}{C}}\overset{\overset{O}{\parallel}}{C}Co(CO)_3 \longrightarrow \boxed{C_3H_7\underset{\underset{H_3C}{|}}{C}DCH\underset{\underset{CH_3}{|}}{C}\overset{\overset{O}{\parallel}}{C}H}$$

$$+ HCo(CO)_3$$

$$C_3H_7\underset{\underset{CH_3}{|}}{C}DCH_2CH_2\overset{\overset{O}{\parallel}}{C}Co(CO)_3 + H_2 \rightleftharpoons C_3H_7\underset{\underset{CH_3}{|}}{C}DCH_2CH_2\overset{\overset{OH}{|}}{\underset{\underset{H}{|}}{C}}Co(CO)_3 \longrightarrow$$

$$\boxed{C_3H_7\underset{\underset{CH_3}{|}}{C}DCH_2CH_2\overset{\overset{O}{\parallel}}{C}H} \quad + HCo(CO)_3$$

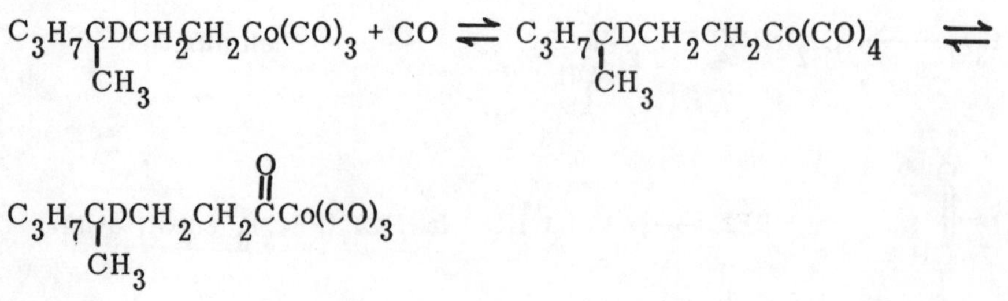

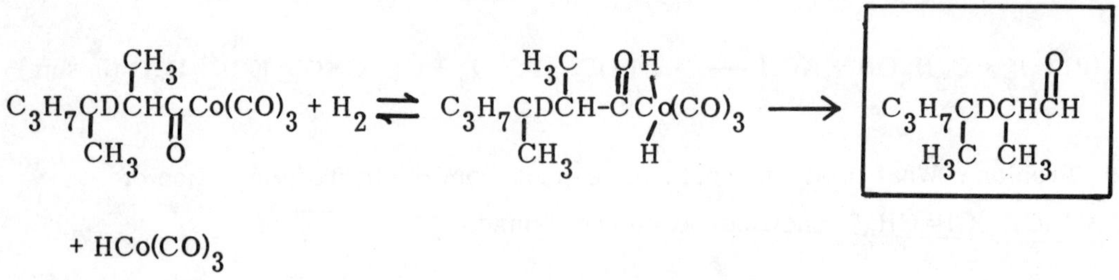

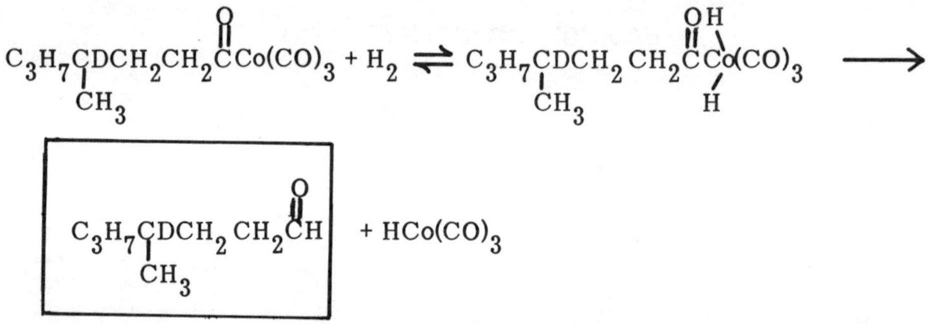

β-elimination and olefin insertion can occur in the opposite sense to move the D to the next C:

$$C_3H_7\underset{\underset{H_3C}{|}}{C}DCHCH_3 \rightleftharpoons DCo(CO)_3 \longleftarrow \underset{\underset{Co(CO)_3}{|}}{|} \underset{CH_3 \quad C_3H_7}{\overset{CH_3 \quad H}{C}} \rightleftharpoons C_3H_7\overset{\overset{Co(CO)_3}{|}}{\underset{\underset{CH_3}{|}}{C}}CDHCH_3$$

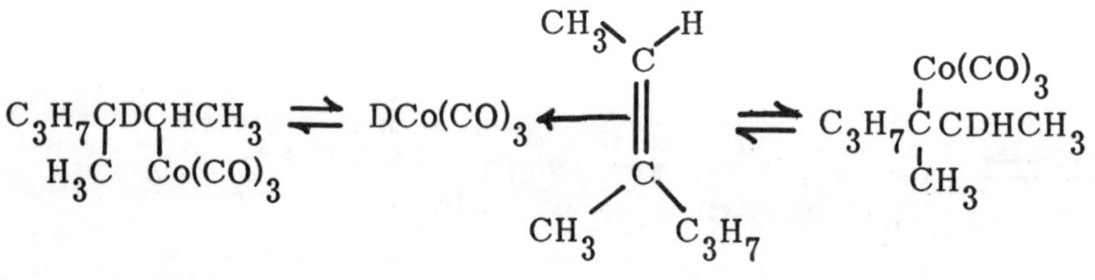

$$\underset{\underset{\displaystyle CDHCH_3}{|}}{\overset{\overset{\displaystyle CH_3}{|}}{C_3H_7C}}\text{-Co(CO)}_3 + CO \;\rightleftharpoons\; \underset{\underset{\displaystyle CDHCH_3}{|}}{\overset{\overset{\displaystyle CH_3}{|}}{C_3H_7C}}\text{Co(CO)}_4 \;\rightleftharpoons\; \underset{\underset{\displaystyle CDHCH_3}{|}}{\overset{\overset{\displaystyle H_3C\ O}{|\ \ ||}}{C_3H_7C}}\text{-}\overset{}{C}\text{Co(CO)}_3$$

$$\underset{\underset{\displaystyle CDHCH_3}{|}}{\overset{\overset{\displaystyle H_3C\ O}{|\ \ ||}}{C_3H_7C}}\text{-CCo(CO)}_3 + H_2 \;\rightleftharpoons\; \underset{\underset{\displaystyle CH_3HDC}{|}}{\overset{\overset{\displaystyle H_3C\ \ O\ H}{|\ \ ||\ /}}{C_3H_7C}}\text{ - }CCo(CO)_3 \;\longrightarrow\; \boxed{\underset{\underset{\displaystyle CDHCH_3}{|}}{\overset{\overset{\displaystyle H_3C\ O}{|\ \ ||}}{C_3H_7C}}\text{-CH}} \quad +$$

$$HCo(CO)_3$$

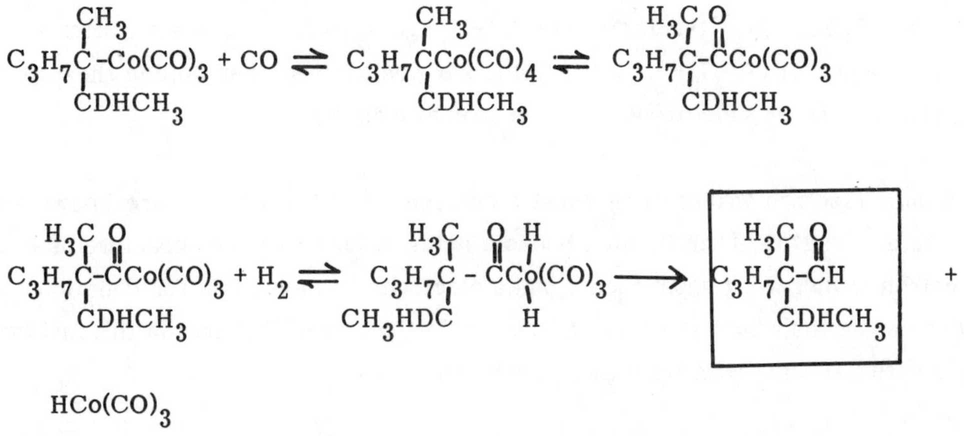

This product, which has the aldehyde function coordinated to a tertiary C, is not actually observed. (R. Heck and D.S. Breslow, J. Am. Chem. Soc. 1961, 88, 4023.) The likely reason is the instability of the tertiary alkyl precursor which can undergo β-elimination and reinsertion leading to a primary alkyl:

$$\underset{\underset{\displaystyle CDHCH_3}{|}}{\overset{\overset{\displaystyle CH_3}{|}}{C_3H_7C}}\text{ - Co(CO)}_3 \;\rightleftharpoons\; HCo(CO)_3 \leftarrow \underset{\underset{\displaystyle C_3H_7\ \ CHDCH_3}{\diagup \ \diagdown}}{\overset{\overset{\displaystyle CH_2}{||}}{\underset{\displaystyle C}{}}} \;\rightleftharpoons\; \underset{\underset{\displaystyle CHDCH_3}{|}}{C_3H_7CHCH_2}Co(CO)_3$$

$$\underset{\underset{\displaystyle CHDCH_3}{|}}{C_3H_7CHCH_2}Co(CO)_3 + CO \;\rightleftharpoons\; \underset{\underset{\displaystyle CHDCH_3}{|}}{C_3H_7CHCH_2}Co(CO)_4 \;\rightleftharpoons\; \underset{\underset{\displaystyle CHDCH_3}{|}}{C_3H_7CHCH_2}\overset{\overset{\displaystyle O}{||}}{C}Co(CO)_3$$

$$\underset{\underset{\displaystyle CHDCH_3}{|}}{C_3H_7CHCH_2}\overset{\overset{\displaystyle O}{||}}{C}Co(CO)_3 + H_2 \;\rightleftharpoons\; \underset{\underset{\displaystyle CHDCH_3}{|}}{C_3H_7CHCH_2}\overset{\overset{\displaystyle O\ H}{||\ \backslash}}{C}\underset{\displaystyle H}{Co(CO)_3} \;\longrightarrow\; \boxed{\underset{\underset{\displaystyle CHDCH_3}{|}}{C_3H_7CHCH_2}\overset{\overset{\displaystyle O}{||}}{CH}} \;+$$

$$HCo(CO)_3$$

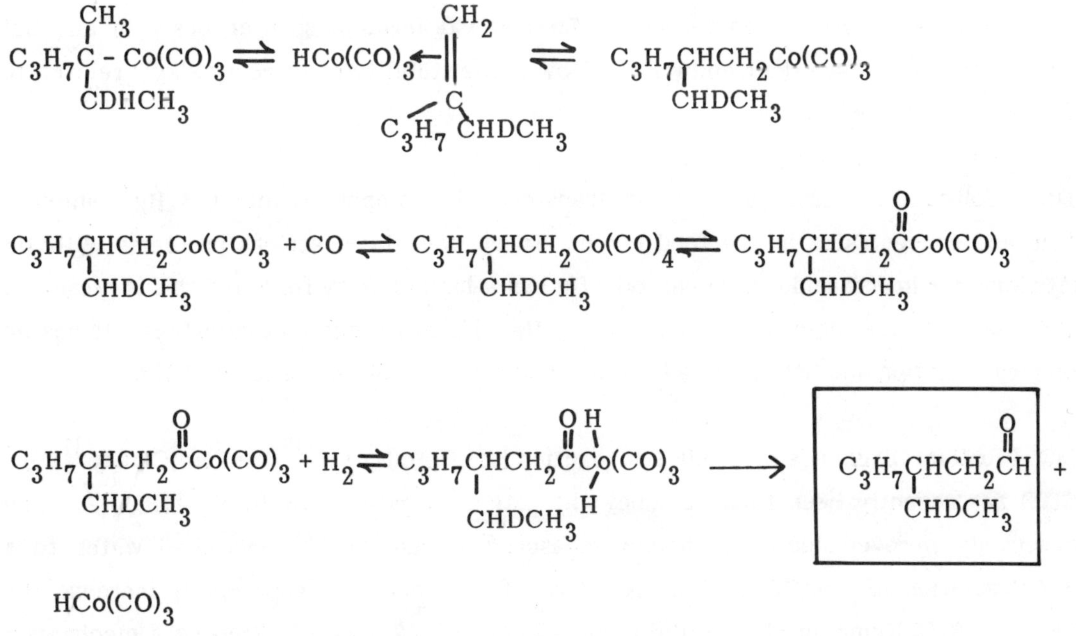

10.30 Problem: Explain how the facts that $(\eta^6\text{-}C_6H_5CO_2H)Cr(CO)_3$ is a stronger acid than benzoic acid and that $(\eta^6\text{-}C_6H_5NH_2)Cr(CO)_3$ is a weaker base than aniline show that the $Cr(CO)_3$ group withdraws electrons from the aromatic rings.

10.30 Solution: Electron-withdrawing substituents on the benzene ring are known to increase the acid strengths of substituted benzoic acids by assisting the departure of H^+. Similarly electron-withdrawing substituents make electrons less available for donation by bases thus decreasing their base strength. Since the $Cr(CO)_3$ group behaves in this fashion it must act to withdraw electrons from the benzene ring.

10.31 Problem: To what point group does the Li_4Me_4 tetramer belong? It consists of a tetrahedron of four Li with a methyl group triply bridging each face.

10.31 Solution: Since the methyl groups lie on the threefold axes of the Li_4 tetrahedron, these axes are preserved in Li_4Me_4 and the symmetry is T_d.

10.32 Problem: Alkyl compounds of Zn and Cd inflame spontaneously in air, but $Hg(CH_3)_2$ is stable toward air and water. How is the stability of mercury alkyls related to the problem of Hg pollution?

10.32 Solution: Microorganisms can transform Hg compounds into CH_3Hg^+ which is stable and concentrates in the food chain. There have been serious problems with Hg contamination in fish and other seafood. Hg has a high affinity for S and this is presumed to be related to the inhibition of enzymes by Hg. Hg has a cumulative effect. It has no biological function and hence there is no mechanism to regulate the level of Hg.

10.33 Problem: Exchange of ^{17}O in the reaction $[Re(CO)_6]^+ + H_2{}^{17}O \rightarrow [Re(CO)_5(C^{17}O)]^+ + H_2O$ has recently been found to obey the rate law, rate = $k[Re(CO)_6{}^+][H_2O]^2$. The experiments were conducted by adding measured amounts of ^{17}O-enriched water to a CH_3CN solution of $[Re(CO)_6]PF_6$. Addition of OH^- does not appreciably increase the rate. (See R.L. Kump and L.J. Todd, Inorg. Chem. 1981, 20, 3715.) Propose a mechanism which accounts for these results.

10.33 Solution: Ligands coordinated to positively charged metals are subject to nucleophilic attack. The results on addition of OH^- indicate that hydroxide is not the principal attacking species. A possible mechanism is:

$$Re(CO)_6^+ + H_2{}^{17}O \rightleftharpoons (CO)_5ReC(=O){-}^{17}OH_2^+ \qquad K$$

$$(CO)_5ReC(=O){-}^{17}OH_2^+ + H_2O \longrightarrow (CO)_5ReC(=O){-}^{17}OH + H_3O^+ \qquad k$$

$$(CO)_5ReC(=O){-}^{17}OH \longrightarrow (CO)_5ReC(OH){=}^{17}O \qquad \text{fast}$$

$$(CO)_5ReC(OH){=}^{17}O + H_3O^+ \longrightarrow (CO)_5Re(C^{17}O)^+ + 2H_2O \qquad \text{fast}$$

$$\text{rate} = k[Re(CO)_5C(=O){-}^{17}OH_2^+][H_2O]$$

$$K = \frac{[Re(CO)_5C(=O){-}^{17}OH_2]}{[Re(CO)_6^+][H_2{}^{17}O]}$$

$$\text{rate} = k\,K\,[Re(CO)_6^+][H_2{}^{17}O][H_2O]$$

But $[H_2{}^{17}O] = f[H_2O]$ where f is the fraction containing ^{17}O, so

$$\text{rate} = kKf\,[Re(CO)_6^+][H_2O]^2$$

10.34 Problem: Why are heat and/or light often necessary for catalytic behavior of transition metal complexes? What types of substances are most likely to poison transition metal complexes? Why?

10.34 Solution: Catalysis involves coordinating the reacting species to the transition metal, and this must involve creation of a vacant coordination position. The energy supplied by heat or light can dissociate an already coordinated ligand to create a vacant position. Any substance which coordinates to a metal forming a bond which cannot be broken by available thermal or light energy acts as a catalyst poison.

10.35 Problem: Discuss the addition of H_2 to Vaska's compound from the viewpoint of orbital symmetry.

10.35 Solution: The oxidative additions involve transfer of electron density from the highest occupied molecular orbital (HOMO) of Vaska's compound to the lowest unoccupied MO (LUMO) of the adding molecule. The LUMO of H_2 is σ_u^* having two lobes of opposite sign. The activated complex for H_2 addition must involve overlap of orbitals having like signs. If we imagine H_2 to approach $Ir(CO)(PPh_3)_2Cl$ along the z-axis, in-phase overlap could occur between lobes of the filled d_{xz} or d_{yz} orbital of Ir and the σ_u^* of H_2. This arrangement would lead to cis addition of H_2. On the other hand approach along the x or y axis by H_2 could lead to favorable overlap of σ_u^* of H_2 with Ir d_{xz} or d_{yz} resulting in addition of one H above and one below the molecular plane (trans addition), but this pathway is not as attractive from steric considerations. Reversable cis addition of H_2 to Vaska's compound is observed.

A third possibility would be electron transfer via overlap of the Ir d_{z^2} orbital with the + lobe of the σ_u^* orbital of H_2. This would lead to breaking of the H_2 bond with bonding of H^+ to Ir and departure of H^-. This is energetically very unfavorable for H_2, but possible in solution for alkyl halides, which are found to undergo trans addition.

XI Prediction and Correlation of Oxidation-Reduction Reactions

11.1 Problem: Write balanced ionic half-reactions for

(a) $HgO \xrightarrow{\text{base}} Hg$

(b) $VO^+ \xrightarrow{\text{acid}} HV_2O_5^-$

(c) $HGeO_3^- \xrightarrow{\text{base}} Ge$

(d) $S_2O_3^{2-} \xrightarrow{\text{acid}} S$

(e) $F_2O \xrightarrow{\text{acid}} HF$

(f) $C_{12}H_{22}O_{11} \xrightarrow{\text{acid}} CO_2$

11.1 Solution: The sequence of steps used to balance a half-reaction is: 1. Select those species which undergo a change and balance with respect to atoms other than H and O. Thus in part (e) we write $F_2O \rightarrow 2HF$; in part (f) $C_{12}H_{22}O_{11} \rightarrow 12CO_2$. Where two or more species are produced from a single species, two half-reactions will usually be written. 2. Balance the oxygen atoms by adding H_2O to the appropriate side of the equation. In part (f) we obtain $13H_2O + C_{12}H_{22}O_{11} \rightarrow 12CO_2$. 3. Balance the hydrogen atoms by adding H^+ to the appropriate side of the equation. In part (f) we obtain $13H_2O + C_{12}H_{22}O_{11} \rightarrow 12CO_2 + 48H^+$. 4. Balance the equation with respect to charge by adding electrons to the appropriate side. Again, to conclude part (f) we have $13H_2O + C_{12}H_{22}O_{11} \rightarrow 12CO_2 + 48H^+ + 48e$. 5. For reactions occurring in base, we add a number of OH^- equal to that of the H^+ in the half-reaction to each side. The $n(H^+ + OH^-)$ are treated as nH_2O and water appearing on both sides of the equation cancelled.

The half-reactions are thus

(a) $2e + HgO + 2H^+ \rightarrow Hg + H_2O$

$$\begin{array}{l} 2e + HgO + 2H^+ \rightarrow Hg + H_2O \\ \quad\quad + 2OH^- \quad\quad\quad + 2OH^- \\ \hline 2e + HgO + H_2O \rightarrow Hg + 2OH^- \end{array}$$

(b) $2VO^+ + 3H_2O \rightarrow HV_2O_5^- + 5H^+ + 2e$

(c)
$$\begin{array}{l} 4e + 5H^+ + HGeO_3^- \rightarrow Ge + 3H_2O \\ \quad 5OH^- \quad\quad\quad\quad\quad + 5OH^- \\ \hline 4e + 2H_2O + HGeO_3^- \rightarrow Ge + 5OH^- \end{array}$$

(d) $4e + 6H^+ + S_2O_3^{2-} \rightarrow 2S + 3H_2O$

(e) $4e + 4H^+ + F_2O \rightarrow 2HF + H_2O$

(f) $13H_2O + C_{12}H_{22}O_{11} \rightarrow 12CO_2 + 48H^+ + 48e$

11.2 Problem: The following are somewhat more challenging equations to balance.

(a) $H_2O + P_2I_4 + P_4 \rightarrow PH_4I + H_3PO_2$

(b) $ReCl_5 + H_2O \rightarrow Re_2Cl_9^{2-} + ReO_4^- + Cl^- + H^+$

(c) $B_{10}H_{12}CNH_3 + NiCl_2 + NaOH \rightarrow Na_4[(B_{10}H_{10}CNH_2)_2Ni] + NaCl + H_2O$

(d) $ICl + H_2S_2O_7 \rightarrow I_2^+ + I(HSO_4)_3 + HS_3O_{10}^- + HSO_3Cl + H_2SO_4$

11.2 Solution:

(a)
$$\begin{array}{l} (32e + 32H^+ + P_4 + 2P_2I_4 \rightarrow 8PH_4I) \\ (8H_2O + P_4 \rightarrow 4H_3PO_2 + 4H^+ + 4e)8 \\ \hline 9P_4 + 2P_2I_4 + 64H_2O \rightarrow 8PH_4I + 32H_3PO_2 \end{array}$$

(b)
$$\begin{array}{l} (3e + 2ReCl_5 \rightarrow Re_2Cl_9^{2-} + Cl^-)2 \\ (4H_2O + ReCl_5 \rightarrow ReO_4^- + 8H^+ + 5Cl^- + 2e)3 \\ \hline 7ReCl_5 + 12H_2O \rightarrow 2Re_2Cl_9^{2-} + 3ReO_4^- + 24H^+ + 17Cl^- \end{array}$$

(c) $2(B_{10}H_{12}CNH_3 \rightarrow B_{10}H_{10}CNH_2^{2-} + 3H^+ + e)$

$2e + NiCl_2 \rightarrow Ni^o + 2Cl^-$

$6NaOH \rightarrow 6Na^+ + 6OH^-$

$$2B_{10}H_{12}CNH_3 + NiCl_2 + 6NaOH \rightarrow Na_4(B_{10}H_{10}CNH_2)_2Ni + 6H_2O + 2NaCl$$

(d) $(e + 2ICl \rightarrow I_2^+ + 2Cl^-)2$

$(3HSO_4^- + ICl \rightarrow I(HSO_4)_3 + Cl^- + 2e)$

$5ICl + 3HSO_4^- \rightarrow 2I_2^+ + I(HSO_4)_3 + 5Cl^-$

$(H_2S_2O_7 + Cl^- \rightarrow HSO_3Cl + HSO_4^-)\ 5$

(1) $5ICl + 5H_2S_2O_7 \rightarrow 2I_2^+ + I(HSO_4)_3 + 5HSO_3Cl + 2HSO_4^-$

(2) $2H_2S_2O_7 \rightarrow H_2S_3O_{10} + H_2SO_4$

(3) $2H_2S_3O_{10} + 2HSO_4^- \rightarrow 2HS_3O_{10}^- + 2H_2SO_4$

(1) + 2 x (2) + (3) =

$$5ICl + 9H_2S_2O_7 \rightarrow 2I_2^+ + 2H_2S_3O_{10}^- + 5HSO_3Cl + 4H_2SO_4 + I(HSO_4)_3$$

11.3 Problem: Should the value of the heat of a reaction calculated from the variation of K_p with temperature be affected by the units in which K_p is expressed? Under what circumstances would ΔH be unaffected and under what circumstances affected by the selection of units for the K_{eq}?

11.3 Solution:

(a) The units in which K_p is expressed will affect the ΔG^o (since the numerical value of K_p is affected) but will not affect ΔH^o if ideal gas behavior is observed. (The slope of $\log K_p$ vs 1/T will remain constant.) Note that ΔS^o will be affected and the change in ΔG^o will be the same as the $-T\Delta S$ term.

(b) ΔH^o will be affected if the change in the units of K involves a change in standard states which cannot be converted from one to another by a simple multiplicative factor, that is, K_N, K_C, K_p, etc.

11.4 Problem: Calculate E^o values for the following cells and write the reactions for which these apply.

(a) Pt, H_2 | HCl || KCl | Hg_2Cl_2(s), Hg

(b) Cu | Cu^{2+} || I^-, CuI(s) | Cu

(c) Pt, CuI(s) | Cu^{2+} || I^- | CuI(s), Cu

From parts (b) and (c), comment on the necessity of knowing the half-cells involved in making predictions of (1) the spontaneity of a reaction and (2) the free energy change of a reaction. (E^o $Hg_2Cl_2 \xrightarrow{0.2682}$ Hg + Cl^-; E^o CuI $\xrightarrow{-0.1852}$ Cu + I^-)

11.4 Solution:

(a) Pt, H_2 | HCl || Hg_2Cl_2 (s), Hg

$$H_2 \rightarrow 2H^+ + 2e \qquad E^o = 0.00$$
$$2e + Hg_2Cl_2 \rightarrow 2Hg + 2Cl^- \qquad E^o = 0.2682$$
$$H_2 + Hg_2Cl_2 \rightarrow 2H^+ + 2Cl^- + 2Hg \qquad E^o = 0.2682 \text{ V}$$

(b) Cu | Cu^{2+} || I^-, CuI(s), Cu

$$Cu \rightarrow Cu^{2+} + 2e \qquad E^o = -0.337$$
$$2(e + CuI \rightarrow Cu + I^-) \qquad E^o = -0.1852$$
$$2CuI \rightarrow Cu^{2+} + Cu + 2I^- \qquad E^o = -0.523 \text{ V}$$

(c) Pt, CuI(s) | Cu^{2+} || I^- CuI(s), Cu

$$CuI \rightarrow Cu^{2+} + I^- + e \qquad E^o = -0.86$$
$$e + CuI \rightarrow Cu + I^- \qquad E^o = -0.1852$$
$$2CuI \rightarrow Cu^{2+} + Cu + 2I^- \qquad E^o = -1.045$$

The spontaneity depends on the sign of E^o which is independent of the reaction path. ΔG^o is a function of nFE where n depends on the reaction path but ΔG^o does not.

11.5 Problem: Calculate the emf values for the following cells.

(a) Pt, H_2 | H^+ (a = 0.1) || H^+ (a = 10^{-7})| H_2, Pt

(b) Zn | Zn^{2+} (a = 1) || Cu^{2+} (a = 10^{-4}) | Cu

(c) Fe, $Fe(OH)_2$ | OH^- (a = 0.1) || Fe^{2+} (a = 1) | Fe

11.5 Solution:

(a) $H_2 \rightarrow 2H^+(10^{-1}) + 2e$ $\quad E = E^o - \frac{.059}{n}\log(H^+)^2$

$E = 0.00 - \frac{.059}{2}\log(0.1)^2 = 0.06$

$2H^+(10^{-7}) \rightarrow H_2$ $\quad E = 0.00 - \frac{.059}{2}\log\frac{1}{(10^{-7})^2} = -0.42$

$2H^+(10^{-7}) \rightarrow 2H^+(10^{-1})$ $\quad E = -0.36$

(b) $Zn \rightarrow Zn^{2+} + 2e$ $\quad E^o = 0.76$

$2e + Cu^{2+}(10^{-4}) \rightarrow Cu$ $\quad E = 0.337 - \frac{.059}{2}\log\frac{1}{10^{-4}} = 0.22$

$Zn + Cu^{2+}(10^{-4}) \rightarrow Cu + Zn^{2+}$ $\quad E = 0.98$

(c) $Fe^{2+} + 2e \rightarrow Fe$ $\quad E^o = -0.44$

$2(OH^-)(10^{-1}) + Fe \rightarrow Fe(OH)_2 + 2e$ $\quad E = 0.887 - \frac{.059}{2}\log\frac{1}{10^{-1}} = 0.86$

$Fe^{2+} + 2OH^-(10^{-1}) \rightarrow Fe(OH)_2(s)$ $\quad E = 0.42$

11.6 Problem: Calculate the approximate half-cell emf values for the following reactions by using the equation $E^o \simeq -(\chi_X - \chi_H)^2$ which applies to the oxidative coupling reaction

$$2HX \rightarrow X_2 + 2H^+ + 2e$$

Use Huggin's assignment of 2.1 for the electronegativity of H.

(a) $2H_2O \rightarrow H_2O_2 + 2H^+ + 2e$

(b) $2HF \rightarrow F_2 + 2H^+ + 2e$

(c) $2PH_3 \rightarrow H_2PPH_2 + 2H^+ + 2e$

11.6 Solution: The electronegativity values of O, F, and P are 3.5, 4.0, and 2.3, respectively. Using the equation given above, we find the oxidative coupling half-cell emf values to be

(a) -1.7 v

(b) -3.2 v

(c) -0.01 v

The experimental values for the half-cell reactions are -1.78, -2.87, and -0.006 volts, respectively.

11.7 Problem: Describe the conditions of acidity most appropriate for the following processes.

(a) $Mn^{2+} \rightarrow MnO_4^-$

(b) $CrO_4^{2-} \rightarrow Cr_2O_7^{2-}$

(e) $Fe^{3+} \rightarrow FeO_4^{2-}$

(d) $ClO_4^- \rightarrow ClO_3^-$

(e) $C_2O_4^{2-} \rightarrow 2CO_2$

(f) $H_2O_2 \rightarrow H_2O$

(g) $H_2O_2 \rightarrow O_2$

11.7 Solution: The reaction or half-reaction is first balanced by the procedure outlined in Problem 11.1, and then the conditions are selected by applying Le Chatelier's principle.

(a) $Mn^{2+} + 4H_2O \rightarrow MnO_4^- + 8H^+ + 5e$	Reaction favored in base
(b) $2H^+ + 2CrO_4^{2-} \rightarrow Cr_2O_7^{2-} + H_2O$	Reaction favored in acid
(c) $4H_2O + Fe^{3+} \rightarrow FeO_4^{2-} + 8H^+ + 3e$	Reaction favored in base
(d) $2e + 2H^+ + ClO_4^- \rightarrow ClO_3^- + H_2O$	Reaction favored in acid
(e) $C_2O_4^{2-} \rightarrow 2CO_2 + 2e$	Independent of acidity except for the removal of oxalate ion by protonation
(f) $2e + 2H^+ + H_2O_2 \rightarrow 2H_2O$	Reaction favored in acid
(g) $H_2O_2 \rightarrow O_2 + 2H^+ + 2e$	Reaction favored in base

11.8 Problem: By extrapolation from known data for Cr, W, and Mo, Seaborg estimated the following emf values for the couples involving possible species for the element 106.

$$E^o_A \quad MO_3 \xrightarrow{0.5} M_2O_5 \xrightarrow{0.2} MO_2 \xrightarrow{0.7} M^{3+} \xrightarrow{0.0} M$$

Predict the results of the following

(a) M is placed in 1M HCl

(b) MCl_3 is placed in an acidic solution containing $FeSO_4$

(c) MO_2 and M are in contact in acid solution

(d) M^{3+} and MO_3 are in contact in acid solution

(e) M is treated with excess concentrated HNO_3.

11.8 Solution: For each set of reactants you should consider the possibility of hydrogen evolution, oxygen evolution, disproportionation, and oxidation or reduction by the anions and cations present with the M species. For each possible reaction an E^o value should be calculated. The reaction with the largest E value will be favored.

(a) No reaction occurs. $E^o = 0.0$ for the evolution of H_2 with the formation of M^{3+}, therefore, overvoltage effects would be sufficient to prevent this reaction.

(b) $3H_2O + 2M^{3+} + SO_4^{2-} \rightarrow 2MO_2 + H_2SO_3 + 4H^+$ $E^o = 0.87v$

(c) No reaction. ($E^o = -0.7$ for $12H^+ + M + 3MO_2 \rightarrow 4M^{3+} + 6H_2O$)

(d) $3H_2O + 2M^{3+} + MO_3 \rightarrow 3MO_2 + 6H^+$ $E^o = 0.35$

(e) $M + 2HNO_3 \rightarrow MO_3 + 2NO + H_2O$ $E^o = 0.72$

11.9 Problem: Given the following half-cell emf diagram for osmium

$$E^o_A \quad OsO_4(s) \xrightarrow{1.0} OsCl_6^{2-} \xrightarrow{0.85} OsCl_6^{3-} \xrightarrow{0.4} Os^{2+} \xrightarrow{0.85} Os$$

(a) Which of the above species, if any, would be unstable in 1M HCl? Give balanced equations for any reactions that occur.

(b) Which couple(s) would remain unchanged in their emf value on altering the pH?

(c) Which couple(s) would remain unchanged in their emf value on altering the chloride ion concentration?

(d) Calculate the value of E^o_A for $OsO_4(s) \longrightarrow Os$

(e) Predict the results of mixing excess osmium with solid OsO_4 in contact with 1M HCl.

11.9 Solution: This problem should be attacked in the same fashion as Problem 11.8.

(a) $3Os^{2+} + 12Cl^- \longrightarrow Os + 2OsCl_6^{3-}$ $\quad E^o = 0.45$ v

(b) To be unaffected by the hydrogen ion concentration, the half-reaction must not involve H^+ ion. Thus all couples except those involving OsO_4 would be unaffected by a change in pH.

(c) To be unaffected by the chloride ion concentration, the half-reaction must not involve free Cl^- ion. Such couples would be Os/Os^{2+}; Os/OsO_4; Os^{2+}/OsO_4; and $OsCl_6^{3-}/OsCl_6^{2-}$

(d)

$$OsO_4 \xrightarrow{1.0} OsCl_6^{2-} \xrightarrow{0.85} OsCl_6^{3-} \xrightarrow{0.4} Os^{2+} \xrightarrow{0.85} Os$$

$$\frac{4(1.0) + 1(0.85) + 1(0.4) + 2(0.85)}{8} = 0.87$$

(e) $4Os + 3OsO_4 + 42Cl^- + 24H^+ \longrightarrow 4OsCl_6^{3-} + 3OsCl_6^{2-} + 12H_2O$ $\quad E^o = 0.3V$

11.10 Problem:

(a) Reconstruct the Latimer diagrams given below so that only species which are stable with respect to disproportionation remain.

$$E^o_A \quad ClO_4^- \xrightarrow{1.19} ClO_3^- \xrightarrow{1.21} HClO_2 \xrightarrow{1.645} HClO \xrightarrow{1.63} Cl_2 \xrightarrow{1.36} Cl^-$$

$$E^o_B \quad ClO_4^- \xrightarrow{0.36} ClO_3^- \xrightarrow{0.33} ClO_2^- \xrightarrow{0.66} ClO^- \xrightarrow{0.40} Cl_2 \xrightarrow{1.36} Cl^-$$

(b) Construct a Pourbaix diagram for Cl_2 with unit activity for predominant species.

11.10 Solution:

(a) A species will disproportionate if the sum of its oxidation half-cell emf and reduction half-cell emf is positive. In base solution we find that Cl_2 will spontaneously yield Cl^- and ClO^- ($E^o = +0.96$). Calculating a new E^o for $ClO^- \longrightarrow Cl^-$ as +0.98 we find that ClO^- is also unstable with respect to disproportionation. (The emf for the intermediate step,

$ClO^- \xrightarrow{-0.40} Cl_2$, however, suggests an activation energy which may cause ClO^- to undergo such slow autooxidation that it may be metastable, that is, the reaction is slow although the species is thermodynamically unstable. Such is the actual case here.) The following diagrams result from this procedure.

$$E^o_A \quad ClO_4^- \xrightarrow{1.39} Cl_2 \xrightarrow{1.36} Cl^-$$

$$E^o_B \quad ClO_4^- \xrightarrow{0.56} Cl^-$$

(b) A Pourbaix diagram shows the E vs. pH dependence of stable (or sometimes metastable) species. The Cl_2/Cl^- couple has no pH dependence so a plot of E vs. pH for unit activities of Cl_2 and Cl^- is a horizontal line with $E^o = 1.36$. The ClO_4^-/Cl_2 couple shows the expected pH dependence given by the Nernst equation for

$$7e + 8H^+ + ClO_4^- \longrightarrow 1/2\ Cl_2 + 4H_2O$$

$$E = E^o - \frac{RT}{n} \ln Q = 1.39 - \frac{0.059}{7} \log \frac{1}{[H^+]^8}$$

$$E = 1.39 - 0.067\ \text{pH}$$

In a similar fashion, for ClO_4^-/Cl^- we have

$$8e + 8H^+ + ClO_4^- \longrightarrow Cl^- + 4H_2O$$

$$E = E^o - 0.059 \log \frac{1}{[H^+]^8} = 1.39 - 0.059\ \text{pH}$$

The Pourbaix diagram is thus:

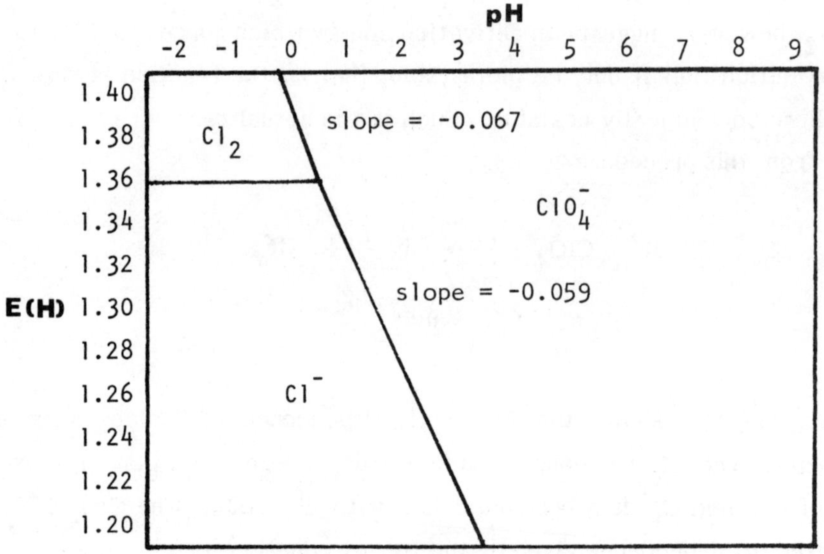

Pourbaix diagram constructed for Cl species in the low pH range

11.11 Problem: A Pt electrode maintained at a potential of 0.525 V relative to the standard hydrogen electrode is immersed in an aqueous solution of Fe^{2+} and Fe^{3+} salts. What is the value of $[Fe^{3+}]/[Fe^{2+}]$?

$$Fe^{3+} \xrightarrow{0.771} Fe^{2+}$$

11.11 Solution: The Nernst equation for the Fe^{3+}/Fe^{2+} couple is

$$E = E^{o} - \frac{0.059}{1} \log \frac{[Fe^{2+}]}{[Fe^{3+}]}$$

If E is arbitrarily fixed at 0.525 V:

$$0.525 = 0.771 - 0.59 \log [Fe^{2+}]/[Fe^{3+}]$$

From which we obtain $[Fe^{2+}]/[Fe^{3+}] = 1.4 \times 10^{4}$

11.12 Problem: Given the following estimated half-cell emf values (A. Stiller, Ph.D. Dissertation, University of Cincinnati, 1973)

$$E^o_A \quad SeO_2 \xrightarrow{0.72} Se_4^{2+} \xrightarrow{1.0} Se_8^{2+} \xrightarrow{0.8} Se_8$$

colorless yellow green red

Write equations to describe the following reactions:

(a) Se_8 dissolves in $H_2S_2O_7$ to give a yellow solution.

(b) On diluting the solution from part (a) to about 98% H_2SO_4, the solution turns green.

(c) On long standing in H_2SO_4 having a concentration above 80%, Se_8 dissolves to give green solutions from which red Se_8 can be recovered almost quantitatively by dilution with water.

11.12 Solution: It should be noted that the emf values are those appropriate for a hydrogen activity of one; in fuming sulfuric acid the hydrogen ion activity would be much greater than one. Accordingly, the $SeO_2 \rightarrow Se_4^{2+}$ half reaction would show a more positive value than the E^o_A value, but the other couples above would not be affected since hydrogen ion is not involved in the remaining half-reactions. The reactions are thus written

(a) $8H_2S_2O_7 + Se_8 \rightarrow 2Se_4^{2+} + 4HS_2O_7^- + 6H_2SO_4 + 2SO_2$

It may be noted that SO_3 becomes a strong oxidizing agent at high acidities. The $HS_2O_7^-$ and H_2SO_4 products arise from H balance and from the reaction of $H_2S_2O_7$ with water produced simultaneously with SO_2.

(b) $8H_2O + Se_4^{2+} \rightarrow 4SeO_2 + 16H^+ + 14e$

$7[2e + 2Se_4^{2+} \rightarrow Se_8^{2+}]$

$8H_2O + 15Se_4^{2+} \rightarrow 7Se_8^{2+} + 4SeO_2 + 16H^+$

Reducing the hydrogen ion activity allows this disproportionation reaction to occur. The hydrogen ion would be taken up by the anions present.

(c) $3H_2SO_4 + Se_8 \rightarrow Se_8^{2+} + 2HSO_4^- + 2H_2O + SO_2$

At the lower acid concentration, the H_2SO_4 oxidizes the Se_8 only to Se_8^{2+}. Again, reducing the hydrogen ion activity allows a disproportionation reaction to occur.

$$16H_2O + Se_8^{2+} \longrightarrow 8SeO_2 + 32H^+ + 30e$$
$$15[2e + Se_8^{2+} \longrightarrow Se_8]$$
$$16H_2O + 16Se_8^{2+} \longrightarrow 15Se_8 + 8SeO_2 + 32H^+$$

The disproportionation reaction yields 15/16th of the Se_8 initially dissolved.

11.13 Problem: Calculate ΔG^o for the following reactions:

(a) $2Al + Fe_2O_3 \longrightarrow Al_2O_3 + 2Fe$

(b) $3Cs_2O + 2Al \longrightarrow Al_2O_3 + 6Cs(g)$

ΔG_f^o Al_2O_3, 1580 kJ/mole; Fe_2O_3, 740 kJ/mole;

Cs_2O, 280 kJ/mole; Cs(g) 51 kJ/mole

11.13. Solution: The ΔG^o for a reaction is the sum of the G_f^o of the products minus the sum of the G_f^o of the reactants. Elements in their standard states have a G_f^o value of zero. For the above reactions we obtain:

(a) ΔG^o = -1580 -(-740) = -840 kJ at 25°C. This reaction, known as the thermite reaction, is thermodynamically favored even at room temperature. However, its reaction rate is not perceptible at room temperature, so it is initiated at high temperature (usually by an Mg fuse), after which the reaction sustains itself.

(b) ΔG^o = 6(51) + (-1580) -3(-280) = -434 kJ at 25°C. This reaction is also carried out at high temperature, the Cs vapor being removed, the positive entropy change making the free energy change even more negative than the above calculation for 25°C.

XII Acids and Bases

12.1 Problem:

(a) Indicate the conjugate bases of the following: NH_3, NH_2^-, NH^{2-}, H_2O, HI.

(b) Indicate the conjugate acids of the above species.

(c) What relationship exists between the strength of a conjugate acid and a conjugate base for a neutral substance, such as NH_3 in parts (a) and (b) above.

12.1 Solution: An acid becomes a "conjugate" base when it loses a proton; a base becomes a "conjugate" acid when it gains a proton.

(a)

acid	→	conjugate base	+ H^+
NH_3	→	NH_2^-	+ H^+
NH_2^-	→	NH^{2-}	+ H^+
H_2O	→	OH^-	+ H^+
HI	→	I^-	+ H^+

(b)

base	+	H^+	→	conjugate acid
NH_3	+	H^+	→	NH_4^+
NH_2^-	+	H^+	→	NH_3
H_2O	+	H^+	→	H_3O^+
HI	+	H^+	→	H_2I^+

(c) This question can be answered on several different levels and from several different perspectives. The simplest would be to consider behavior in aqueous solution where we know that HI(aq) contains mainly I^-(aq) and no detectable H_2I^+(aq), H_2O contains equal, but small, amounts of H_3O^+(aq) and OH^-(aq), while NH_3(aq) contains some NH_4^+(aq) but no detectable NH_2^-. From these observations we might conclude that substances with strong conjugate acids have weak conjugate bases and vice versa. As a first order approximation, the generalization would be correct, but extensions are very tenuous. Thus in the gas phase some substituents enhance (or decrease) both the acid strength and the base strength. If one makes the assumption (which is not too good for second period elements) that the homolytic bond energy in H_2X^+ is the same as that in HX, then difference in PA's for X^- and HX will be the sum of the EA of X and the IE of HX. Returning to a solution view and putting water in as the solvent we may rewrite (a) and (b) for NH_3 as follows:

$$NH_3 + H_2O \rightarrow NH_2^- + H_3O^+ \qquad K_{a(NH_3)}$$

$$NH_3 + H_2O \rightarrow NH_4^+ + OH^- \qquad K_{b(NH_3)}$$

from which

$$K_aK_b = \frac{(NH_2^-)(NH_4^+)}{(NH_3)^2} K_w$$

For gross approximations $(NH_2^-)(NH_4^+)/(NH_3)^2$ may be replaced with the K_s for $NH_3(\ell)$, i.e. 10^{-30}. Using $K_b = 10^{-9.5}$ for NH_3 and our approximation $K_aK_b = K_sK_w$ gives a value of 10^{-35} for K_a for NH_3 in aqueous solutions.

12.2 Problem: Into a low pressure system containing an excess of HI(g) the following species are introduced—(1) H_3O^+(g); (2) OH^-(g); (3) HCl(g); (4) PH_3(g); and (5) H_2Cl^+(g).

(a) Use the proton affinities given below to indicate any proton transfer reactions which occur.

(b) Indicate the minimun energy "cost" of a proton in such a system, that is, the effective PA of the system.

(c) Would you characterize this "solvent" as giving a more acidic or more basic system than H_2O(g)? Explain.

PA values in kJ/mole

HI 611	H_2O 711	HCl 565	PH_3 774
I^- 1315	OH^- 1632	Cl^- 1395	PH_2^- 1548

12.2 Solution: HI will lose a proton to any species whose proton affinity is greater than 1315 kJ/mole. HI will gain a proton from any species whose conjugate base has a PA of less than 611 kJ/mole. The "cost" of a proton in the resulting system will be the lowest value of the PA for any species remaining, including the excess HI.

(a) Reaction	(b) Lowest Cost of Proton	Source
(1) $H_3O^+ + HI \not\rightarrow$	711 kJ/mole	H_3O^+
(2) $OH^- + HI \rightarrow H_2O + I^-$	1315 kJ/mole	excess HI
(3) $HCl + HI \not\rightarrow$	1315 kJ/mole	HI
(4) $PH_3 + HI \not\rightarrow$	1315 kJ/mole	HI
(5) $H_2Cl^+ + HI \rightarrow H_2I^+ + HCl$	611 kJ/mole	H_2I^+

(c) The proton cost in our HI(g) system must always fall between the PA of HI (611 kJ/mole) and the PA of I^- (1315 kJ/mole). For a system containing excess monomolecular $H_2O(g)$, the proton cost would fall between the PA of H_2O (711 kJ/mole) and the PA of OH^- (1632 kJ/mole). Thus the HI system would be more acidic than the H_2O system.

12.3 Problem: The heat of neutralization of $H^+(aq)$ and $OH^-(aq)$ is considerably higher in concentrated salt solution (ΔH_{neut} = 85.4 kJ/mole at μ = 16) than in dilute aqueous solution ($\Delta H_{neut.}$ = 56.5 kJ/mole as $\mu \rightarrow 0$). Explain what might cause the increase in the heat of neutralization when a high concentration of NaCl is present.

12.3 Solution: The water activity in concentrated salt solutions is decreased because of the large amount of water tied up in ion hydration. This will cause a decrease in the average number of water molecules associated with the H^+, thereby increasing the availability of the proton, that is, less energy is lost by the breakup of $(H_2O)_nH^+$ when the proton is transferred. The same reasoning applies to the solvation of the OH^- ion. Lowering the water activity of strong acid or base solutions thus increases the effective strength of the acid or base.

12.4 Problem: Write a thermochemical cycle from which you can obtain the hydride affinity of a positive ion from the bond dissociation energy of EH and other suitable data.

12.4 Solution:

$$\begin{array}{ccc} E^+ + H^- & \xrightarrow{1} & EH \\ \uparrow 4 & & \downarrow 2 \\ E^+ + H + e & \xleftarrow{3} & E + H \end{array}$$

$$\text{Hydride affinity} = -\Delta H_1$$

$$\Delta H_2 = D_{E\text{-}H}$$

$$\Delta H_3 = IE(E)$$

$$\Delta H_4 = -EA(H)$$

$$\Delta H_1 + \Delta H_2 + \Delta H_3 + \Delta H_4 = 0$$

$$H^-A(E^+) = D_{E\text{-}H} + IE_E - EA_H$$

12.5 Problem: Give the approximate pK_a values for the following acids:

(a) H_3PO_3 (b) HNO_3 (c) $HClO_4$ (d) H_5IO_6

12.5 Solution: Pauling's rules for estimating the strength of acids of the general formula $XO_n(OH)_m$ are

$$pK_1 = 7 - 5n$$

and successive ionization steps give pK_2, etc., which are successively 5 units greater. In the following we have rewritten the formulas of the acids to more readily apply Pauling's rules. The values of n and pK_a are thus:

(a) H_3PO_3	$HPO(OH)_2$	$n = 1$	$pK_1 = 2$	$pK_2 = 7$
(b) HNO_3	$(HO)NO_2$	$n = 2$	$pK_1 = -3$	
(c) $HClO_4$	$HOClO_3$	$n = 3$	$pK_1 = -8$	
(d) H_5IO_6	$(HO)_5IO$	$n = 1$	$pK_1 = 2$	$pK_2 = 7$

12.6 Problem: Select the best answer and give the basis for your selection.

(a) Thermally most stable: PH_4Cl PH_4Br PH_4I

(b) Strongest acid: H_2O H_2S H_2Se H_2Te

(c) Acidic oxide: Ag_2O V_2O_5 CO Ce_2O_3

(d) Strongest acid: MgF_2 $MgCl_2$ $MgBr_2$

(e) Stronger base (toward a proton): PH_2^- NH_2^-

12.6 Solution:

(a) $PH_4X \xrightarrow[\Delta]{} PH_3 + HX$

HI is the strongest acid (or I^- has the lowest proton affinity) in the series so PH_4I would be the most stable. (Lattice energy effects, which would be in the opposite direction, do not override the primary acid strength effects here.)

(b) H_2Te is the strongest protonic acid in this sequence. The major factor (as in the case of the hydrogen halides) is the decreasing bond X–H energy as we descend in the group. (This changes much more than the EA's of SH, SeH, etc.)

(c) V_2O_5 is the most acidic oxide here, the charge on the metal ion being the major consideration.

(d) $MgBr_2$ is the strongest acid here. The lattice energy of MgF_2 makes acidic behavior difficult to detect. The factors are probably the same as with the BX_3 series although far less experimental work has been done here.

(e) NH_2^- is the stronger base. Important factors are similar to those in part (b) of this problem.

12.7 Problem: Give equations to explain why adding ammonium acetate to either zinc amide(s) in liquid ammonia or zinc acetate(s) in acetic acid causes the solid to dissolve.

12.7 Solution: Zinc salts tend to show amphoteric behavior. In liquid ammonia, $NH_4CH_3CO_2$ is an acid since it increases the positive ion involved in the autoionization of the solvent. In glacial acetic acid, $NH_4CH_3CO_2$ is a base since it increases the concentration of the negative ion involved in the autoionization reaction. Consequently, the zinc salt dissolves in the acidic or basic medium produced. Equations are:

$$Zn(NH_2)_2(s) + 2NH_4^+ \xrightarrow{NH_3(\ell)} Zn^{2+}\ (solvated) + 4NH_3$$

or

$$\xrightarrow{NH_3(\ell)} Zn(NH_3)_4^{2+}$$

$$Zn(OAc)_2(s) + 2OAc^- \xrightarrow{CH_3CO_2H} Zn(OAc)_4^{2-}$$

12.8 Problem: Would the following increase, decrease, or have no effect on the acidity of the solution?

(a) Addition of Li_3N to liquid NH_3.
(b) Addition of HgO to an aqueous KI solution.
(c) Addition of SiO_2 to molten Fe + FeO.
(d) Addition of $CuSO_4$ to aqueous $(NH_4)_2SO_4$.
(e) Addition of $Al(OH)_3$ to aqueous NaOH.
(f) Addition of $KHSO_4$ to H_2SO_4
(g) Addition of CH_3CO_2K to liquid NH_3.

12.8 Solution: See Problem 12.7 for guide.

(a) $N^{3-} + 2NH_3 \rightarrow 3NH_2^-$
Increases NH_2^- concentration, behaves as base, decreases acidity

(b) $HgO + 4KI + H_2O \rightarrow HgI_4^{2-} + 4K^+ + 2OH^-$
Increases OH^- ion concentration, behaves as base, decreases acidity

(c) $FeO + SiO_2 \rightarrow FeSiO_3$
Oxide ion acceptor, behaves as a Lux-Flood acid, increases acidity

(d) $Cu^{2+}(aq) + 4NH_4^+(aq) \rightarrow Cu(NH_3)_4^{2+} + 4H^+(aq)$
Increases H^+ ion concentration, therefore acidity increases

(e) $Al(OH)_3(s) + OH^-(aq) \rightarrow Al(OH)_4^-(aq)$
Decreases OH^- ion concentration, therefore acidity increases

(f) $KHSO_4$ behaves as a base in H_2SO_4 since it increases the HSO_4^- ion concentration. Therefore the acidity of the solution decreases.

(g) CH_3CO_2K does not solvolyze since acetic acid is a strong acid in $NH_3(\ell)$. Since neither K^+ nor OAc^- are ions common to the autoionization of liquid ammonia this salt will be neutral in $NH_3(\ell)$.

12.9 Problem: Select the best response within each horizontal group and indicate the major factor governing your choice.

Strongest protonic acid

(a)	SnH_4	SbH_3	H_2Te
(b)	NH_3	PH_3	SbH_3
(c)	H_5IO_6	H_6TeO_6	HIO
(d)	$Fe(H_2O)_6^{3+}$	$Fe(H_2O)_6^{2+}$	H_2O
(e)	$Na(H_2O)_x^+$	$K(H_2O)_x^+$	

Strongest Lewis Acid

(f)	BF_3	BCl_3	BI_3
(g)	$BeCl_2$	BCl_3	
(h)	B(<u>n</u>Bu)$_3$	B(<u>t</u>Bu)$_3$	

More basic toward BMe_3

(i)	Me_3N	Et_3N		
(j)	2-MePy	4-MePy	Py	(Py = pyridine)
(k)	2-MeC_6H_4CN	C_6H_5CN		

12.9 Solution: For (a) and (b) you may give an answer based on your knowledge of periodic trends in acidity of binary hydrides. A more complete understanding can be obtained by analyzing an appropriate thermodynamic cycle similar to the type written for Problem 12.4.

(a)	H_2Te	Changes in EA of central atom
(b)	SbH_3	Changes in X—H bond energy
(c)	H_5IO_6	Based on Pauling's rules (See Problem 12.5)
(d)	$Fe(H_2O)_6^{3+}$	Charge on central atom repels protons most strongly

(e)	$Na(H_2O)_x^+$	Smaller ionic radius of Na^+ leads to greater electrostatic repulsion of proton
(f)	BI_3	Least π-conjugation
(g)	BCl_3	More highly developed positive charge on central atom
(h)	$B(\underline{n}\text{-Bu})_3$	Least B-strain in adducts
(i)	Me_3N	Least F-strain in adducts
(j)	4-MePy	Combination of electron releasing inductive effect of Me group and least steric effect when group is in para-position
(k)	2-MeC_6H_4CN	The steric effect of the 2-Me is not appreciable here since the reactive center is farther removed than in part (j) above

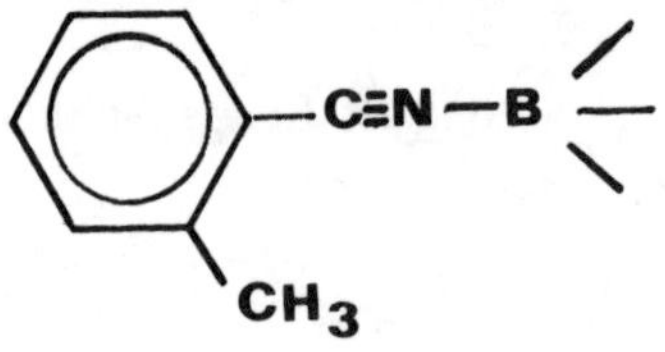

12.10 Problem: In general, the anhydrous nitrites of a given metal ion have a lower thermal stability than the nitrates, and similarly for the sulfites and the sulfates. How may this be explained? (For a more general survey of such data, and generalizations about the area, see R.T. Sanderson, Chemical Periodicity, Chap. 7, Reinhold, New York, 1960.)

12.10 Solution: This problem is similar to Problem 12.6(a) and 12.6(c). Nitrites and nitrates may be considered as oxide complexes of N_2O_3 and N_2O_5; sulfites and sulfates may be considered to be oxide complexes of SO_2 and SO_3. The oxide in the higher oxidation state is the better oxide ion acceptor (or stronger acid) and hence the nitrates and sulfates will show greater thermal stability. It should be noted that whereas the sulfites and sulfates will generally decompose on strong heating to give SO_2 and SO_3 respectively, decomposition of the nitrites and nitrates is more complex.

12.11 Problem: Use the data given in the table below to sketch a field distribution diagram for a 1 M solution of V(V) as a function of pH. What pH range is suitable for precipitation of V(V) from solution? (Hint: First determine the logarithm of the concentration ratios as a function of pH.)

Some vanadium(V) equilibria in aqueous solution at 25°C[a]

		pK	$pH_{50\%}$[b]
$2VO_4^{3-} + 2H^+$	$\rightleftharpoons V_2O_7^{4-}$	-27.0	13.5
$2V_2O_7^{4-} + 4H^+$	$\rightleftharpoons V_4O_{12}^{4-} + 2H_2O$	-40.0	10.0
$5V_4O_{12}^{4-} + 8H^+$	$\rightleftharpoons 2V_{10}O_{28}^{6-} + 4H_2O$	-58.0	7.2
$V_{10}O_{28}^{6-} + H^+$	$\rightleftharpoons HV_{10}O_{28}^{5-}$	- 5.8	5.8
$HV_{10}O_{28}^{5-} + H^+$	$\rightleftharpoons H_2V_{10}O_{28}^{4-}$	- 3.6	3.6
$H_2V_{10}O_{28}^{4-} + 4H^+$	$\rightleftharpoons 5V_2O_5nH_2O + (3 - n)H_2)$	-12.0	3.0
$V_2O_5 \cdot nH_2O + 2H^+$	$\rightleftharpoons 2VO_2^+ + (1 + n)H_2O$	+ 2.4	-1.2

[a]From C.S.G. Phillips and R.J.P. Williams, Inorganic Chemistry, Oxford, Oxford, 1965.

[b]$pH_{50\%}$ is the pH at which the two species are present in equal amounts at molar concentration and pK is - log(equilibrium constant of the reaction).

12.11 Solution: From the table above we may write the following equilibrium relationships:

$$-\log \frac{(V_2O_7^{4-})}{(VO_4^{3-})^2(H^+)^2} = -27 \quad \text{or} \quad 0.5 \log \frac{(V_2O_7^{4-})}{(VO_4^{3-})} = 13.5 - pH$$

$$0.25 \log \frac{(V_4O_{12}^{4-})}{(V_2O_7^{4-})^2} = 10 - pH$$

$$0.4 \log \frac{(V_{10}O_{28}^{6-})}{(V_4O_{12}^{4-})^5} = 7.2 - pH$$

$$\log \frac{(HV_{10}O_{28}^{5-})}{(V_{10}O_{28}^{6-})} = 5.8 - pH$$

$$\log \frac{(H_2V_{10}O_{28}{}^{4-})}{(HV_{10}O_{28}{}^{5-})} = 3.6 - pH$$

$$0.25\log \frac{1}{(H_2V_{10}O_{28}{}^{4-})} = 3.0 - pH$$

$$\log(VO_2{}^{+}) = -1.2 - pH$$

To obtain the field distribution diagram for a total initial vanadium concentration of 1 M, we add the condition that the sum of the vanadium ion concentrations total 1.0 M, that is

$$[VO_4{}^{3-}] + 2[V_2O_7{}^{4-}] + 4[V_4O_{12}{}^{4-}] + \ldots = 1.0$$

In the immediate vicinity of the $pH_{50\%}$ only the species involved in that particular equilibrium need be considered. In general, no more than two equilibria (those whose $pH_{50\%}$ bracket the pH chosen) need be considered. When solid V_2O_5 is present, the <u>initial</u> solution concentrations are taken as 1.0 M. A solution to the resulting simultaneous equations can be rapidly obtained by approximation methods, and plotted to give the figure below.

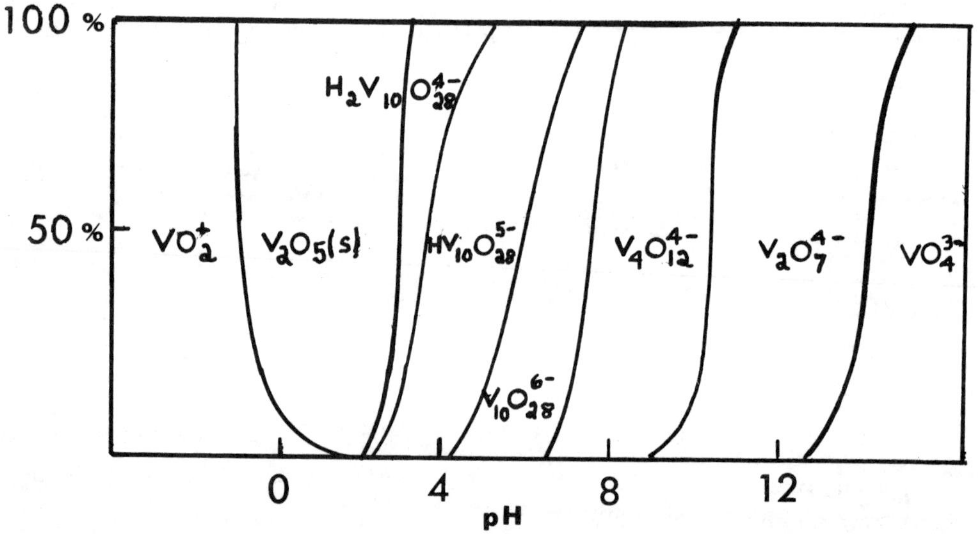

Field distribution diagram for vanadium(V) species.

Normally solid species are omitted from field distribution diagrams, but we were interested in the pH range suitable for the precipitation of V_2O_5. From the above diagram we find that optimum recovery of V_2O_5 occurs between 1.50 and 1.75 pH, but the pH range from 2 to -1 provides practical recovery.

12.12 Problem: Compared with the spectra of the halopentaamminecobalt(III) complexes, the spectra of the comparable rhodium compounds show shifts in the d-d bands toward the uv end of the spectrum, and in the charge transfer bands toward the visible. Offer an explanation for these shifts.

12.12 Solution: The second and third transition metals exhibit progressively greater ligand field splitting of the d-orbital energy levels than do comparable complexes of the first transition series. Thus the shift in the d-d bands to greater energy is expected. Charge transfer bands are generally low energy for soft-soft interactions and high energy for hard-hard interactions. The softness of a metal ion of fixed oxidation state increases as we descend a particular group of transition metals, so we expect the charge transfer band for the soft halogen ligands to move to lower energy on going from Co(III) to Rh(III) complexes.

12.13 Problem: Tetrahydrofuran has an ionization energy of 11.1 eV and diethyl ether has an ionization energy of 9.6 eV. How can we account for this difference? What difference is expected in their PA's? Which is higher? THF is much more basic toward $MgCl_2$ and Grignard reagents than Et_2O; how might this observation be rationalized?

12.13 Solution: The large difference in ionization energy of THF and Et_2O cannot readily be rationalized on the basis of the electronic effects of a ring vs. an open chain. This suggests that rehybridization takes place on ionization—going from essentially sp^3 for the neutral to sp^2 for the ion (this is known to occur with NH_3). The THF would suffer from ring strain in the ion, which would be reflected in its higher ionization energy. The higher ionization energy of THF is reflected in a lower proton affinity for this compound compared to Et_2O. The greater basicity of THF, compared to Et_2O, toward Lewis acids may be explained by the lower steric requirements of THF.

12.14 Problem: Indicate which of the following ligands should show ambidentate character, and indicate for such ligands the hard and soft ends: NO_3^-, SO_3^{2-}, $S_2O_3^{2-}$, $S_2O_7^{2-}$, NO_2^-, $(CH_3)_2SO$, CN^-, $SeCN^-$.

12.14 Solution: Ambidentate ligands must have more than one atom capable of donating an electron pair in bond formation. This rules out the nitrate ion, since here only the O atoms have unshared pairs. Chelating agents are not classified as ambidentate since these ligands generally do not show preferential ligation through a single atom. In the following ligands the soft end is the one of lower electronegativity.

SO_3^{2-}, $S_2O_3^{2-}$	S - soft	NO_2^-	N - soft
	O - hard		O - hard
CN^-	C - soft	$SeCN^-$	Se - soft
	N - hard		N - hard

12.15 Problem: Given the following C and E parameters for the Drago-Wayland equation, select the stronger base in each pair.

		E_B	C_B
(a)	Acetone	0.937	2.33
	Dimethylsulfoxide	1.34	2.85
(b)	Dimethylsulfide	0.343	7.46
	Dimethylsulfoxide	1.34	2.85

Select from among the following acids to support your conclusions.

	E_A	C_A
I_2	1.00	1.00
$HCCl_3$	3.02	0.159
SO_2	0.92	0.808
BF_3	9.88	1.62

12.15 Solution: The enthalpy of interaction of an acid and a base is given by

$$-\Delta H = E_A E_B + C_A C_B \text{ (kcal/mole)}$$

(a) Since both E_B and C_B are greater for dmso than for acetone, dmso will be the stronger base in all cases.

(b) The situation here is ambiguous. Thus $(CH_3)_2S$ is more basic than dmso towards acids with $C_A \simeq E_A$ but is a weaker base for acids where $E_A \gg C_A$. Thus with iodine, dimethyl sulfide releases

$$Q = -\Delta H = (1.00)(0.343) + (1.00)(7.46) = 7.8 \text{ kcal}$$

Whereas dmso releases only

$$Q = -\Delta H = (1.00)(1.34) + (1.00)(2.85) = 4.2 \text{ kcal}$$

On the other hand, with BF_3, dimethylsulfide is expected to release

$$(9.88)(0.343) + (1.62)(7.46) = 15.5 \text{ kcal}$$

whereas dmso releases

$$(9.88)(1.34) + (1.62)(2.85) = 17.9 \text{ kcal.}$$

12.16 Problem: Neglecting the metals with partially filled p orbitals, in what regions of the periodic table are the class (a) metal acceptors (those that prefer N, O, or F donors) and the class (b) metal acceptors (those that prefer ligand atoms from the third period or a later period)? How does this classification relate to size, charge, electronegativity, and the hard-soft acid-base classification?

12.16 Solution: The class (a) metal acceptors are the large metal atoms with low electronegativity found on the left side of the periodic table. The class (b) metals are near Au in the periodic table—these are the highly electronegative and highly polarizing (also highly polarizable) metals, particularly in low oxidation states. The class (a) metals are soft acids, preferring soft bases.

12.17 Problem: Use values for the E and C parameters given below to rank the basicities of NH_3, $(CH_3)_2NH$, $(CH_3)_3P$ and $(CH_3)_2S$ toward ICl and H_2O.

Base	E	C		Acid	E	C
NH_3	1.15	4.75		ICl	5.10	0.830
$(CH_3)_2NH$	1.09	8.73		H_2O	1.64	0.571
$(CH_3)_3P$	0.838	6.55				
$(CH_3)_2S$	0.343	7.46				

12.17 Solution: The Drago-Wayland equation allows you to predict heats of acid-base reactions:

$$-\Delta H = E_A E_B + C_A C_B$$

For any given acid, the stronger the base the more negative ΔH. The following values of ΔH are in kcal/mole:

Base	ICl	H_2O
NH_3	- 9.80	-4.60
$(CH_3)_2NH$	-12.8	-6.77
$(CH_3)_3P$	- 9.71	-5.11
$(CH_3)_2S$	- 7.94	-4.82

Hence the rankings are:

toward ICl $(CH_3)_2S < (CH_3)_3P \sim NH_3 < (CH_3)_2NH$

toward H_2O $NH_3 < (CH_3)_2S < (CH_3)_3P < (CH_3)_2NH$

ICl is a surprisingly hard acid when compared with I_2 (both E and C set at 1.00) which displays the order $NH_3 < (CH_3)_3P < (CH_3)_2S < (CH_3)_2NH$

12.18 Problem: Indicate the order of increasing values of the proton affinity expected for the following species: NH_2^-, NH_3, NH^{2-}, C_5H_5N, and CH_3CN.

12.18 Solution: In Problem 2.23 we found the order of the electronegativities of the conjugate acids of the above compounds to be $NH_2^- < NH_3 < NH_4^+ < C_5H_5NH^+ < CH_3CNH^+$.

The attraction for protons by atoms of the same element would be expected to be just the opposite of the attraction for electrons, so the order of PA's would be expected to be

$$NH^{2-} > NH_2^- > NH_3 > C_5H_5N > CH_3CN.$$

Experimentally, the PA of C_5H_5N is greater than that of NH_3 but the other compounds fall in the predicted order. The inversion of the NH_3 and C_5H_5N order has been rationalized on the basis of the greater polarizability of pyridine than that of ammonia—in aqueous solution this "self-solvation" becomes much less important and NH_3 is more basic than C_5H_5N.

12.19 Problem: Transmission of signals by optical waveguides has recently become practical. One of the factors used in calculating the wavelength for minimum signal loss is the "average electronic excitation gap". Rationalize the following data on the band gap (in eV) of various materials (from K. Nassau, The Bell System Technical Journal 1981, 60, 327).

LiF 16.5	LiCl 11.0	LiBr 9.5	BeO 13.7	
KF 14.7	KCl 10.5	KBr 9.2	MgO 11.4	
ZnF_2 13	$ZnCl_2$ 9	$ZnBr_2$ 7.5	CaO 9.9	
HgF_2 9	—	--	ZnO 6.1	ZnTe 4.4
			B_2O_3 12.4	BN 10.6

12.19 Solution: The above data may be rationalized on the basis of Jorgensen's optical electronegativities (DMA p. 549). The band gap is greatest for hard-hard interactions and least for soft-soft interactions.

XIII The Chemistry of Some Nonmetals

13.1 Problem: The electron affinity of N ($\leq$0) seems anomalous. Explain the order of electronegativities for Group VB and why one can conclude that N is really more "regular" than oxygen.

Electron affinities (kJ) for Group VB and VIB elements

N	P	As	Sb	Bi
0	71.7	77	101	91.4
O	S	Se	Te	Po
141.1	200.4	195	190	183

13.1 Solution: The ionization energies change by small amounts from one element to the next for P $\rightarrow$ Bi, but the ionization energy of N is much higher than that of P. The electron affinities in general for the Group VB elements are lower than those of the neighboring elements in the same period because an electron must be added to the half-filled p^3 configuration. Unlike the trend for most main group families, the electron affinities for Group VB increase with increasing atomic number, except for Bi. The increase might be expected because of the decrease in pairing energy with increasing atomic radius. The electron affinity of N is zero or slightly negative. This results from the combination of three trends: (1) It is expected to be lower than that of P because of the high pairing energy of N, (2) It is expected to be lower than the values of C or O because of the half-filled configuration of N, and (3) The first members of Groups VB-VIIB are anomalous with respect to further increase in electron density for the small compact atoms, as seen in their low electron affinities and single bond energies. The electron affinity of N appears to deviate more from the family trend than in the case of O or F. However, because of the opposite trend in electron affinities for Groups VB and VIB, plots of ionization energy vs. electron affinity reveal that N comes closer to the straight line determined by the rest of the family than is the case for O.

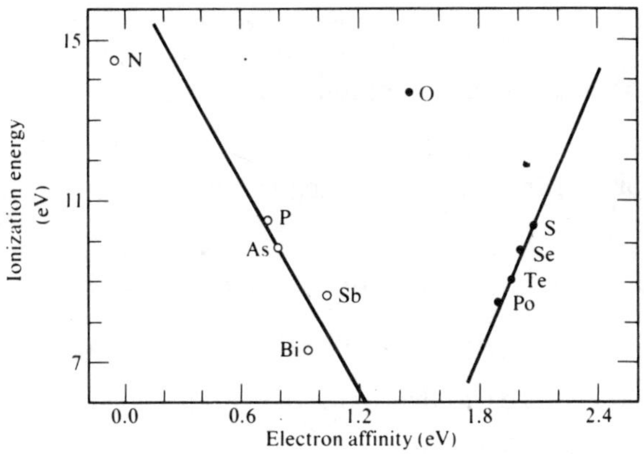

Ionization energy versus electron affinity plots for Groups VB and VIB. (Adapted from P. Politzer in *Homoatomic Rings, Chains and Macromolecules of Main Group Elements*, A. L. Rheingold, Ed., Elsevier, Amsterdam, 1977, p. 95.)

13.2 Problem: What compound is used for inflating airbags in cars?

13.2 Solution: NaN_3, which decomposes thermally to give N_2.

13.3 Problem: Write balanced equations for the preparation of HNO_3, starting with N_2.

13.3 Solution:

$$N_2 + 3H_2 \underset{\substack{500^\circ C \\ 500 \text{ atm}}}{\overset{\text{Fe catalyst}}{\rightleftharpoons}} 2NH_3 \qquad \text{Haber process}$$

$$4NH_3 + 5O_2 \xrightarrow{\text{Pt catalyst}} 4NO + 6H_2O \qquad \text{Ostwald process}$$

$$2NO + O_2 \rightarrow 2NO_2$$

$$2NO_2 + H_2O \rightarrow HNO_3 + HNO_2$$

HNO_2 disproportionates to give more HNO_3 and NO, which is recycled.

13.4 Problem: When dilute nitric acid reacts with Cu turnings in a test tube, a colorless gas is formed that turns brown near the mouth of the tube. Explain the observations and write equations for the reactions involved.

13.4 Solution: $3Cu + 8HNO_3 \rightarrow 3\ Cu(NO_3)_2 + 2NO + 4H_2O$ colorless

$2NO + O_2 \rightarrow 2NO_2$ brown

The brown color appears as NO mixes with O_2 (air) and is oxidized to NO_2.

13.5 Problem: Write equations for the preparation of five nitrogen oxides.

13.5 Solution: NH_4NO_3 (aqueous) $\xrightarrow[\text{(Cl}^-)]{\text{heat}} N_2O + 2\ H_2O$ Heat the solution containing some Cl^-.

NO is obtained in the Ostwald process for the catalytic oxidation of NH_3 (See Problem 13.3). Other oxides can be obtained from NO as follows:

$$
\begin{array}{ccccc}
& & \text{ONON} & & \\
& & \uparrow \text{acid} & & \\
\text{ONNO} \underset{\overleftarrow{}}{\overset{\text{cooling}}{\rightleftharpoons}} & \text{NO} \xrightarrow{O_2} & \text{NO}_2 & \overset{\text{cooling}}{\rightleftharpoons} & \text{N}_2\text{O}_4 \\
\swarrow \text{NO}_2 & & O_3 \downarrow & & \downarrow O_3 \\
\text{O}_2\text{NNO} & & \text{NO}_3 & \xrightarrow{\text{NO}_2} & \text{N}_2\text{O}_5 \\
\end{array}
$$

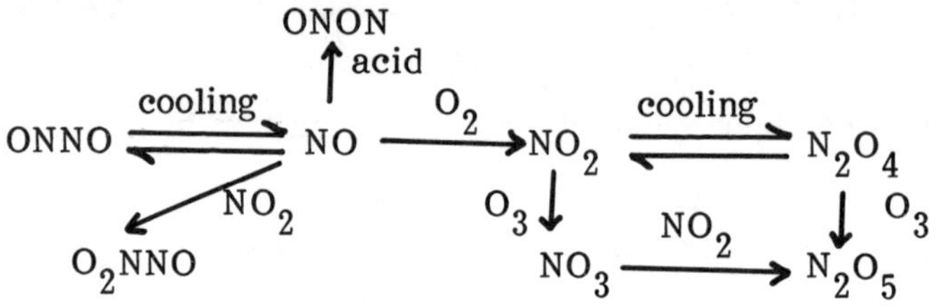

(See DMA p. 561)

HNO_3 may be used for the laboratory preparation of the common nitrogen oxides other than N_2O. At low temperatures, 100% HNO_3 may be dehydrated with P_4O_{10} to give N_2O_5 (at elevated temperatures the N_2O_5 dissociates to give NO_2 and O_2). The reduction products with glassy As_2O_3 are a function of the acid concentration as indicated by the density of the acid.

density	1.2	1.35	1.45
products	NO	N_2O_3	$NO_2 + 10\%\ N_2O_3$

Dry NO_2 may be prepared by thermal decomposition of predried $Pb(NO_3)_2$.

$$Pb(NO_3)_2 \rightarrow PbO + 1/2\ O_2 + 2NO_2$$

13.6 Problem: Sketch the <u>cis</u>- and <u>trans</u>-isomers of hyponitrous acid.

13.6 Solution:

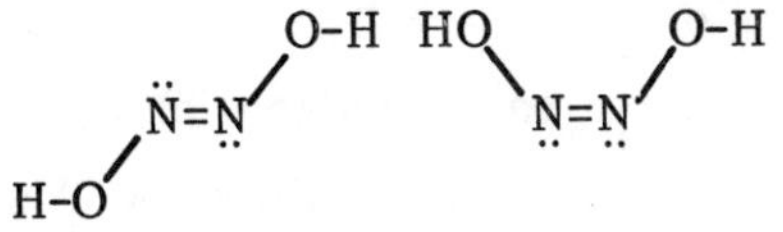

<u>trans</u> <u>cis</u>

13.7 Problem: What properties of polymers of $(SN)_x$ and $(N\text{-}PR_2)_x$ make them of interest for practical uses?

13.7 Solution: $(SN)_x$ forms chains with high electrical conductivity along the chain, sometimes referred to as a "one-dimensional metal" (See DMA p. 574). $(NPR_2)_x$ polymers have high thermal stability, oil resistance, etc. (See DMA p. 561).

13.8 Problem: Give the formula of a biodegradable polyphosphazene polymer that could be used for medical devices.

13.8 Solution:

$$[\text{-N}\overset{\displaystyle NHCH_2CO_2C_2H_5}{\underset{\displaystyle NHCH_2CO_2C_2H_5}{\text{P}}}\text{-}]_x$$

This polymer degrades to give harmless amino acids.

13.9 Problem: PI_5 is not known. What would be its likely structure in the solid?

13.9 Solution: $PI_4{}^+I^-$ P is not expected to show C.N. 5 with atoms as large as I. In the solid PCl_5 exists as $PCl_4{}^+PCl_6{}^-$ and PBr_5 exists as $PBr_4{}^+Br^-$.

13.10 Problem:

(a) What N species are compatible with the H_3PO_4-$PO_4{}^{3-}$ species from the Pourbaix diagrams given by J.A. Campbell and R.A. Whiteker, <u>J. Chem. Educ.</u> 1969, <u>46</u>, 90?

(b) What P species are compatible with $NH_3(aq)$?

13.10 Solution:

(a) Phosphoric acid represents the highest oxidation state of P, so the stable N species are those existing in the same high E^o region, N_2 and NO_3^-. If we superimpose Pourbaix diagrams on one another, species in the same region are compatible.

(b) Ammonia is the reduced form of N existing at high pH. The P species compatible (existing in the same pH-E^o region) are PH_3 and $H_2PO_3^-$.

13.11 Problem: How can white P be separated from red P?

13.11 Solution: White P (P_4) dissolves in diethyl ether or benzene, polymeric red P does not.

13.12 Problem: How are P_4, P_4O_6, and P_4O_{10} related structurally?

13.12 Solution: All have tetrahedral arrangements of 4 P--6 O are inserted in the edges of the tetrahedron to form P_4O_6 and 4 more are added at the apices to form P_4O_{10} (See DMA p. 563).

13.13 Problem: In what way are phosphine ligands in metal complexes similar to CO?

13.13 Solution: PR_3 ligands have vacant d orbitals which makes them π acceptors (back bonding) as in the case of CO (using π*orbitals).

13.14 Problem: Write the formulas for the diethylester of phosphorous acid and the monoethylester of hypophosphorous acid. Would these be protonic acids?

13.14 Solution: In both compounds P has C.N. 4. Neither is protonic, the hydrogens bonded to P are not acidic.

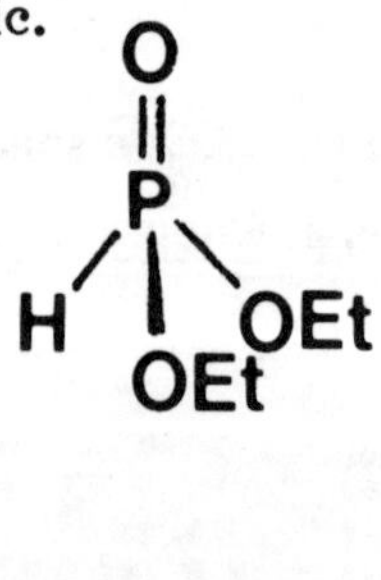

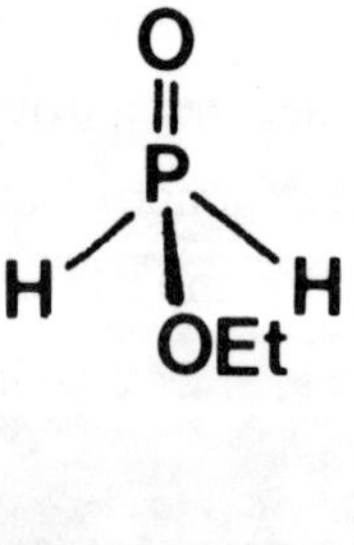

13.15 Problem: Sketch the p-d-π bonding in hexachlorotriphosphazene and in cyclic $(PCl_2N)_4$ assuming it to be planar.

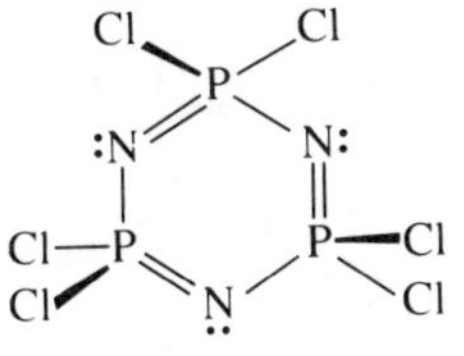

Hexachlorotriphosphazene

13.15 Solution: The signs are mismatched for any odd n in $(PN)_n$. Only lobes of the P 3d orbitals and N 2p orbitals above the plane of the paper are shown.

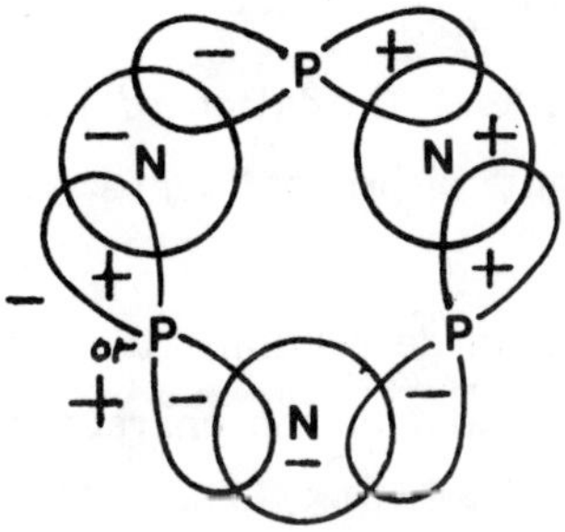

$(PCl_2N)_3$
Signs mismatched

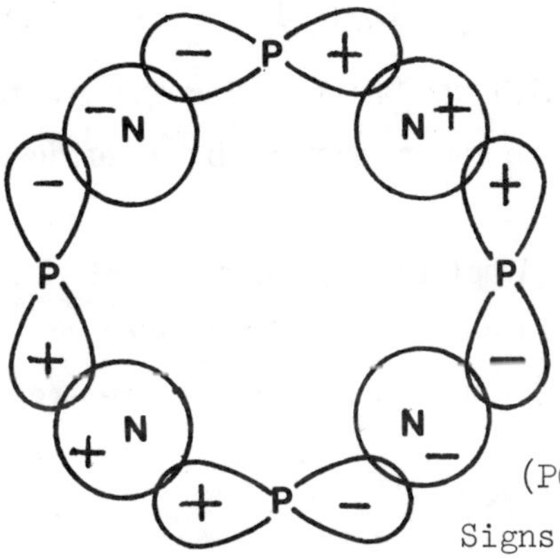

$(PCl_2N)_4$
Signs are matched

13.16 Problem: Write separate balanced equations and calculate E^o_{cell} for the reaction of H_2O_2 and NO_2^- and with MnO_4^- in acid solution.

13.16 Solution: H_2O_2 can be oxidized or reduced, but H_2O_2 cannot oxidize Mn^{2+} in acid solution, so MnO_4^- oxidizes H_2O_2.

$$O_2 \xrightarrow{0.68} H_2O_2 \xrightarrow{1.77} H_2O$$

$$MnO_4^- + 8H^+ + 5e \rightarrow Mn^{2+} + 4H_2O \qquad E^o = 1.51\ V$$

$$H_2O_2 \rightarrow O_2 + 2H^+ + 2e \qquad E^o = -0.68\ V$$

$$2MnO_4^- + 5H_2O_2 + 6H^+ \rightarrow 2Mn^{2+} + 5O_2 + 8H_2O \qquad E^o_{cell} = +0.83\ V$$

$$HNO_2 + H^+ + e \rightarrow NO + H_2O \qquad E^o = 1.0\ V$$

$$HNO_2 + H_2O \rightarrow NO_3^- + 3H^+ + 2e \qquad E^o = -0.94\ V$$

$$H_2O_2 + 2H^+ + 2e \rightarrow 2H_2O \qquad E^o = 1.77\ V$$

The two reactions

$$2HNO_2 + H_2O_2 \rightarrow 2NO + 2H_2O + O_2 \qquad E^o_{cell} = +0.32$$

$$HNO_2 + H_2O_2 \rightarrow NO_3^- + H_2O + H^+ \qquad E^o_{cell} = +0.83$$

are favorable, with the latter having the greater driving force. Oxidation of HNO_2 to NO_3^- should be the major reaction, but in effect, both reactants disproportionate.

13.17 Problem: What thermodynamic factors are important in determining the relative stabilities of solid metal oxides, peroxides, and superoxides? What cation characteristics (size and charge) would you choose to prepare metal oxides, peroxides, superoxides, and ozonides?

13.17 Solution: The oxides are favored by high lattice energies and this is particularly favorable for small cations. Peroxides are favored over oxides because the electron repulsion is lower for O_2^{2-} than for O^{2-} and the fact that it is not necessary to dissociate O_2 to form O_2^{2-}. The lattice energies are lower for the larger O_2^{2-}. Peroxides are favored for cations of intermediate size. Superoxides are favored over peroxides by the favorable electron affinity. The lattice energies are lower because of the lower ionic charge of O_2^-. Superoxides are favored by large cations with low charge (actually +1). Ozonides (O_3^-) are expected only for very large +1 ions where the more favorable lattice energies of MO_2, M_2O_2, and M_2O are less important.

13.18 Problem: Give a description of bonding in O_2F_2 to account for the very short O–O bond and very long O–F bonds.

13.18 Solution: The structure of O_2F_2 is similar to that of H_2O_2, but with O–O–F bond angles of 109.5° and the angle between the planes containing the O-F bonds is 87.5°. The O-O bond distance corresponds to a bond order of 2 and the O-F bond to a fractional bond order. This suggests that the O-O bond is much like that of O_2 with a singly occupied σ orbital of each F atom overlapping with one of the singly occupied π^* orbitals of oxygen to form a weak 3-center electron pair bond. Because of the high electronegativity of F, little electron density is transferred into the π^* orbital to weaken the O-O bond.

Bond Distances for Some Oxygen and Fluorine Compounds

Molecule	O-O distance	O-F distance
OF_2		142 pm
H_2O_2	148 pm	
O_2F_2	122	158
O_2	121	
Sum of covalent radii	148	145

13.19 Problem: Write (separate) equations for the oxidation of $S_2O_3^{2-}$ by I_2 and by H_2O_2.

13.19 Solution:

$$2S_2O_3^{2-} + 4H_2O_2 \rightarrow \underset{\text{trithionate}}{S_3O_6^{2-}} + SO_4^{2-} + 4H_2O$$

$$2S_2O_3^{2-} + I_2 \rightarrow \underset{\text{tetrathionate}}{S_4O_6^{2-}} + 2I^-$$

13.20 Problem: In the Pourbaix diagram for oxygen, H_2O_2 does not appear. Explain. (If you care to check the diagram, see reference in Problem 13.10.)

13.20 Solution: H_2O_2 is thermodynamically unstable (disproportionates) over the entire pH/E° range. The species shown in a Pourbaix diagram are thermodynamically stable.

13.21 Problem: From the Pourbaix diagrams for S and Se,

(a) what S and Se species are stable in contact with O_2?

(b) what S and Se species are stable in contact with H_2?

(c) what S species are stable in contact with Se?

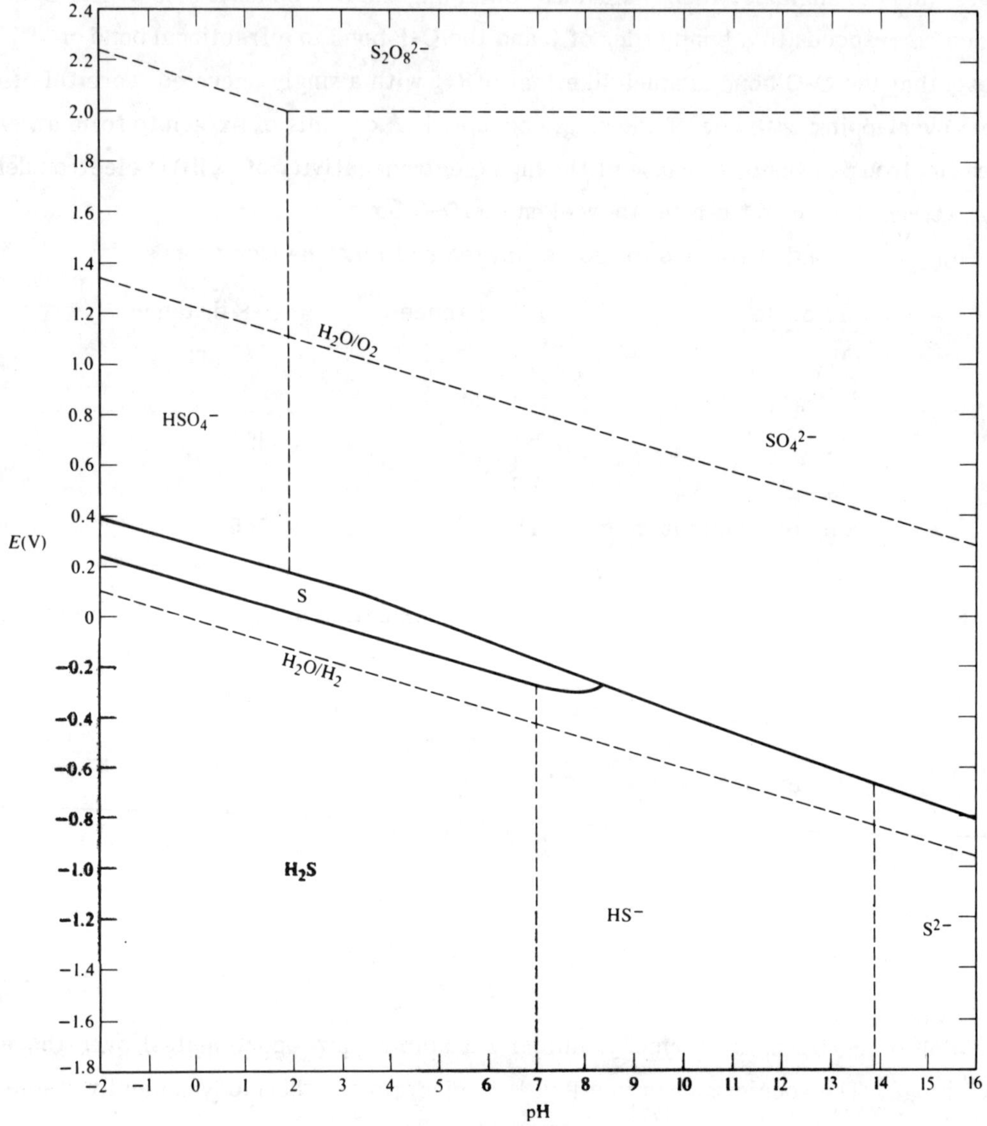

Pourbaix diagram for sulfur.

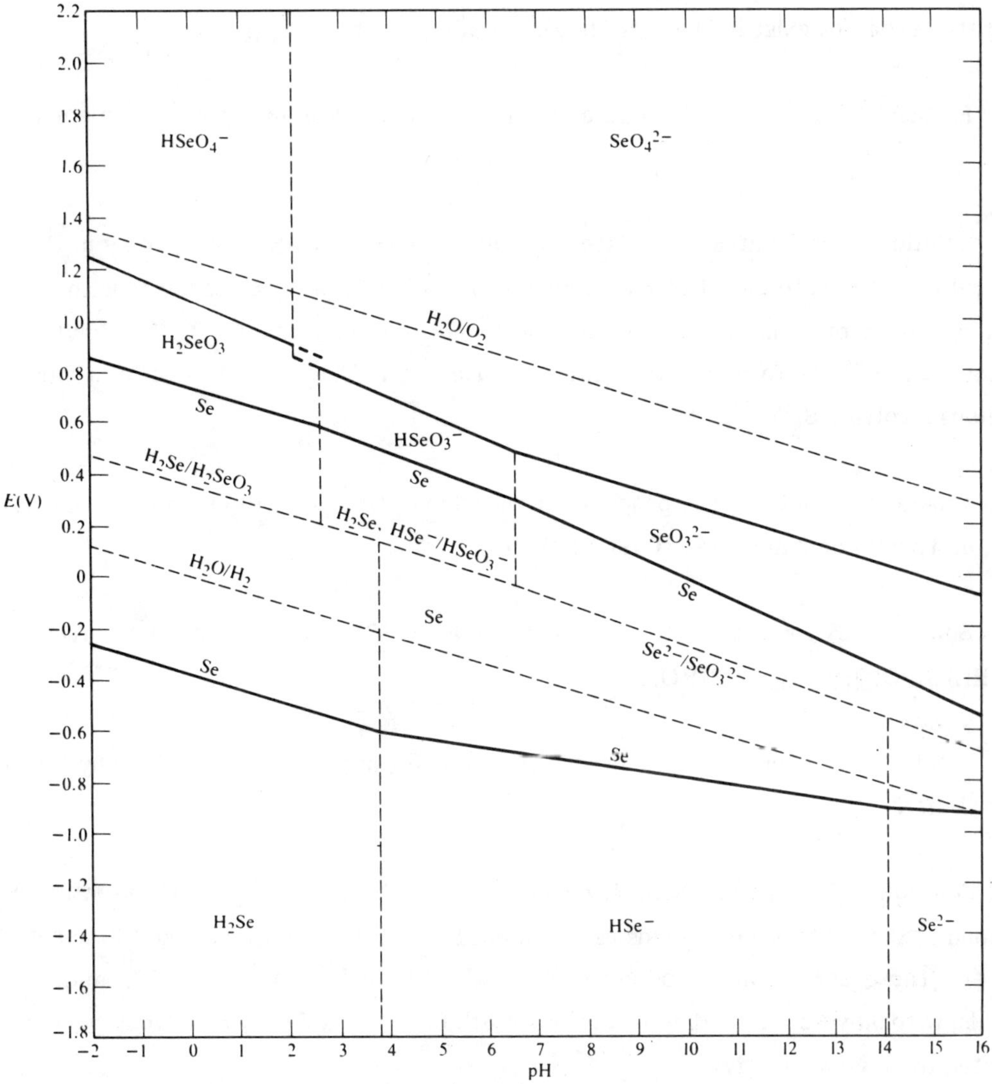

Pourbaix diagram for selenium.

13.21 Solution:

(a) HSO_4^-/SO_4^{2-} and $S_2O_8^{2-}$; $HSeO_4^-/SeO_4^{2-}$. These are the species not oxidized by O_2.

(b) H_2S/S^{2-}, and H_2Se/Se^{2-}. These are the species not reduced by H_2.

(c) H_2S/S^{2-}, S, and HSO_4^-/SO_4^{2-}
Compatible species exist in the same pH/E^o area of Pourbaix diagrams.

13.22 Problem: How are dithionates and polythionates alike, and in what respect do they differ?

13.22 Solution: Dithionates are related to polythionates $(O_3S\text{-}S_n\text{-}SO_3)^{2-}$ only in the sense that n=0 for dithionate ion. Dithionates are stable with respect to reactions with H_2O and resist oxidation or reduction. They are obtained by oxidation of SO_3^{2-}. Polythionates decompose readily to form S and S oxides or oxoacids. They are formed by a variety of reactions involving $S_2O_3^{2-}$.

13.23 Problem: What are the products of reaction of SF_4 with alkylcarbonyls, carboxylic acids, and phosphonic acids [R(RO)PO(OH)] ?

13.23 Solution: SF_4 reacts with RR'C=O to give $RR'CF_2$, with $R\text{-}CO_2H$ to give $R\text{-}CF_3$, with R(RO)PO(OH) to give $R(RO)PF_3$.

13.24 Problem: Give valence bond descriptions for SF_3 and SF_5 and indicate why they have low stability.

13.24 Solution: SF_2 has two ordinary 2-center bonds. As in the case of the interhalogen compounds, the VB treatment leads us to expect SF_4 (promotion of 1 e) and SF_6 (promotion of 2 e). These correspond in bonding descriptions to ICl_2^- and IF_6^-. SF_3 and SF_5 are radicals, with only 1 e for bonding to the 3rd or 5th F. Such radicals, with weak bonding, are expected to be very reactive.

13.25 Problem: Write balanced equations for the following preparations:

(a) Cl_2 (from NaCl).
(b) Br_2 (recovery from seawater).
(c) I_2 (from $NaIO_3$).
(d) HF (from CaF_2).
(e) HCl (from NaCl).

13.25 Solution: (a) Cl_2 can be obtained by oxidation of Cl^- with an inexpensive oxidizing agent, such as MnO_2 in acid solution. On a commercial scale electrolysis is used.

$$2NaCl + MnO_2 + 2H_2SO_4 \rightarrow MnSO_4 + Cl_2 + Na_2SO_4 + 2H_2O \quad \text{or}$$

$$2Na^+ + 2Cl^- + 2H_2O \xrightarrow{\text{electr.}} 2Na^+ + 2OH^- + Cl_2 + H_2$$

(b) Br_2 is recovered by first evaporating seawater partially to concentrate the Br^- and the Br^- is oxidized by Cl_2 in an air stream blown through the solution. The Br_2 in the air stream is absorbed in a solution of a strong base (Na_2CO_3 is used because it is cheap) since the Br_2 disproportionates. The Br_2 is recovered by acidifying the solution since BrO_3^- oxidizes Br^- in acid solution.

$$Cl_2 \text{ (in air stream)} + 2Br^- \rightarrow Br_2 + 2Cl^-$$

$$2Br_2 + 3CO_3^{2-} \text{ (solution)} \rightarrow 5Br^- + BrO_3^- + 3CO_2$$

Add acid for recovery of Br_2 from Na_2CO_3 solution.

$$5Br^- + BrO_3^- + 6H^+ \rightarrow 3Br_2 + 3H_2O$$

(c) $NaIO_3$ can be reduced to I_2 using $NaHSO_3$ as an inexpensive reducing agent.

$$2\,NaIO_3 + 5NaHSO_3(aq) \rightarrow 3\,NaHSO_4 + 2\,Na_2SO_4 + H_2O + I_2$$

(d) CaF_2 is the important F^- ore. It is converted to HF using the cheapest strong acid of low volatility, H_2SO_4.

$$CaF_2 + H_2SO_4 \rightarrow CaSO_4 + 2\,HF$$

(e) NaCl is the important Cl^- ore. As noted for the production of HF, H_2SO_4 is the best choice for displacing the volatile HCl.

$$NaCl + H_2SO_4 \rightarrow NaHSO_4 + HCl$$

13.26 Problem: Draw the structures of the following species, indicating the approximate bond angles: IF_2^+, IF_4^+, and IF_6^+

13.26 Solution: The bond angle for IF_2^+ is smaller than the tetrahedral angle because of lone pair-lone pair repulsion. IF_4^+ has four bonding electron pairs and one lone pair. The lone pair occupies an equatorial position. The F atoms bend away from the lone pair because the lone pair-bonding pair repulsion is most important. IF_6^+ (isoelectronic with

SF_6) has six bonding electron pairs and no lone pairs. The structure is a regular octahedron.

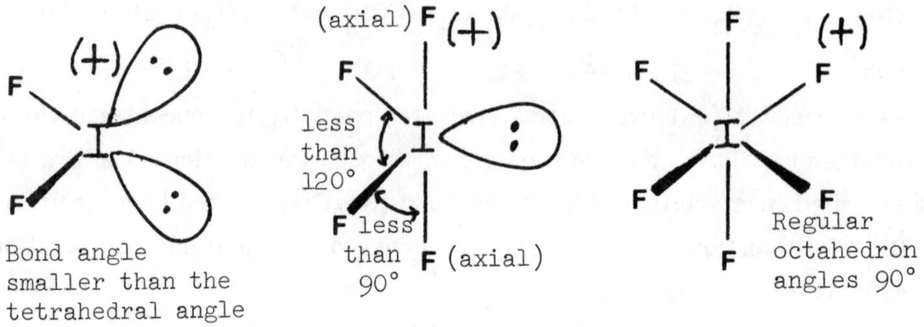

13.27 Problem: Indicate reactions which might be suitable for the preparation of:

(a) Anhydrous tetramethylammonium fluoride.

(b) Aluminum bromide.

(c) Barium iodide.

13.27 Solution:

(a) Tetraalkylammonium chlorides are generally available. The fluoride can be obtained by precipitating AgCl using AgF in CH_3OH.

$$AgF + Me_4NCl \xrightarrow{CH_3OH} Me_4NF + AgCl$$

(b) $AlBr_3$ cannot be obtained readily by metathesis from $AlCl_3$. Direct reaction of Al and Br_2 is preferable to obtain an anhydrous product.

$$2Al + 3Br_2 \rightarrow 2AlBr_3$$

(c) BaI_2, free of other salts, can be obtained in a nonprotonic solvent by reaction of BaH_2 and with NH_4I, since NH_4^+ and H^- give H_2 and NH_3.

$$BaH_2(xs) + 2NH_4I \xrightarrow{\text{pyridine}} BaI_2 + 2NH_3 + H_2$$

13.28 Problem: The ΔH_f^o of BrF(g), $BrF_3(\ell)$, and $BrF_5(\ell)$ are -61.5, -314, and -533 kJ/mole, respectively. Which of these species should be predominate on reacting Br_2 and F_2 under standard conditions?

13.28 Solution: Since ΔH_f^o of BrF_3 is more negative than that of BrF and that of BrF_5 is still more negative, the reaction will proceed to the extent of the availability of F_2. BrF_5 will be formed if sufficient F_2 is available.

13.29 Problem: Bromites and perbromates have been obtained only recently. Give the preparation of each.

13.29 Solution: Anodic oxidation in basic solution of Br^- produces BrO_2^- and anodic oxidation of BrO_3^- produces BrO_4^-. The perbromate is obtained also by oxidation of BrO_3^- by XeF_2 or F_2 in basic solution. Oxoanions in general are obtained more easily in basic solution.

$$Br^- + 4\,OH^- \xrightarrow{\text{anodic oxid.}} BrO_2^- + 2H_2O + 4e$$

$$BrO_3^- + 2\,OH^- \xrightarrow{\text{anodic oxid.}} BrO_4^- + H_2O + 2e$$

13.30 Problem: What are the expected structures of ClO_3^+, $F_2ClO_3^-$, and $F_4ClO_2^-$? Sketch the regular figure and indicate deviations from idealized bond angles.

13.30 Solution: ClO_3^+ with no unshared electrons on Cl is planar with 120° bond angles (SO_3 structure). Pi bonding, using the unhybridized p orbitals on Cl, does not alter the molecular geometry. There are no unshared electrons on Cl in $ClO_3F_2^+$ so a trigonal bipyramid with normal bond angles results. The electroneutrality principle (see Problem 2.14) favors partial Cl-O π bonding, placing the O atoms in equatorial positions, but causing no distortion. In the octahedral arrangement of $F_4ClO_2^-$ (no unshared electrons on Cl) the partially π bonded oxygens would be expected to occupy trans positions with all 90° angles. There is some expectation of preference for the cis-configuration (See K.O. Christe and C.J. Schack, Adv. Inorg. Chem. Radiochem. 1976, 18, 319).

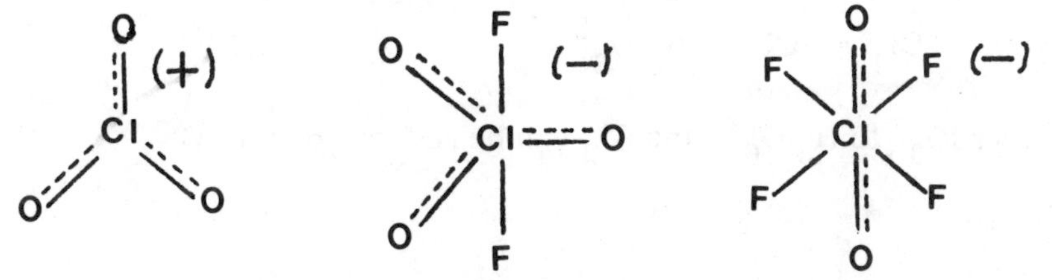

13.31 Problem: From emf data predict the results of mixing the following:

(a) Cl^- and BrO_3^- and 1M acid;

(b) Cl_2 and IO_3^- in 1M base;

(c) At_2 and Cl_2 in 1M base.

$$E^o_{acid} \quad ClO_4^- \xrightarrow{1.19} ClO_3^- \xrightarrow{1.21} HClO_2 \xrightarrow{1.645} HClO \xrightarrow{1.63} Cl_2 \xrightarrow{1.36} Cl^-$$

$$BrO_4^- \xrightarrow{1.743} BrO_3^- \xrightarrow{1.49} HBrO \xrightarrow{1.59} Br_2(\ell) \xrightarrow{1.07} Br^-$$

$$E^o_{base} \quad ClO_4^- \xrightarrow{0.36} ClO_3^- \xrightarrow{0.33} ClO_2^- \xrightarrow{0.66} ClO^- \xrightarrow{0.40} Cl_2 \xrightarrow{1.36} Cl^-$$

$$H_3IO_6^{2-} \xrightarrow{0.7} IO_3^- \xrightarrow{0.14} IO^- \xrightarrow{0.45} I_2 \xrightarrow{0.535} I^-$$

$$AtO_3^- \xrightarrow{0.6} AtO^- \xrightarrow{0.3} At_2 \xrightarrow{0.3} At^-$$

(emf data from DMA Appendix C)

13.31 Solution:

(a)

$Cl_2 + 2e \rightarrow 2Cl^-$	$E^o = 1.36$
$2BrO_3^- + 12H^+ + 10e \rightarrow Br_2 + 6H_2O$	1.51
$2BrO_3^- + 12H^+ + 10Cl^- \rightarrow 5Cl_2 + Br_2 + 6H_2O$	0.15 V

Br_2 does not oxidize Cl^-, so the reaction stops with Br_2. BrO_3^- cannot oxidize Cl_2 to HClO.

(b)

$Cl_2 + 2e \rightarrow 2Cl^-$	1.36
$H_3IO_6^{2-} + 2e \rightarrow IO_3^- + 3\,OH^-$	0.7
$IO_3^- + 3OH^- + Cl_2 \rightarrow 2Cl^- + H_3IO_6^{2-}$	0.7 V

Cl_2 easily oxidizes IO_3^- to $H_3IO_6^{2-}$ and Cl_2 can be reduced only to Cl^-.

(c)

$$Cl_2 + 2e \rightarrow 2Cl^- \qquad 1.36$$

$$2AtO_3^- + 6H_2O + 10e \rightarrow At_2 + 12\,OH^- \qquad 0.5$$

$$5Cl_2 + At_2 + 12\,OH^- \rightarrow 10Cl^- + 2\,AtO_3^- + 6H_2O \qquad 0.9V$$

Cl_2 is the stronger oxidizing agent, oxidizing At_2 to AtO_3^- in basic solution.

13.32 Problem: Compare the expected rate of reaction of AtO_3^- and IO_3^- as oxidizing agents. What formula is expected for perastatinate? Why?

13.32 Solution: AtO_3^- should be a faster oxidizing agent than IO_3^-, since the order is

$$IO_3^- > BrO_3^- > ClO_3^-$$

Perastinate ion is expected to be $H_3AtO_6^{2-}$. C.N. 6 is expected since At has a larger radius than I.

13.33 Problem: Give practical uses of He, Ne, and Ar, and give the sources of each.

13.33 Solution: Helium is used for cryogenic work because of its very low boiling point, and for lighter-than-air craft because of its low density and non-combustability. He is obtained from deposits of natural gas such as those in the southwestern U.S. Ne is used in Ne signs and some lamps. It is recovered from air. Ar is used to provide an inert atmosphere and in light bulbs. It is recovered from the fractional distillation of liquid air.

13.34 Problem: What known compounds related in bonding and structure to the Xe halides should have prompted a search for the Xe halides earlier?

13.34 Solution: The interhalogen halides, for example, ICl_2^- and ICl_4^-, are isoelectronic (considering valence electrons) with XeF_2 and XeF_4. The existence of such compounds of halogens with positive oxidation states should have prompted the search for fluorides, at least, of Xe. The ionization of Xe is about the same as that of O_2 and the discovery of compounds of O_2^+ led to the discovery of Xe compounds.

13.35 Problem: Give the expected shape and approximate bond angles for ClF_3O, considering the effects of the lone pair on Cl and the directional effects of the Cl==O bond. (See K.O. Christe and H. Oberhammer, Inorg. Chem. 1981, 20, 296.)

13.35 Solution: Both the lone pair and the Cl–O bond with double bond character (see Problem 13.30) cause the axial F to bend away, but the effect of the π bonding to O is slightly greater. Both the lone pair and double bonding to O decrease the $OClF_{eq}$ angle to much less than 120°. The $OClF_{ax}$ angle is influenced primarily by Cl–O double bonding. The rather large increase (above 90°) suggests that the π bond is perpendicular to the equatorial plane, using d_{xz} (or d_{yz}). See K.O. Christe and H. Oberhammer for a discussion of directional effects of double bonds.

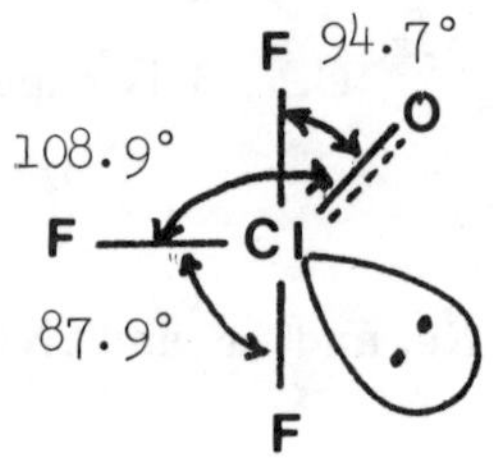

13.36 Problem: Draw the structure expected for $XeOF_4$ and indicate the approximate bond lengths (the Xe–F bond length in XeF_4 is 195 pm).

13.36 Solution: A square pyramid is expected to accommodate one lone pair. The Xe–F bond lengths should be about the same as for XeF_4. The Xe–O bond should be shorter because of some double bond character, perhaps about 170–180 pm. The bond lengths are 190 pm for Xe–F and 170 pm for Xe–O.

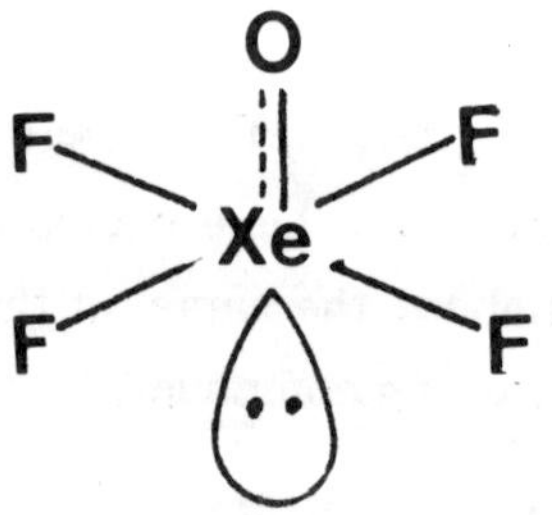

13.37 Problem: XeF_6 gives solutions in HF which conduct electricity. How might one distinguish between the following possible modes of dissociation:

$$XeF_6 + HF \rightleftharpoons XeF_5^+ + HF_2^-$$

$$XeF_6 + 2HF \rightleftharpoons XeF_7^- + H_2F^+$$

13.37 Solution: One might attempt to obtain an IR spectrum of the solution and look for the bands associated with the known FHF^- species. Alternatively one might attempt to assign bands to the new XeF_5^+ or XeF_7^- species based on their symmetry. Addition of a strong fluoride ion acceptor, such as SbF_5, would increase the concentration of XeF_5^+ (if this were present) or decrease the concentration of XeF_7^- (if this were present) and thus increase or decrease the intensity of the IR absorption. If fluoride exchange were not too rapid, the fluorine NMR spectrum might be used to distinguish between these possibilities.

13.38 Problem: How might one distinguish between Xe^+ ions and Xe_2^{2+} ions in the compound $XePtF_6$?

13.38 Solution: A distinction between Xe^+ and Xe_2^{2+} in $XePtF_6$ could be made by a determination of the Xe-Xe separation or possibly by a magnetic moment determination, Xe^+ would be paramagnetic and Xe_2^{2+} should be diamagnetic.

13.39 Problem: Heating B_2S_3 in a glass tube gave the compound B_8S_{16} with a planar porphine-like structure. On the basis of the chemistry of B and S, sketch the formula for B_8S_{16}.

13.39 Solution: B tends to form three bonds and S forms two bonds. Thus B is best suited for the bridge-heads of the five-membered rings and these are joined to one another by S. (See B. Krebs and H.-U. Hürter, Angew. Chem. Int. Ed. Engl. 1980, 19, 482)

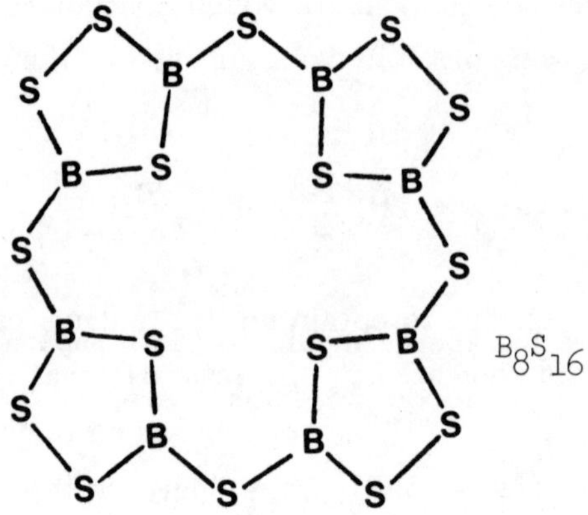

13.40 Problem: Ozone is used in place of Cl_2 for water purification in many European countries. Explain the advantages and disadvantages of the use of ozone for water treatment.

13.40 Solution: Ozone, like chlorine, is a strong oxidizing agent that will oxidize bacteria. Unlike Cl_2, ozone decomposes fairly rapidly, leaving no residual taste, and no potentially carcinogenic chlorocarbons. It must be generated on site by a process much more expensive than for chlorination (i.e. by an electric discharge in O_2).

13.41 Problem: Explain why the ozone molecule has a dipole moment.

13.41 Solution: In the contributing structures to the resonance hybrid for O_3 the "single bond" must be a dative bond giving rise to a positive formal charge on the central oxygen atom and a formal negative charge for the terminal oxygen.

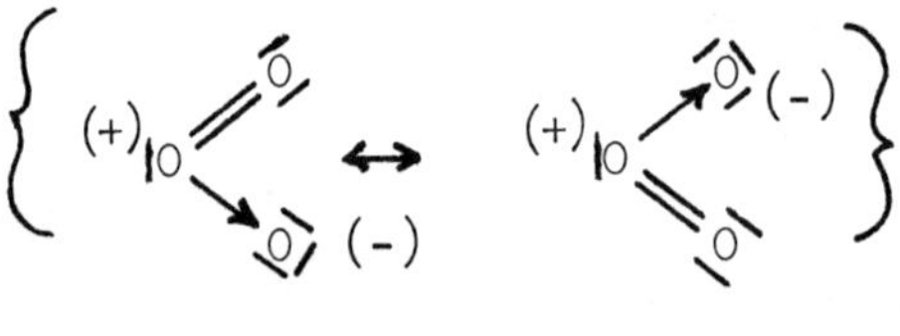

13.42 Problem: Explain the function of SiO_2 and C in the production of elemental phosphorus from phosphates. What other substances could be substituted for these? Why are these other substances not used?

13.42 Solution: The SiO_2 functions as a nonvolatile acid in the Lux-Flood sense—that is as an oxide ion acceptor—and thus displaces the more volatile P_4O_{10}, which is reduced to P_2 by the C. The P_2(g) condenses to white P_4. Other nonvolatile Lux-Flood acids are TiO_2, V_2O_5, MoO_3, etc.; other reducing agents should be easily separable from the product, and further the reduction of the Lux-Flood acid should not occur readily. A ferrosilicon alloy or hydrogen might function satisfactorily. The expense of the substitute reagents rules them out for commercial production of phosphorus.

13.43 Problem: Although the pentahalides of phosphorus are known for all but the iodide, only recently has a "pentahalide" been found for nitrogen, NF_5. Do you expect NCl_5 will soon be found? Explain.

13.43 Solution: Both PF_5 and PCl_5 exist in the vapor as discrete molecules, for which sp^3d hybridization may be assumed. The NF_4^+ ion has been prepared through the reaction of NF_3 with F_2 in the presence of a good F^- ion acceptor (AsF_5, SbF_5, BF_3). Metathesis reactions with a suitable metal fluoride (e.g., CsF) then give NF_4F. Since N has no low lying d orbitals, a molecular NF_5 should not exist. NF_3 is the only stable nitrogen trihalide known, the others being shock-sensitive compounds undergoing exothermic, explosive, decomposition. If NCl_4Cl could be put together, coordination of the Cl^- ion with an empty d orbital of a chlorine attached to nitrogen would provide a low energy path for decomposition to NCl_3 and Cl_2. The greater bond energy of Cl_2 than that of F_2, coupled with the lower lattice energy expected for NCl_4Cl than that of NF_4F, and the unimportance of resonance stabilization of NCl_4 compared to NF_4 all militate against the possible existence of NCl_4Cl.

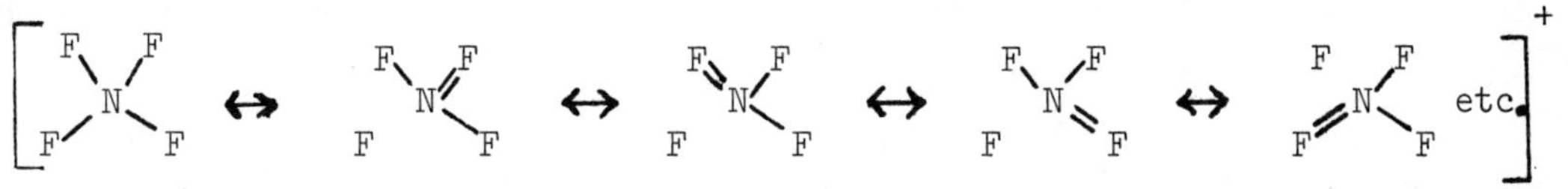

13.44 Problem: Explain the trends in the melting and boiling points of the fluorides of the third period:

	NaF	MgF_2	AlF_3	SiF_4	PF_5	SF_6
M.p. (^oC)	992	1263	1270	-90	-83	-51
B.p. (^oC)	1704	-	1270 (subl.)	-86	-75	-64 (subl.)

13.44 Solution: Boiling points and melting points depend on forces between constituent particles. NaF and MgF_2 are both ionic solids. Their increasing melting points parallel increasing lattice energies. Boiling points are expected to follow the same trend because of the greater coulombic attraction for 2+ ions. For AlF_3 the lattice energy and melting points are still higher. In the solid, Al^{3+} has six F^- neighbors. AlF_3 sublimes near the boiling point, indicating that once the AlF_6^{3-} units are broken down on melting, AlF_3 vaporizes. We know that $AlCl_3$ vaporizes as Al_2Cl_6, two tetrahedra sharing an edge. The sublimation of AlF_3 suggests the formation of a molecular species in the vapor.

The small Si(IV) is shielded by four F^- to form discrete, polar covalent SiF_4 molecules with T_d symmetry. Only van der Waals forces bind the molecules together so the melting point and boiling point are very low. The trigonal bipyramidal PF_5 molecule has two long P-F bonds in the vapor state. Insofar as this persists in the solid, the very small change in melting point of PF_5 (compared to SiF_4) could reflect weak van der Waals interaction resulting from greater distances between molecular centers and less efficient packing. The normal increase in melting point for molecules with more electrons is seen for the highly symmetrical octahedral SF_6 molecule. The trend is often described as paralleling the change in molecular weight, but it is not dependent on mass. The boiling points show the expected trend, increasing in the series from SiF_4 to PF_5 to SF_6. The boiling points being near the melting points indicate weak attraction between the compact, nonpolarizable fluoride molecules once the solid structure breaks down. No great significance can be associated with the fact that the boiling point of SF_6 is below the melting point. We choose arbitrarily to define the boiling point at 1 atmosphere. Variations could occur at another pressure.

XIV Chemistry and Periodic Trends of Compounds of Metals

14.1 Problem: Write balanced equations for the preparation of

(a) Na_2CO_3 (d) $ZrCl_3$

(b) MgF (e) $VOCl_3$

(c) $TiBr_3$ (f) WCl_6

14.1 Solution: (a) Na_2CO_3 is the cheapest strong base. It is made efficiently from inexpensive starting materials (CO_2 and NaCl). The reaction of CO_2, NH_3 and NaCl depends on the low solubility of $NaHCO_3$. Limestone ($CaCO_3$) provides CaO and CO_2.

$$CO_2 + NH_3 + H_2O + NaCl \rightarrow NaHCO_3(s) + NH_4Cl(aq)$$

$$2NaHCO_3(s) \xrightarrow{heat} Na_2CO_3 + H_2O + CO_2 \text{ (recycle)}$$

$$2NH_4Cl(aq) + Ca(OH)_2 \rightarrow CaCl_2 + 2H_2O + 2\,NH_3 \text{ (recycle)}$$

(b) The direct reaction of Mg and a limited supply of F_2 should produce some MgF in the gas phase at high temperature. It disproportionates to form Mg + MgF_2(s) on cooling.

$$Mg + 1/2\,F_2 \rightarrow MgF(g) \text{ stable at high T}$$

(c) Zn can reduce $TiBr_4$ to $TiBr_3$.

$$2TiBr_4(aq) + 2Zn + 2HBr \rightarrow 2TiBr_3 + 2ZnBr_2 + H_2$$

(d) $ZrCl_4$ can be reduced by Zr at elevated temperature.

$$3ZrCl_4(s) + Zr \rightarrow 4ZrCl_3(s)$$

(e) Vanadium is oxidized by a strong oxidizing agent to VO^{3+} and, in the presence of HCl, $VOCl_3$ is formed.

$$3V + 5HNO_3(dil) + 9HCl \rightarrow 3VOCl_3 + 5NO + 7H_2O$$

(f) The free elements combine to form WCl_6

$$W + 3Cl_2 \rightarrow WCl_6$$

14.2 Problem: The bond dissociation energies for the alkali metal M_2 molecules decrease regularly from 100.9 kJ/mole for Li_2 to 38.0 kJ/mole for Cs_2. The bond dissociation energies are greater for the M_2^+ ions, decreasing from 138.9 kJ/mole for Li_2^+ to 59 kJ/mole for Cs_2^+. Explain why the dissociation energies are greater for the M_2^+ ions. What is the bond order for M_2^+?

14.2 Solution: The bond dissociation energies are greater for M_2^+ ions, even though they have a bond order of 1/2, because of polarization. M^+ polarizes M, with the effect increasing from Li → Cs, so that the percentage increase (M_2^+ compared to M_2) is greatest for Cs_2^+. This effect is not important for H_2^+ because the H atom has low polarizability.

14.3 Problem: Predict the following for Fr.

(a) The product of the burning of Fr in air.

(b) An insoluble compound of Fr.

(c) The structure of FrCl.

(d) The relative heats of formation of FrF and FrI.

14.3 Solution: (a) FrO_2 — The larger alkali metals stabilize O_2^-.

(b) $FrClO_4$ and Fr_2PtCl_6 — These are the salts of low solubility for the larger alkali metal ions.

(c) CsCl structure — C.N. 8 is favored by the large cation size.

(d) ΔH_f is more negative for FrF — the heats of formation of all alkali metal fluorides are more negative than those of the other halides.

14.4 Problem: How could you remove unreacted Na (metal) in liquid ammonia safely?

14.4 Solution: Excess Na in liquid ammonia can be removed safely by adding NH_4Cl (a strong acid in NH_3) gradually to form NaCl and H_2.

14.5 Problem: Write equations for the preparation of the following metals
(a) Na (b) K (c) Cs (d) Mg

14.5 Solution: (a) Electrolysis is used to obtain Na since ordinary reducing agents do not reduce alkali metal ions.

$$2NaCl(fused) \xrightarrow{electr.} 2Na + Cl_2$$

(b) CaC_2 reduces KF at high temperature because CaC_2 decomposes and CaF_2 has very high lattice energy.

$$2KF + CaC_2 \xrightarrow{heat} CaF_2 + 2K + 2C$$

(c) Al reduces Cs_2O because Cs is volatile and Al_2O_3 has very high lattice energy.

$$3Cs_2O + 2Al \rightarrow Al_2O_3 + 6\ Cs(g)$$

(d) The electrolysis of fused $MgCl_2$ is made more energy efficient by adding KCl to lower the melting point of $MgCl_2$.

$$KCl\text{-}MgCl_2(fused) \xrightarrow{electr.} Mg + Cl_2$$

14.6 Problem: How can you stabilize solutions containing Na^-?

14.6 Solution: Na^- can be stabilized by dissolving Na in ethylenediamine and adding a cryptand ligand (see DMA p. 612). This increases the solubility of Na by coordinating Na^+, shifting the equilibrium to the right:

$$2\ Na \xrightleftharpoons{crypt.} Na(crypt.)^+ + Na^-$$

14.7 Problem: What are laboratory or everyday uses of Li, K and Cs?

14.7 Solution: Li is used in batteries and alloys. K is used for analytical reagents because K^+ salts are not so highly solvated as those of Na^+, making it easier to obtain anhydrous reagents. Cs is used in photocells because of its low ionization energy.

14.8 Problem: Cite several properties that show the diagonal relationships between Li and Mg, and between Be and Al.

14.8 Solution: The ionic potentials (charge/size ratios) are very similar for the pairs Li,

Mg and Be, Al. The solubilities of the fluorides, carbonates, and phosphates of Li and Mg are relatively low, unlike these salts of other members of the families. Li^+ and Mg^{2+} hydrolyze appreciably. Li alkyls and aryls are used in organic syntheses as are Grignard reagents (RMgX). Li and Mg form nitrides by reaction with N_2. Fused $BeCl_2$ and $AlCl_3$ are poor electrolytes. BeO and Al_2O_3 are hard and refractory. $BeCO_3$ and $Al_2(CO_3)_3$ are unstable except in the presence of CO_2. Be_2C and Al_4C are unusual as metal carbides, hydrolyzing to form methane. Be^{2+} and Al^{3+} hydrolyze extensively. Be^{2+} and Al^{3+} have a great tendency to form complexes with F^- and oxo anions.

14.9 Problem: The elements Li-Mg, Na-Ca, and Be-Al are closely related because of the diagonal relationship. Would you expect Mg^{2+} to be more closely related to Sc^{3+} or to Ga^{3+}? Why?

14.9 Solution: Mg^{2+} should be more closely related to Sc^{3+} than to Ga^{3+} because of comparable ionic potentials and Mg^{2+} and Sc^{3+} both have noble-gas electron configurations, while Ga^{3+} has the 18-electron configuration. Ga^{3+} is more polarizing and more polarizable.

14.10 Problem: Why is Au expected to form Au^-?

14.10 Solution: The electron affinity of Au is higher than that of any element other than the halogens. It should form an anion if any metal does.

14.11 Problem: Write equations for the reduction of Cu^{2+} with a limited amount of CN^- and with an excess of CN^-.

14.11 Solution: Cyanide ion reduces Cu^{2+} to precipitate CuCN which forms a soluble complex ion in the presence of excess CN^-.

$$2Cu^{2+} + 4CN^- \rightarrow 2CuCN(s) + (CN)_2$$

$$CuCN(s) + CN^-(xs.) \rightarrow Cu(CN)_2^-$$

14.12 Problem: Why is aqua regia (HCl-HNO_3) effective in oxidizing noble metals when neither HCl nor HNO_3 is effective alone?

14.12 Solution: Formation of chloro complexes stabilizes the positive oxidation states and lowers the emf required for oxidation to the useful range for HNO_3.

14.13 Problem: Why are large anions effective in stabilizing unusually low oxidation states, such as Cd(I)?

14.13 Solution: Unusual low oxidation states, such as Cd(I), are often stabilized using $AlCl_4^-$ or some other large anion that has little tendency to form stable complexes with the metal in its higher oxidation state. This avoids stabilization of the disproportionation products. (See DMA p 229f.)

14.14 Problem: Sketch a linear combination of s, p_z and d_{z^2} that would give very favorable overlap for bonding in a linear MX_2 molecule.

14.14 Solution:

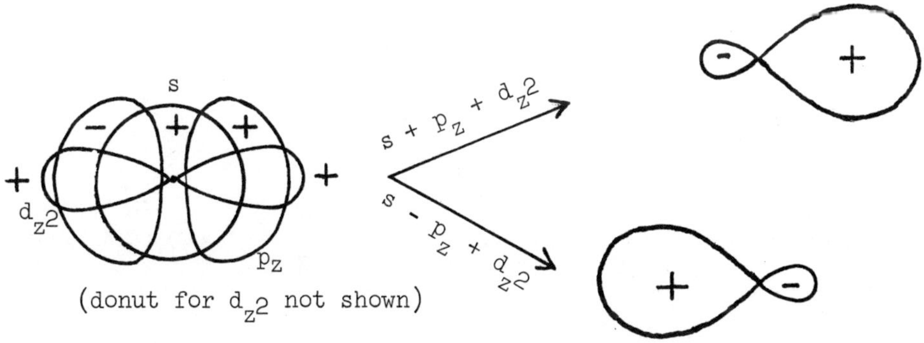

A third linear combination has a large donut in the xy plane (nonbonding).

14.15 Problem: The structures of ZnO, CdO, and HgO are quite different. Describe the structures.

14.15 Solution: ZnO has the zinc blende (6PT) or wurtzite (4PT) structure with C.N. 4.

CdO has the NaCl (6PO) structure with C.N. 6. HgO (orthorhombic) has zigzag chains of essentially linear O-Hg-O, giving C.N. 2. (See DMA p. 629.)

14.16 Problem: The structures of HgF_2, $HgCl_2$, $HgBr_2$, and HgI_2 show interesting variations. Describe the structures.

14.16 Solution: HgF_2 has the fluorite structure (9PTT) with C.N. 8 for Hg^{2+}. $HgCl_2$ contains discrete linear $HgCl_2$ molecules - a molecular lattice. Red HgI_2 has a layer lattice and, at 126°C it goes to a yellow form containing linear HgI_2 molecules. $HgBr_2$ has a layer structure, but the presence of two Br closer than the other two indicates a transition between layer and molecular structures.

14.17 Problem: Sketch the structure of the silicate anion in beryl.

14.17 Solution: Beryl, $Al_2Be_3[Si_6O_{18}]$, contains a cyclic "discrete" silicate anion.

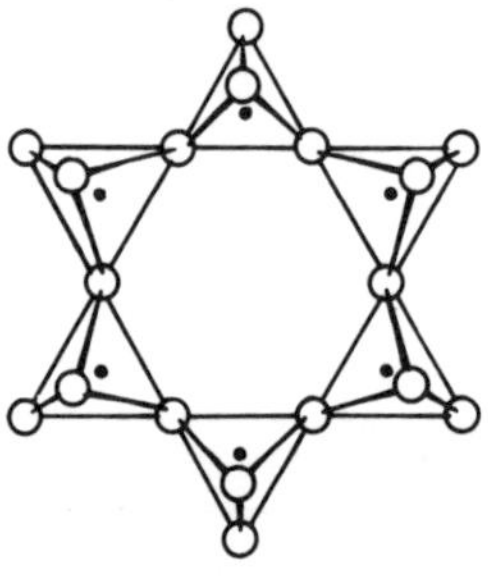

$[Si_6O_{18}]^{12-}$

14.18 Problem: Give equations for reactions that could be used to separate Zn^{2+}, Cd^{2+}, and Hg^{2+} present in solution.

14.18 Solution: Zn^{2+} can be separated easily since it is the only amphoteric ion of the group.

$$Zn^{2+} + Cd^{2+} + Hg^{2+} \xrightarrow[OH^-]{excess} Zn(OH)_4^{2-} + CdO(s) + HgO(s)$$

$$CdO(s) + HgO(s) \xrightarrow{H^+} Cd^{2+} + Hg^{2+} \xrightarrow{H_2S} CdS(s) + HgS(s)$$

$$3CdS(s) + 3\ HNO_3 \longrightarrow 3Cd(NO_3)_2 + 3S + 2NO + 4H_2O$$

HgS does not dissolve in HNO_3.

14.19 Problem: Can one obtain
(a) Hg^{2+} salts free of Hg_2^{2+}?

(b) Hg_2^{2+} salts free of Hg^{2+}?

14.19 Solution: (a) Yes, Hg_2^{2+} can be oxidized completely to Hg^{2+}.
(b) Hg_2^{2+} salts cannot be obtained free of Hg^{2+} because Hg_2^{2+} is in equilibrium with Hg^{2+} and Hg (K = 166).

14.20 Problem: Discuss the factors involved in determining the following solubility patterns: LiF is much less soluble than LiCl, but AgF is much more soluble than AgCl.

14.20 Solution: LiF has such a high U_o that the high solvation energy of F^- is not sufficient to make it more soluble than LiCl. The difference in U_o is not so great for AgF and AgCl and the high solvation energy of F^- determines the relative solubilities.

14.21 Problem: The metal perchlorates have been referred to as "universal solutes". What properties are important in causing most metal perchlorates to be quite soluble in water and several other solvents?

14.21 Solution: U_o is fairly low for salts of the large ClO_4^- and the solvation energy is fairly high in a wide range of solvents. There is little tendency for formation of perchlorate complexes in most solvents.

14.22 Problem: For which elements in the rare earth series are M(II) and M(IV) oxidation numbers expected?

14.22 Solution: M(IV) is expected for f^1 (Ce) and for f^8 (Tb) since removal of one more electron gives empty (Ce^{IV}) or half-filled (Tb^{IV}) f orbitals. M(II) is expected just before Gd (Eu) and Lu (Yb) since these ions achieve half-filled (Eu^{II}) and filled (Yb^{II}) f orbitals.

14.23 Problem: Why should it have been expected that there would be more uncertainty concerning the identity of the first member of the second inner-transition series compared with the lanthanide series?

14.23 Solution: The energy levels of the valence electrons are closer together for the larger actinides and greater variation of oxidation states is expected.

14.24 Problem: One of the common M_2O_3 structures is that of α-alumina (Al_2O_3). Another is the La_2O_3 structure (C.N. 7). Check the La_2O_3 structure in Wells and describe the coordination about La^{3+}.

14.24 Solution: In the La_2O_3 structure (A-M_2O_3 type) La^{3+} has C.N. 7, with 3 O^{2-} at 238 pm, 1 at 245 pm, and 3 at 272 pm. Approximately octahedral LaO_6 units share edges, with the oxygen of another octahedron also shared through one of the octahedral faces. (See A. F. Wells, Structural Inorganic Chemistry, 4th edition, Oxford Press, Oxford, 1975.)

14.25 Problem: Ti is the ninth most abundant element in the earth's crust, and its minerals are reasonably concentrated in nature. Why is it less commonly used than rarer metals?

14.25 Solution: Ti is a rather active metal with a great affinity for oxygen. Its compounds are not easily reduced. Advanced technology is required for obtaining and fabricating the metal.

14.26 Problem: Some elements are known as dispersed elements, forming no independent minerals, even though their abundances are not exceptionally low; whereas others of comparable or lesser abundance are highly concentrated in nature. Explain the following cases:

Dispersed: Rb, Ga, Ge, Hf.

Concentrated: Li, Be, Au.

14.26 Solution: Rb^+ has about the same size and follows the more abundant K^+. Similarly Ga^{3+} follows the very abundant Al^{3+} (almost same size). Ge^{4+} follows the very abundant Si^{4+}. Hf^{4+} follows Zr^{4+} (same size). Li^+ and Be^{2+} differ so much from other members of their families that they concentrate in nature in spite of their low abundances. Au is much more noble than other metals and concentrates as the dense free element.

14.27 Problem: The stabilities of the oxidation states of the lanthanides other than +III can be explained in terms of empty, half-filled, and filled f orbitals. Attempt to apply a similar approach to the series Sc-Zn. Where does it work well? Why is it much less useful? Is this approach effective for the actinide elements?

14.27 Solution: The "characteristic" oxidation number for the first transition series is M^{2+} since the "regular" configuration is $3d^x4s^2$. Expected "other" oxidation states are Sc^{3+} (d^0), Fe^{3+} (d^5), and Cu^+ (d^{10}). Mn^{2+} (d^5) and Zn^{2+}(d^{10}) are expected to be the stable oxidation states for these metals. The situation is complicated by the importance of ligand field effects, so the half-filled t_{2g} orbitals (Cr^{3+}) and filled t_{2g} orbitals (low-spin Fe^{2+} and Co^{3+}) need to be considered also. It does not work as well for the transition metals becuase the d electrons are valence electrons (low ionization energies) and the f electrons are not. It does not work as well for the actinides as for the lanthanides because the energy levels of the valence electrons are closer together for the larger actinides.

14.28 Problem: Compare the syntheses of the highest and lowest stable oxidation states of Mn and Re? Which halide ions can be oxidized by MnO_4^- and by ReO_4^-?

14.28 Solution: Mn(II) is obtained by dissolution of Mn in acid or by reduction of higher oxidation states by most strong reducing agents in acid solution. Re(IV) is obtained by strong reduction of Re_2O_7 with H_2 to form ReO_2. MnO_4^- is produced by electrolytic oxidation of lower oxidation states or by Cl_2 oxidation of K_2MnO_4. ReO_4^- is obtained by mild oxidation of lower oxidation states using O_2 or milder oxidizing agents.

MnO_4^- in acid solution will oxidize Cl^- (slowly), Br^- and I^-. ReO_4^- will not oxidize any halide.

14.29 Problem: Would the removal of Hf from Zr be important in most applications of zirconium compounds? In the use of Zr metal in flash bulbs?

14.29 Solution: For most applications the removal of Hf from Zr compounds would be unimportant because they are so similar. This is certainly true for use in flash bulbs where only the ease of oxidation of the metal is important.

14.30 Problem: What properties of tungsten make it so suitable for filaments for light bulbs?

14.30 Solution: The very low volatility of W and its suitable resistance are most important.

14.31 Problem: The isopoly anions containing MO_6 can be considered as fragments of oxide structures. Why are the arrangements usually ccp?

14.31 Solution: The ccp structure involves sharing octahedral edges only, while hcp requires sharing faces also. Sharing faces brings the metal ions at the center of the octahedra closer together, increasing cation-cation repulsion.

14.32 Problem: The simple anions CrO_4^{2-}, MoO_4^{2-} and WO_4^{2-} are tetrahedral. Why do the polyacids and polyanions of Cr differ structurally from those of Mo and W?

14.32 Solution: The small Cr^{VI} cannot expand its C.N., while Mo^{VI} and W^{VI} become 6-coordinate in polyacids and polyanions. The electroneutrality principle (see Problem 2.14) leads us to expect higher C.N. when the oxygens are shared.

14.33 Problem: Give an example of

(a) An acidic oxide of a metal.

(b) An amphoteric oxide of a transition metal.

(c) A diamagnetic rare earth metal ion.

(d) A compound of a metal in the +8 oxidation state.

(e) A liquid metal chloride.

(f) A compound of a transition metal in a negative oxidation state.

14.33 Solution:

(a) CrO_3, WO_3, V_2O_5 — Oxides of metals in very high oxidation states are acidic.

(b) Cr_2O_3 -- Oxides of transition metals in intermediate oxidation states are amphoteric.

(c) Ce^{4+}, La^{3+}, Lu^{3+}, Yb^{2+} — Rare earth d^0f^0 and d^0f^{14} ions are diamagnetic.

(d) OsO_4 — The +8 oxidation state is expected for larger Group VIII metals in combination with O or possibly F.

(e) $TiCl_4$ — Only chlorides of metals in oxidation states of +4 or higher are expected to be liquids.

(f) $H_2Fe(CO)_4$, $HCo(CO)_4$ — Metal carbonyls usually have zero oxidation state; metals in the carbonyl "hydrides" and their anions are assigned negative oxidation states.

14.34 Problem: Describe the quadruple bonding in $Re_2Cl_8^{2-}$ in terms of the bond types (σ, π, etc.) and the atomic orbitals involved. What is significant about the eclipsed configuration?

14.34 Solution: The planar $ReCl_4$ units use dsp^2 (σ) hybrids involving $d_{x^2-y^2}$ and p_x, p_y. The M-M bond involves a p_z-d_{z^2} hybrid. Two π bonds can be formed using the d_{xz} and d_{yz} orbitals on each Re. The δ bond uses the d_{xy} orbitals on each Re. The eclipsed configuration is required for overlap of the d_{xy} orbitals to form the δ bond.

14.35 Problem: Determine the bond order and oxidation number for $Mo_2(O_2CCH_3)_4$ (bridging acetate ions), $W_2Cl_4(PR_3)_4$ (no bridging ligands), and $Mo_2(O\text{-}t\text{-}Bu)_6CO$.

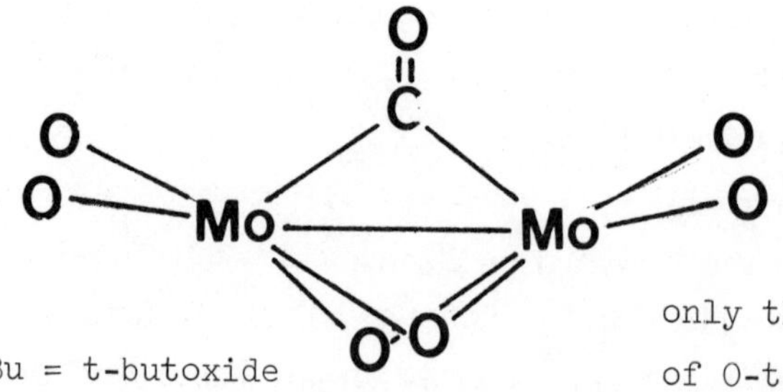

O-t-Bu = t-butoxide

only the coordinated oxygens of O-t-Bu are shown

14.35 Solution:

	B.O.	Oxidation Number
$Mo_2(O_2CCH_3)_4$	4	II
$W_2Cl_4(PR_3)_4$	4	II
Mo_2(O-t-Bu)$_6$CO	2	III

For $Mo_2(O_2CCH_3)_4$ and $W_2Cl_4(PR_3)_4$ there are two square planar MX_4 units joined by quadruple bonds, as in $Re_2Cl_8^{2-}$. For each Mo(III) of Mo_2(O-t-Bu)$_6$CO, one electron is used for bonding to the bridging CO, and four electron pairs are provided by the carboxylate ions. This leaves two electrons per Mo for forming an Mo–Mo double bond in the diamagnetic complex.

14.36 Problem: Give an example of each type of Co complex.

(a) Co(II) tetrahedral.

(b) Co(II) square planar

(c) Co(III) octahedral, high-spin.

(d) Co(III) optically active.

14.36 Solution:

(a) $CoCl_4^{2-}$ — Co^{II} (d^7) with anionic ligands

(b) Co(dimethylglyoximate)$_2$ — Co^{II} (d^7) with a ligand favoring a square planar arrangement

(c) CoF_6^{3-} — only very weak field ligands give high spin for Co^{III} (d^6)

(d) $[Co(en)_3]^{3+}$ — a complex with chelate ligands and no S_n axis.

14.37 Problem: Give one example of a nickel complex illustrating square planar, tetrahedral, and octahedral coordination. What type of ligands favor each of these cases?

14.37 Solution: Square planar Ni^{II} complexes are formed by very strong field ligands, especially π acceptor ligands -- $[Ni(CN)_4]^{2-}$. Tetrahedral complexes of Ni^{II} are formed by anionic weak field ligands -- $[NiCl_4]^{2-}$. Those of Ni^0 are formed by π acceptor ligands -- $Ni(CO)_4$. Octahedral complexes are formed in all cases not favoring C.N. 4 -- $[Ni(NH_3)_6]^{2+}$.

14.38 Problem: The structures of complexes with high coordination number involving bidentate groups can be described in terms of the "average" positions of the bidentate groups. Describe $Ce(NO_3)_5^{2-}$ and $Ce(NO_3)_6^{3-}$ in this way.

14.38 Solution: The structure of $Ce(NO_3)_5^{2-}$ (C.N. 10) is a bicapped trigonal antiprism. Three of the bidentate NO_3^- are coordinated along the edges of the trigonal antiprism (a trigonal prism with one triangular face twisted relative to the other - the octahedron is a trigonal antiprism with the twist angle 60° and all edges of equal length). One of the other two bidentate NO_3^- is coordinated above the top triangular face and one below the bottom triangular face. The "average" positions of the NO_3^- ligands correspond to the positions of the N atoms - three in a trigonal plane, with one above and one below - describing a trigonal bipyramid.

The structure of $Ce(NO_3)_6^{3-}$ (C.N. 12) is an icosahedron. With the icosahedron oriented so that edges at the top and bottom are horizontal, the N of the NO_3^- ions spanning these edges are along the z axis of an octahedron. For the two NO_3^- spanning opposite vertical edges, the N atoms are along the x axis of an octahedron. The remaining pair of NO_3^- span opposite horizontal edges with the N atoms along the y axis of an octahedron. A model of an icosahedron (constructed in Problem 3.8) is helpful. (See DMA p. 325 for simplified shapes of complexes of bidentate ligands giving high C.N. and M.G.B. Drew, Coord. Chem. Rev. 1977, 24, 179.)

14.39 Problem: Give syntheses for cis- and trans-$[Pt(NH_3)_2Cl_2]$, starting with $[Pt(NH_3)_4]^{2+}$ and $[PtCl_4]^{2-}$.

14.39 Solution:

$$Pt(NH_3)_4^{2+} + HCl \rightarrow [PtCl(NH_3)_3]^+ + NH_4^+$$

$$[PtCl(NH_3)_3]^+ + HCl \rightarrow \underline{trans}\text{-}[PtCl_2(NH_3)_2] + NH_4^+$$

$$PtCl_4^{2-} + NH_3 \rightarrow [PtCl_3NH_3]^- + Cl^-$$

$$[PtCl_3NH_3]^- + NH_3 \rightarrow \underline{cis}\text{-}[PtCl_2(NH_3)_2] + Cl^-$$

In each case a ligand trans to Cl^- is displaced more readily than one trans to NH_3.

14.40 Problem: How can one account for the color of the following:

(a) Fe_3O_4.
(b) Ag_2S.
(c) $KFeFe(CN)_6$.
(d) $KMnO_4$.
(e) $Ti(H_2O)_6^{3+}$.
(f) $Cu(NH_3)_4^{2+}$.

14.40 Solution:

(a) Deep color, Fe is present in two oxidation states. This is a defect structure involving charge delocalization between Fe^{II} and Fe^{III} (see DMA p. 230).

(b) Black, there is an intense charge transfer absorption band involving $S^{2-} \rightarrow Ag^+$ donation (see DMA p. 288).

(c) Intensely colored, Fe is in two oxidation states.

(d) The charge transfer band, O → Mn, is more intense than the usual d → d bands observed for transition metal complexes.

(e) and (f) These involve d → d transitions. The number of bands depends on the number of d electrons and the symmetry of the complex. The band energies depend on the ligand field strength (see DMA p. 277f).

14.41 Problem: The colors of lanthanide ions arise primarily from f-electron transitions. Which of the lanthanide ions might be expected to be colorless? Which of the following might be expected to be the same color: Ce^{3+}, Pr^{3+}, Pm^{3+}, Tm^{3+}, Tb^{3+}? Explain.

14.41 Solution: Ions with f^0 (La^{3+}), f^{14} (Lu^{3+}), and half-filled f^7 (Gd^{3+}) configurations are expected to be colorless. One expects from the hole formalism that ions of the same charge for which there are n electrons or n holes will have similar absorption bands (all are expected to be weak field complexes since the f orbitals are rather well shielded from their environment). The number of f electrons of the ions are Ce^{3+} 1, Pr^{3+} 2, Pm^{3+} 4, Tm^{3+} 12, and Tb^{3+} 8. Praseodymium and thulium salts are both green, cerium(III) colorless, terbium(III) pale pink, and promethium(III) pink, thus Pr^{3+} and Tm^{3+} fulfill our expectations.

14.42 Problem: Metals such as Ti and Al are useful structural materials even though they are easily oxidized. Na and K (among other metals) have reduction potentials comparable in magnitude to those of Ti and Al, yet are useless for applications involving contact with air. Explain.

14.42 Solution: The molar volumes of TiO_2 and Al_2O_3 are larger than the molar volumes of Ti and Al, respectively. Hence, the oxides occupy larger volumes than the metals from which they were formed and thus cover the metal surface with an impervious oxide layer. Sometimes oxide layers are deposited electrochemically, a process known as anodizing the metal, for protection. Actually these oxide layers represent defect structures varying in composition from the stoichiometric oxide on the surface (in contact with O_2) to the metal structure below. Because of this there is no sharp phase boundary that would result in flaking. For Na and K, the oxides have smaller volumes than the metals. Hence, O_2 can continue to penetrate the structure and oxidize the underlying metal at a rate governed by the O_2 diffusion rate.

14.43 Problem: The structure of ReO_3 is similar to that of perovskite, $CaTiO_3$, without the Ca. Describe the structure in terms of the roles of the ions in a close-packed structure. Give the PTOT notation (See Problem 6.5).

14.43 Solution: The oxide ions are ccp with 1/4 of the sites (those for Ca^{2+} in perovskite) vacant and with Re in 1/4 of the octahedral holes (there are as many octahedral holes as packing positions). A beautiful model can be built using octahedra (ReO_6) centered at each corner of a cube and sharing the octahedral apices (each O^{2-} is shared by two Re). The PTOT notation is $6P_{3/4}O_{1/4}$, the 6 refers to the 6 layers (3P and 3O) in the repeating ccp arrangement.

14.44 Problem: Generally abundances of elements decrease with increasing atomic number. However, the terrestial and cosmic abundances of the light elements Li, Be, and B are very low. Explain the low abundances of these elements.

14.44 Solution: The elements Li, Be, and B have low atomic numbers and undergo thermonuclear reactions readily because of low barriers for proton or alpha capture. These elements did not accumulate since they were used up as formed.

XV Boron Hydrides, Cluster and Cage Compounds

15.1 Problem: Classify the following species as closo, nido, arachno, or hypho.

C_8H_8 B_6H_{12} B_9H_{15} B_4H_{10} B_4H_8 $B_6H_{11}^+$

15.1 Solution: Clusters having p vertices are classified by skeletal electron count as follows:

Skeletal e pairs	Classification
p + 1	closo
p + 2	nido
p + 3	arachno
p + 4	hypho

Each BH group is considered to contribute two electrons (one pair) to skeletal bonding. Groups isoelectronic with BH (e.g., CH^+, BeH^-) also contribute two electrons. Others contribute a number of electrons given by the formula (v + x - 2) where v is the number of valence electrons of the atom in the cluster and x is the number contributed by exo bonded groups (e.g., for CH, v + x - 2 = 4 + 1 - 2 = 3). "Extra" H's each contribute one electron. Account must also be taken of the charge. The numbers of electron pairs for each compound and the classifications are:

C_8H_8	(8 x 3) = 24e = p + 4 pairs; hypho.
B_6H_{12}	(6 x 2) + 6 = 18e = p + 3 pairs; arachno.
B_9H_{15}	(9 x 2) + 6 = p + 3 pairs; arachno.
B_4H_{10}	(4 x 2) + 6 = p + 3 pairs; arachno.

B_4H_8 $(4 \times 2) + 4 = p + 2$ pairs; nido.

$B_6H_{11}^+$ $(6 \times 2) + 5 - 1 = p + 2$ pairs; nido.

15.2 Problem: What structure do you expect for each species in Problem 15.1?

15.2 Solution: Expected frameworks are as shown below (R.W. Rudolph, Acc. Chem. Res. 1976, 9, 446). The first column contains drawings of the closo frameworks for various numbers (p) of vertices. Excision of the most highly connected group at a vertex leads to a nido structure given in the second column. The p-vertex nido structure is related to the (p + 1)-vertex closo structure by removal of one group. Similarly, excision of a highly connected group from the (p + 1)-vertex nido structure produces the p-vertex arachno structure. Hence, each p-vertex arachno framework is related to the (p + 2)-vertex closo framework by successive excision of two groups. Hypho structures contain one less vertex than arachno structures. Bridging H's are not shown. (See DMA Sec. 15.3)

For C_8H_8 we expect the structure derived from excision of three BH vertices from the closo $B_{11}H_{11}^{2-}$ structure (octadecahedron). This provides a distorted version of the actual structure which is the boat form of a C_8H_8 ring. For B_6H_{12}, the arachno structure for p = 6 is shown in the third column of the figure. Other frameworks may be similarly located.

15.3 Problem: How may $(CH_3)_2B_2H_4$ be prepared? Draw structural formulas for all isomers expected for $(CH_3)_2B_2H_4$. How might one identify these isomers if they were all separated?

15.3 Solution: Mixing $B(CH_3)_3$ and B_2H_6 would lead to a redistribution of the methyl groups through hydrogen-bridged intermediates. Other possible preparative routes are:

$$4\ Li[CH_3AlH_3] + 4\ BCl_3 \xrightarrow{Et_2O} 2(CH_3)_2B_2H_4 + 4\ Li[AlCl_3H]$$

$$2\ [\underline{n}\text{-}Bu_4N][CH_3BH_3] + 2\ BCl_3 \xrightarrow{25^{\circ}C} (CH_3)_2B_2H_4 + 2\ [\underline{n}\text{-}Bu_4N][HBCl_3]$$

Possible isomers are:

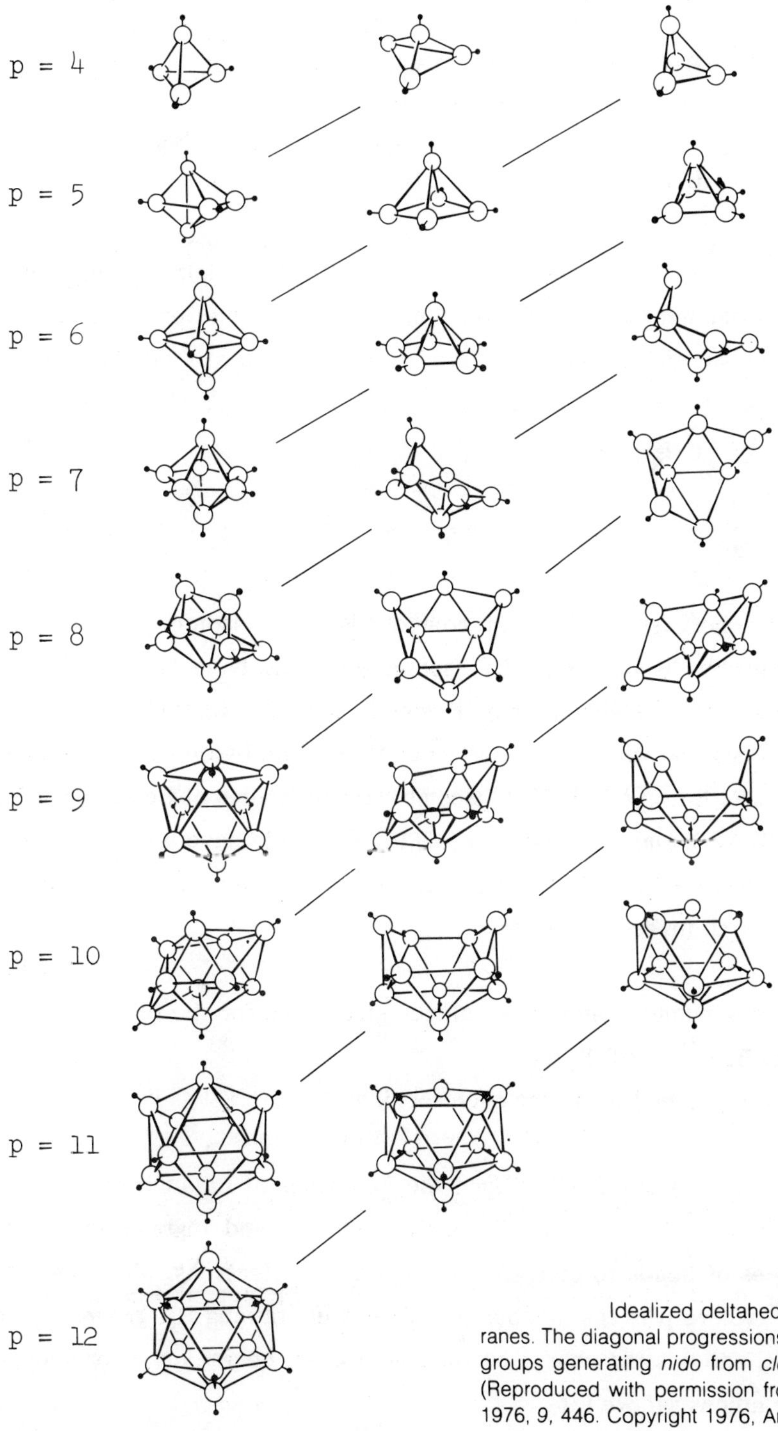

Idealized deltahedra and deltahedral fragments for boranes. The diagonal progressions represent excision of successive BH groups generating *nido* from *closo* and *arachno* from *nido* species. (Reproduced with permission from R. W. Rudolph, *Acc. Chem. Res.* 1976, *9*, 446. Copyright 1976, American Chemical Society.)

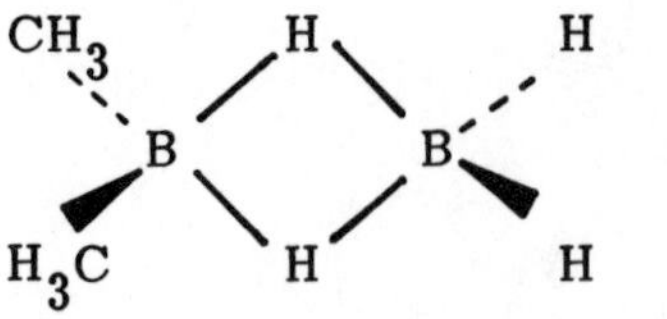

(a) (C_{2v})

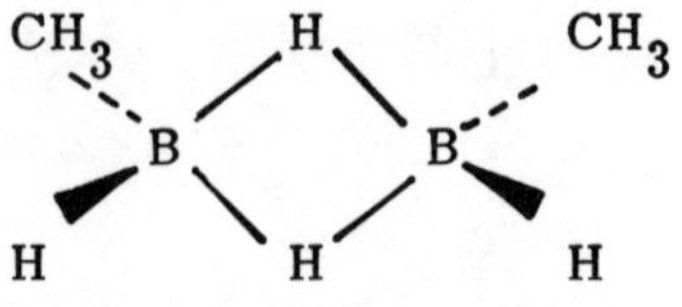

(b) (C_{2v})

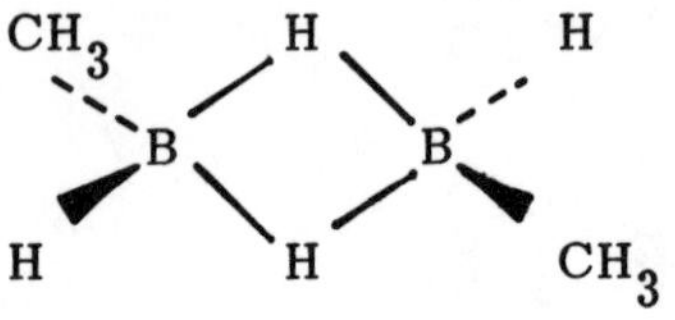

(c) (C_{2h})

Bridging methyl groups are not found in boron compounds. The isomer (a) could be distinguished from (b) and (c) by ^{11}B and ^{1}H NMR. Isomer (a) would have two ^{11}B signals. Both (b) and (c) would have one ^{11}B signal, one ^{1}H signal each for methyl H, μ-H and terminal B-H. In all cases B signals will be split by coupling to H and vice versa. (A good reference for ^{11}B NMR is G.R. Eaton and W.N. Lipscomb, "NMR Studies of Boron Hydrides and Related Compounds," W.A. Benjamin, New York, 1969). (b) would have a (small) dipole moment whereas (c) would have a zero dipole moment. (c) would have no coincident IR and Raman bands since it has an inversion center.

15.4 Problem: Calculate styx numbers and draw valence structures for the following:
(a) B_5H_{11} (b) B_6H_{10} (c) $B_5H_5^{2-}$ (d) B_8H_{12}
Which of the structures is the preferred one for each species?

15.4 Solution: Lipscomb and co-workers developed the Equations of Balance as a way of counting available orbitals and electrons in the boron hydrides and distributing them among various possible types of bonds to give each B an octet of electrons. Formulas are written as $[(BH)_pH_{q+c}]^{c+}$ where p is the number of vertices containing BH groups, q + c is the number of "extra" H, and c is the charge on the species. Of course c = 0 for neutral species and is negative for anions. If we let s = number of B-H-B bonds

t = number of B-B-B three-center bonds

y = number of B-B bonds

x = number of BH_2 groups

the Equations of Balance are (See DMA Sec. 15.22)

$$q + c = s + x$$

$$p + c = s + t$$

$$p - \frac{q}{2} - c = t + y$$

Sets of styx numbers may be calculated for any formula and used to draw valence bond structures. (See DMA p. 663.)

(a) $B_5H_{11} = (BH)_5H_6$: p=5, q=6, c=0. The Equations of Balance are:

$$6 = s + x$$

$$5 = s + t$$

$$5 - 3 = 2 = t + y$$

These are diophantine equations, that is the solutions are whole, positive, numbers. Since the sum of t and y is 2, the possible solutions for t and y respectively are 0,2; 1,1; and 2,0. Since the sum of s and t is 5, and of s and x is 6, the complete set of possible styx numbers would be 5021, 4112, and 3203. This is an arachno species (see Problem 15.1). Arachno structures are derived from closo structures by successive excision of B-H groups. Closo structures have all BH groups at vertices of triangular polyhedral faces. Hence, each BH group is attached to at least four others by B-B bonds. If we start from the seven-vertex closo structure and excise two B-H groups, we must still have at least one B bonded to four other B's by B-B bonds. Only the 3203 structure incorporates this feature and so is preferred.

3203

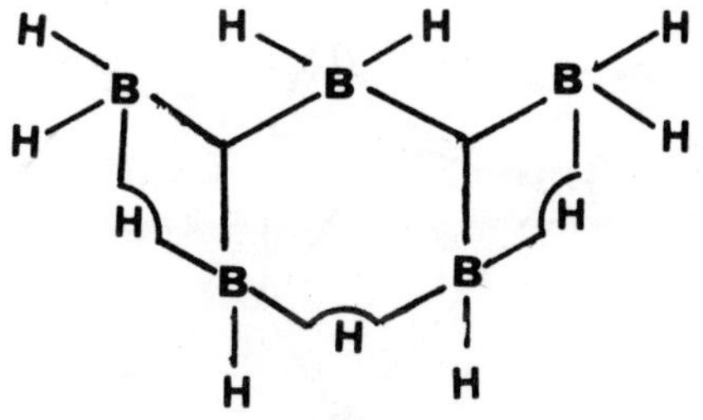

(b) B_6H_{10}: Possible styx numbers are 4220, 3311 and 2402. In this nido structure, the apical BH must be joined to the five basal ones. This will require at least two three-center bonds. Hence, we must have $t \geq 2$. However, t cannot be >3. Otherwise, at least one pair of basal B's would be joined by two three-center B-B-B bonds. Hence, 4220 and 3311 are possibilities. However, 3311 involves an open B-B-B bond. 4220 is preferred.

4220

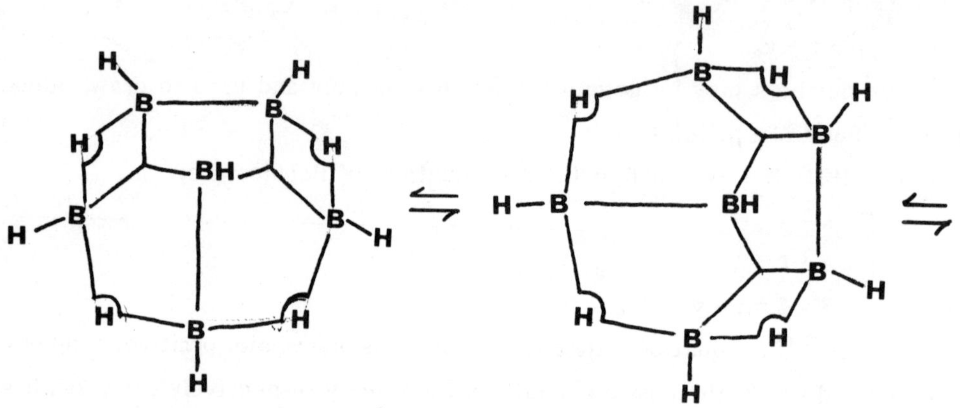

\+ 3 more structures

(c) $B_5H_5^{2-}$:

$q + c = \#$ "extra" $H = 0 = s + x$ Hence, $q = 2$ since $c = -2$.

$p + c = 5 - 2 = 3 = s + t$

$p - \frac{q}{2} - c = 5 - 1 - (-2) = 6 = t + y$

The only solution is 0330 leading to a structure with open B-B-B bonds or the two structures with closed B-B-B bonds.

0330

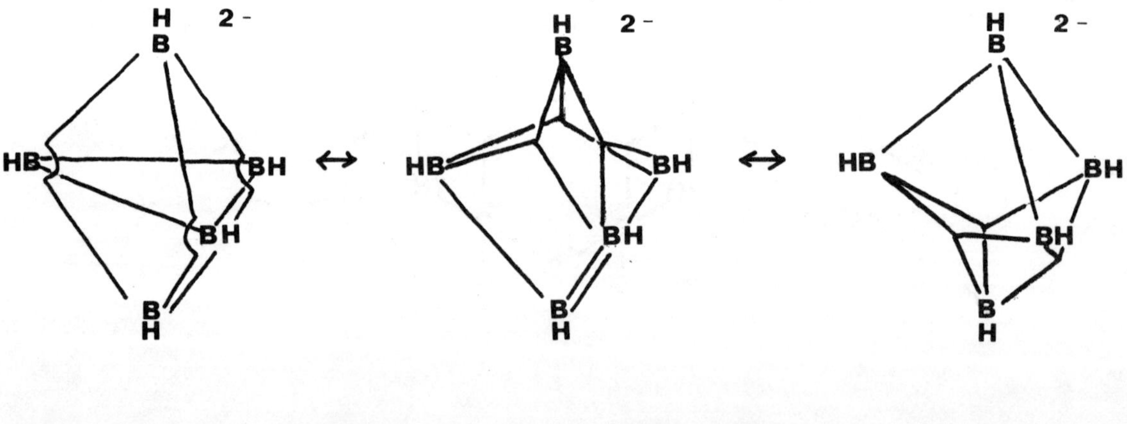

(d) B_8H_{12}: Solutions to the equations of balance are 4420, 3511, and 2602. Both 3511 and 2602 involve open B-B-B bonds. One 4420 structure need not have open B-B-B bonds and is preferred.

4420

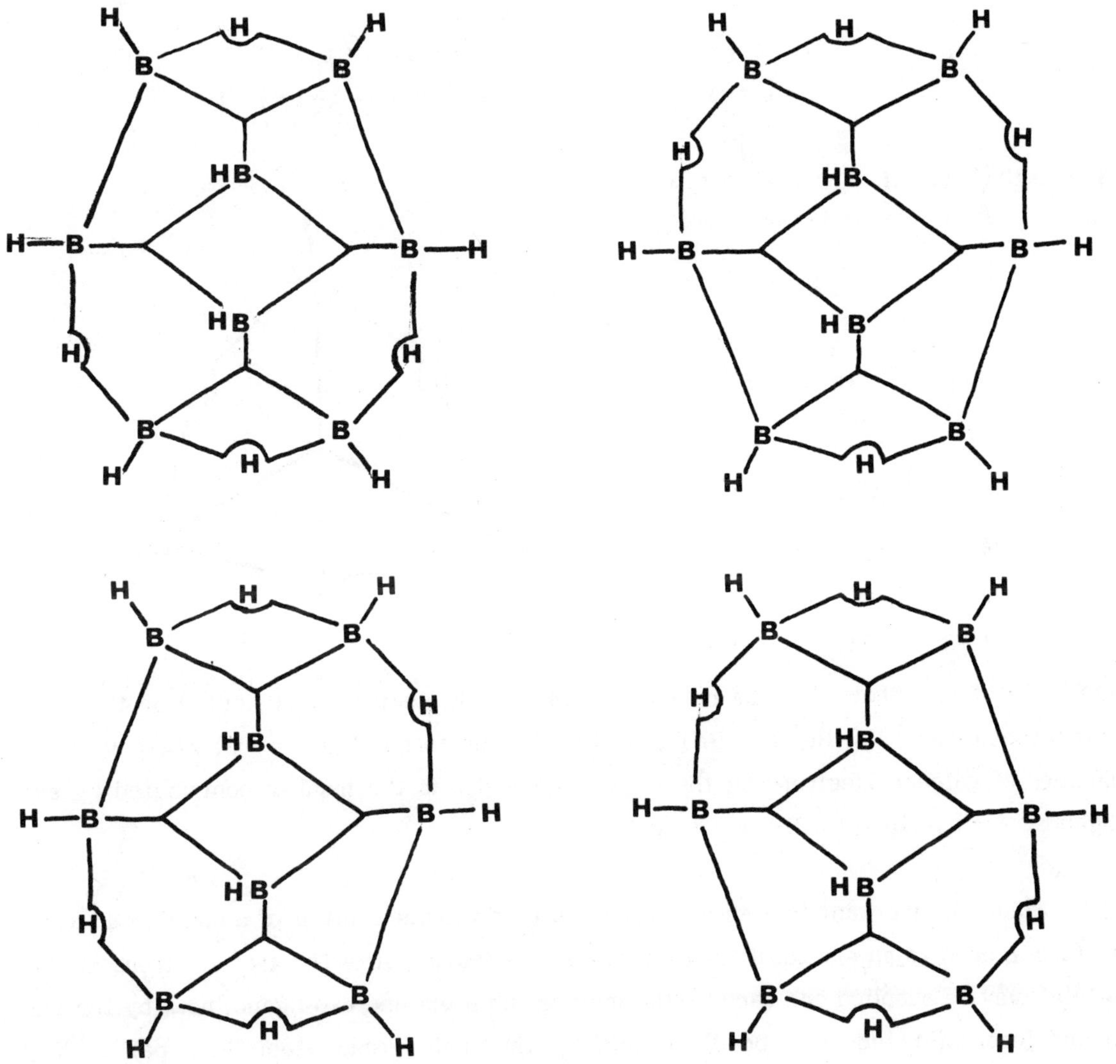

Other 4420 structures with open B-B-B bonds can be drawn.

15.5 Problem: Solve the Equations of Balance for $B_3H_6^+$ (not known) and write a "reasonable" structure for such a hydride.

15.5 Solution: See Problem 15.4 for the procedure used in solving the Equations of Balance. We formulate $B_3H_6^+$ as $(BH)_3H_3^+$ for which

$$p = 3 \qquad q + c = 3 = s + x$$

$$q + c = 3 \qquad p + c = 4 = s + t$$

$$c = +1 \qquad p - \frac{q}{2} - c = 3 - 1 - 1 = 1 = t + y$$

$$q = 2$$

3100 is the solution. The structure is:

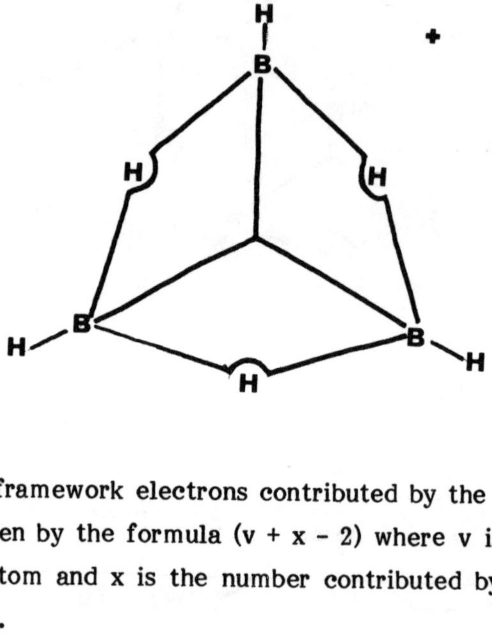

15.6 Problem: Show that the number of framework electrons contributed by the main group species NH^-, Si, Be^+ and BNR_3 is given by the formula (v + x - 2) where v is the number of valence electrons on the cage atom and x is the number contributed by exo ligands. For example, for BH, v = 3 and x = 1.

15.6 Solution: We want to reserve one of the four valence orbitals of a main group atom to form a bond to an exo substituent or to contain a lone pair. The two electrons in this orbital may be supplied one each by the main group atom and exo ligand, both by the exo ligand (coordinate covalent bond) or both by the main group atom (lone pair). The remaining three orbitals and any remaining electrons are contributed to framework bonding.

NH^-: N^- has six valence electrons; one of these plus one from H form the NH bond leaving five framework electrons. $(v + x - 2) = 6 + 1 - 2 = 5$.

Si: Si has four valence electrons; two are used for a lone pair leaving two for framework bonding. $(v + x - 2) = 4 + 0 - 2 = 2$.

Be^+: Be^+ has one valence electron. In order to have a lone pair, it must extract one electron from the framework. $(v + x - 2) = 1 + 0 - 2 = -1$.

BNR_3: Both electrons in the B-N bond are supplied by nitrogen of the amine leaving the three B electrons for donation to the framework. $(v + x - 2) = 3 + 2 - 2 = 3$.

15.7 Problem: Classify the following species as closo, nido, or arachno:

(a) 2-CB_5H_9
(b) 5-CH_3-2,3-$C_2B_4H_7$
(c) 1,2-$C_2B_9H_{11}$
(d) $B_9H_{11}S$
(e) 1,7-$B_{10}CPH_{11}$
(f) $B_{11}SH_{10}Ph$ (phenyl is attached to B)
(g) $B_9H_{12}NH^-$
(h) $C_2B_6H_{10}$
(i) $B_2H_4(PF_3)_2$

Name and sketch the structures of the above species.

15.7 Solution: See Problems 15.1 and 15.6 regarding structure and electron count.

(a) 2-CB_5H_9 $(CH)(BH)_5(H)_3$: 3e + 5 x 2e + 3e = 16e = 8 pr; nido; 2-carbahexaborane(9). The structure is depicted on page 269. It is the p=6 nido framework having a CH group in one of the basal positions. The three μ-H bridge three of the basal BH pairs.

(b) 5-CH_3-2,3-$C_2B_4H_7$ $(CH)_2(BH)_3(BCH_3)_1(H)_2$: 2 x 3e + 3 x 2e + 2e + 2e = 16e = 8 pr; nido; 2,3-dicarba-5-methylhexaborane(7). Same framework as above, except two CH groups are adjacent in the basal plane and a methyl group instead of H is bonded to the B two vertices removed from each CH group.

(c) 1,2-$C_2B_9H_{11}$ $(CH)_2(BH)_9$: 2 x 3e + 9 x 2e = 24e = 12 pr; closo; 1,2-dicarbaundecaborane(11). All frameworks are depicted in the Figure of Problem 15.2. This compound has the eleven-vertex closo framework (octadecahedron) with CH groups at two adjacent vertices in the top and next lower plane. The numbering of vertices (DMA p. 671) is described more fully in R.M. Adams and K.A. Jensen, Pure Appl. Chem. 1972, 30, 681.

(d) $B_9H_{11}S$ $(BH)_9(H)_2S$: 9 x 2e + 2e + 4e = 24e = 12 pr; nido; thiadecaborane(11). The framework is that of $B_{10}H_{14}$. The vertex occupied by S is not specified here, but is probably one of the two in the uppermost plane.

(e) $1,7\text{-}B_{10}CPH_{11}$ $(CH)_1(BH)_{10}P$: $3e + 10 \times 2e + 3e = 26e = 13$ pr; closo; 1-carba-7-phosphadodecaborane(11). The twelve-vertex closo structure is the icosahedron. Here, the CH group is at the apex and P is at one of the five vertices in the lower pentagonal plane.

(f) $B_{11}SH_{10}Ph$ $(BPh)(BH)_{10}S$: $2e + 10 \times 2e + 4e = 26e = 13$ pr; closo; thiaphenyl-dodecaborane(10). The icosahedral framework has two unspecified vertices which contain BPh and S instead of BH.

(g) $B_9H_{12}NH^-$ $(BH)_9(H)_3(NH^-)$: $9 \times 2e + 3e + 5e = 26e = 13$ pr; arachno; imido-dodecahydrodecaborate(1-). The framework is the arachno p=10 one derived by excision of two BH groups from an icosahedron. An NH^- group occupies an unspecified vertex, probably on the open face. The bridging H is between two BH groups. Two BH_2 groups are on the open face.

(h) $C_2B_6H_{10}$ $(CH)_2(BH)_6(H)_2$: $2 \times 3e + 6 \times 2e + 2e = 20e = 10$ pr; nido; dicarba-octaborane(10). The nido eight-vertex framework is derived from the tricapped trigonal prism by removal of a capping vertex. The two CH groups are in non-adjacent positions in the most stable isomer and μ-H bridge BH groups.

(i) $B_2H_4(PF_3)_2$ $(BH)_2(H)_2(PF_3)_2$: $2 \times 2e + 2e + 4e = 10e = 5$ pr; arachno; bis(trifluoro-phosphine)diborane(4). The structure contains two more electrons than B_2H_6 since PF_3 is a two-electron donor.   Enough electrons are available to write an electron-precise structure.

$$F_3P \rightarrow \overset{\overset{\displaystyle H}{|}}{\underset{\underset{\displaystyle H}{|}}{B}} - \overset{\overset{\displaystyle H}{|}}{\underset{\underset{\displaystyle H}{|}}{B}} \leftarrow PF_3$$

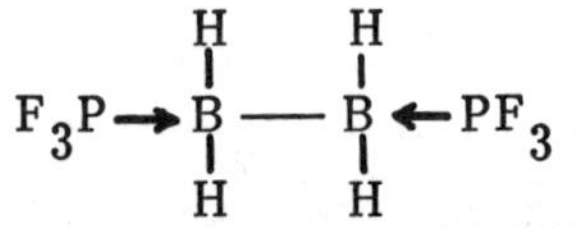

15.8   Problem: Use the Equations of Balance to obtain a reasonable bonding picture of 1,5-dicarba-closo-pentaborane(7).

15.8   Solution: See Problem 15.4 for Equations of Balance. $1,5\text{-}C_2B_3H_7$ is isoelectronic with $B_5H_5^{2-}$. See Solution 15.4. The "extra" H would be expected to bridge two BH groups in the equatorial plane.

15.9   Problem: Predict the products of the following reactions:

(a) B_5H_{11} + KH
(b) B_5H_9 + NMe_3
(c) $B_{10}H_{14}$ + SMe_2
(d) B_5H_9 + HCl(ℓ)
(e) B_6H_{10} + Br_2
(f) $Li_2[\underline{o}\text{-}C_2B_{10}H_{10}]$ + CH_3I
(g) $Li_2[\underline{o}\text{-}C_2B_{10}H_{10}]$ + R_3SiCl
(h) $B_6H_9^-$ + Me_2SiCl_2
(i) 2,3-$C_2B_4H_8$ + NaH
(j) $ZrCl_4$ + 4 $LiBH_4$
(k) $(C_6H_5)_2PCl$ + $LiAlH_4$

15.9 Solution:

(a) $K[B_5H_{10}]$ + $1/2H_2$

(b) $B_5H_9(NMe_3)_2$; B_5H_9 contains no BH_2 groups making fragmentation less likely. $B_5H_9(NMe_3)$ is an acceptable possibility but does not form in fact.

(c) $B_{10}H_{12}(SMe_2)_2$ + H_2

(d) $[B_5H_{10}]^+Cl^-$

(e) B_6H_9Br + HBr

(f) $\underline{o}$-$(CMe)_2B_{10}H_{10}$ + 2 LiI

(g) $\underline{o}$-$[C(SiR_3)]_2B_{10}H_{10}$ + 2 LiI

(h) μ,μ'-$Me_2Si[B_6H_9]_2$ + 2 Cl^-

(i) $Na^+[2,3\text{-}C_2B_4H_7]^-$ + H_2

(j) $Zr(BH_4)_4$ + 4 LiCl

(k) Ph_2PH + LiCl + AlH_3

15.10 Problem: Give a reasonable method for preparing and purifying B_2H_6. How might the purity of the sample be determined? How could one dispose of the diborane?

15.10 Solution: Reduction of boron halides by $LiAlH_4$ in ether, or displacement from BH_4^- salts by a stronger, nonoxidizing acid will produce B_2H_6. Either reaction should be carried out on a vacuum line, where the purity could be checked by the tensiometric homogeneity, i.e. the constancy of the vapor pressure of different fractions. For disposal, produce an amine borane which could then be exposed to air and allowed to react with 2-propanol or n-butanol.

15.11 Problem: (a) Show that the diamond-square-diamond mechanism (shown below) cannot account for the known thermal rearrangement of o-carborane to p-carborane. (b) What experiments might be helpful in shedding light on the o- to p- rearrangement?

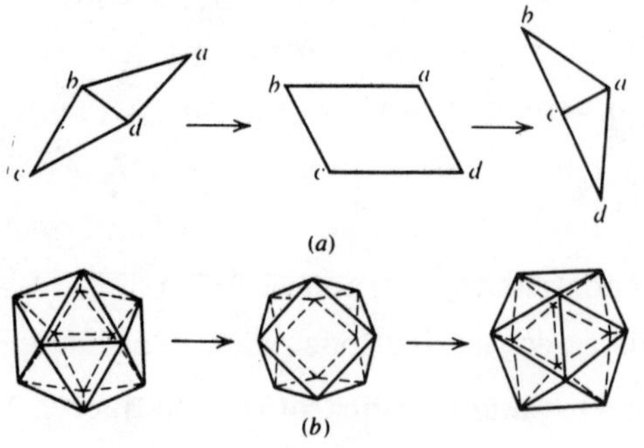

(*a*) The diamond-square-diamond mechanism. (*b*) Rearrangement of *o*- to *m*-carborane via a cuboctahedral intermediate. (From H. Beall, in *Boron Hydride Chemistry*, E. L. Muetterties, Ed., Academic Press, New York, 1975.)

15.11 Solution: (a) The d–s–d path allows a particular C to re-connect only to a B in the plane of atoms immediately above or below it in the starting isomer giving an m arrangement.
(b) One possibility might be rotation of half the icosahedron giving a bicapped pentagonal prism. Another might be rotation of triangular icosahedral faces. Labeling (e.g., starting with a monobromo o-carborane) could distinguish via isomer distribution in the product.

15.12 Problem: Write an essay on the possible relevance of transition cluster compounds to heterogeneous catalysis. (You will want to consult appropriate references).

15.12 Solution: Appropriate references to consult would be:
E.L. Muetterties, Bull. Soc. Chim. Belg. 1975, 84, 959.
E.L. Muetterties and J. Stein, Chem. Rev. 1979, 79, 479.
B.F.G. Johnson, Ed., "Transition Metal Clusters," Wiley, New York, 1980.

15.13 Problem: Classify the following as closo, nido, or arachno.

(a) $[Co_4Ni_2(CO)_{14}]^{2-}$
(b) $[Fe(CO)_3]B_4H_8$
(c) $(CpCo)C_2B_7H_{11}$
(d) $Os_5(CO)_{16}$
(e) $(Et_3P)_2Pt(H)B_9H_{10}S$
(f) $Rh_6(CO)_{16}$
(g) $[Rh_9P(CO)_{21}]^{2-}$ (P is at cage center)

15.13 Solution: Rules formulated by Wade suggest that the relation between framework electron count and molecular geometry should be similar for transition metal clusters and boranes. The number of framework electrons contributed by transition-metal-containing groups is given by (v + x - 12) where v is the number of valence electrons of the metal and x is the number of electrons contributed by all the attached ligands. Transition metals have nine valence orbitals (5d's + 3p's + 1s). By analogy with BH, we wish to allow three orbitals plus whatever electrons they contain to contribute to framework bonding. Hence, six orbitals must be employed in bonding ligands or containing lone pairs. This requires a total of twelve e which may come from the pool of metal valence electrons or from those contributed by ligands. If the number of ligands attached to each metal is not given, the total number of metal valence electrons may be added to the total of electrons donated by ligands (e.g., two for each CO) and the charge. Twelve electrons for each metal are subtracted giving the final framework count. (See DMA Sec. 15.10.4 or K. Wade, Chem. in Britain 1975, 11, 177.)

(a) $[Co_4Ni_2(CO)_{14}]^{2-}$

$4 \times 9e + 2 \times 10e + 14 \times 2e + 2e - 6 \times 12e = 14e = 7$ pr; closo

(b) $[Fe(CO)_3]B_4H_8 = [Fe(CO)_3](BH)_4(H)_4$

$2e + 4 \times 2e + 4e = 14e = 7$ pr; nido

(c) $(CpCo)C_2B_7H_{11} = (CpCo)(CH)_2(BH)_7(H)_2$

$2e + 2 \times 3e + 7 \times 2e + 2e = 24e = 12$ pr; nido

(d) $Os_5(CO)_{16}$

$5 \times 8e + 16 \times 2e - 5 \times 12e = 12e = 6$ pr; closo

(e) $(Et_3P)_2Pt(H)B_9H_{10}S$

The electron-counting rules must be modified for complexes involving Pt(II) which ordinarily forms 16- rather than 18e species. Hence, the number of framework electrons for $(PEt_3)_2Pt(H)^+$ is $v + x - 10 = 4$. For $B_9H_{10}S^-$, the number is 24. Hence, we have 14 electron pairs and 11 vertices for an arachno structure.

(f) $Rh_6(CO)_{16}$

$6 \times 9e + 16 \times 2e - 6 \times 12e = 14e = 7$ pr; closo

(g) $[Rh_9P(CO)_{21}]^{2-}$

$9 \times 9e + 5e + 21 \times 2e + 2e - 9 \times 12e = 22e = 11$ pr; nido

15.14 Problem: Sketch the predicted geometry of the species in Problem 15.13.

15.14 Solution: The only prediction which can be made from Wade's rules is which atoms are at polyhedral vertices. No information is provided about how many carbonyls are bridging and how many terminal nor about the exact placement of bridging CO or H. The descriptions given here are those of actual structures and include details not predictable from Wade's rules.

(a) $[Co_4Ni_2(CO)_{14}]^{2-}$: octahedron; contains μ-CO (V.G. Albano, G. Ciani and P. Chini, J. Chem. Soc. Dalton Trans., 1974, 432).

(b) $[Fe(CO)_3](B_4H_8)$: square pyramidal structure of B_5H_9 with $Fe(CO)_3$ replacing the apical BH (N.N. Greenwood, et al., Chem. Comm., 1974, 718).

(c) $(CpCo)(C_2B_7H_{11})$: not known; would predict replacement by CpCo of one BH group in the $C_2B_8H_{12}$ framework which would be isoelectronic and isostructural with $B_{10}H_{14}$.

(d) $Os_5(CO)_{16}$: Trigonal bipyramid of four $Os(CO)_3$ groups having an $Os(CO)_4$ group in the plane which distorts the TBP framework (B.E. Reichert and G.M. Sheldrick, Acta. Cryst. 1977, B33, 173).

(e) $(Et_3P)_2Pt(H)B_9H_{10}S$: The arachno structure for eleven atoms results from excision of two vertices of a 13-vertex polyhedron shown here. The structure observed results from excision of vertices numbered 1 and 4 with Pt and S at the 2 and 6 positions, respectively. (A.R. Kane, L.J. Guggenberger and E.L. Muetterties, J. Am. Chem. Soc., 1970, 92, 2571.)

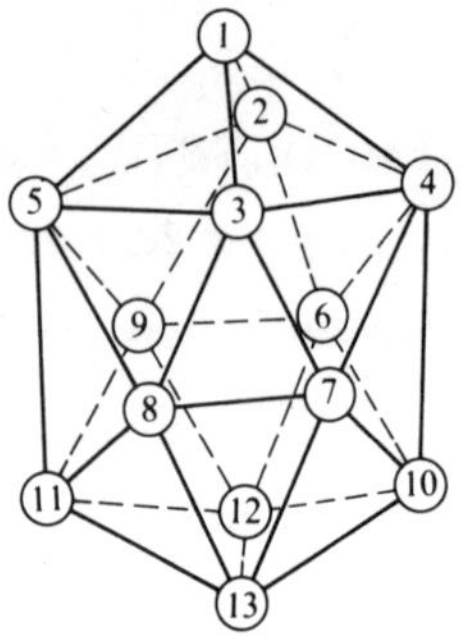

(f) $Rh_6(CO)_{16}$: octahedron of $Rh(CO)_2$ groups with remaining four CO's triply bridging alternate octahedral faces (E.R. Corey, L.F. Dahl and W. Beck, J. Am. Chem. Soc. 1963, 85, 1202).

(g) $[Rh_9P(CO)_{21}]^{2-}$: Archimedian antiprism of Rh(CO)'s with one apex capped by Rh(CO). Three sets of four μ-CO's bridge apical and upper plane Rh's, upper and basal plane Rh's and basal plane Rh's, respectively. P is at center of Rh_9 framework (J.L. Vidal, W.E. Walker, R.L. Pruett and R.C. Schoening, Inorg. Chem. 1979, 18, 129).

15.15 Problem: Some known cluster compounds contain p vertices and p pairs of framework bonding electrons. The structures of these species are often (p-1)-vertex closo structures with one of the triangular faces capped. One example of such a structure is $Os_6(CO)_{18}$ (below). Contrary to the prediction by Wade's rules (See Problem 15.13), this compound is a trigonal bipyramid with one triangular face capped by an $Os(CO)_3$ group. Show that this capping arrangement has the effect of contributing two more electrons and no more orbitals to the framework bonding, thus rationalizing the basic TBP geometry.

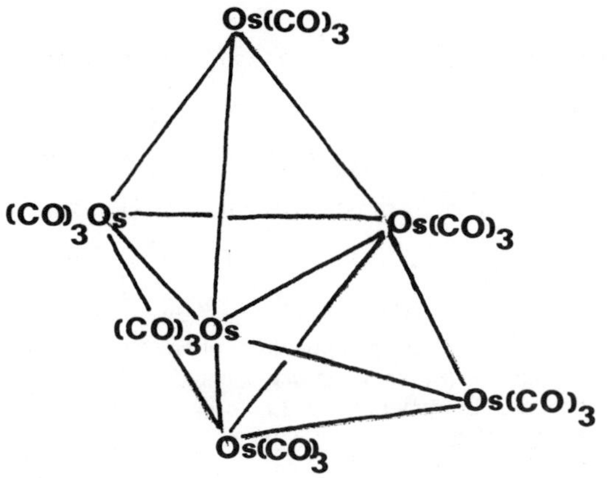

The structure of $Os_6(CO)_{18}$

15.15 Solution: The number of framework electrons contributed by various groups is based on the assumption that three orbitals and their electrons are contributed to framework bonding after six orbitals and their twelve electrons have been used for bonding ligands or as non-bonding orbitals. For $Os(CO)_3$ groups three valence orbitals and six ligand electrons are used for Os-CO bonds. One of the filled non-bonding orbitals on each Os in a triangular face could be used to overlap with the three empty "framework" orbitals of the capping $Os(CO)_3$. The two remaining electrons of the capping group could be considered to be added to the framework pool giving 5 + 1 = 6 electron pairs required for a closo TBP framework. The argument can be applied to other capped structures.

15.16 Problem: The Figure shows the ^{13}C NMR spectrum of $(cot)Fe(CO)_3$ (cot = cyclooctatetraene) at several temperatures. The signals at 214 and 212 ppm are attributed to CO, and the others to cot. Explain the appearance of the spectrum at -134°C, and its change with T.

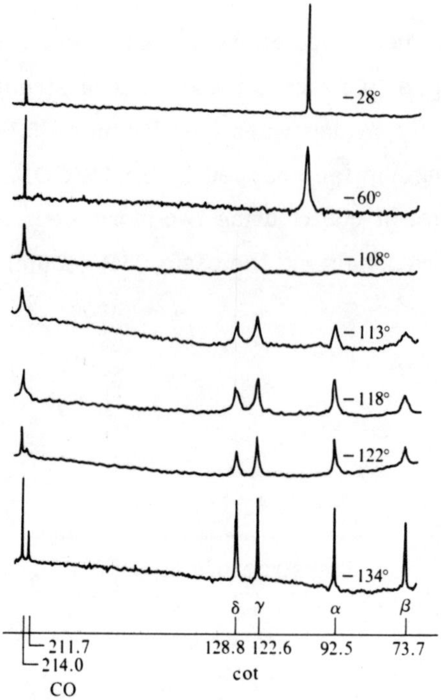

^{13}C NMR spectrum of $Fe(CO)_3$ at several temperatures. (Reproduced with permission from F. A. Cotton, and D. L. Hunter, *J. Am. Chem. Soc.* 1976, *98*, 1413. Copyright 1968, American Chemical Society.)

15.16 Solution: Cyclooctatetraene acts as a four-electron donor in order that Fe conform to the EAN rule. At -134°C the static structure is frozen out. Two different kinds of terminal CO's exist in a 2:1 ratio on Fe. This leads to four non-equivalent pairs of C on the cot ligand. (Compare Figure 10.6 in DMA.) Both CO and cot signals begin to collapse at the same time indicating that the same process achieves equivalence for both CO and cot carbons. Presumably, this is the migration of the $Fe(CO)_3$ group around the cot ring. At -108°C, the speed of the migration process is of the same order as NMR time scale and the signal is broadened almost into the baseline. By -28°C, the migration process is much faster and a single averaged environment is seen for CO and for cot C's.

15.17 Problem: (a) The ^{1}H NMR spectrum $CpFe(CO)_3)(\eta^1-C_5H_5)$ at several temperatures is reproduced in the figure below. Account for the appearance of the temperature changes. (b) (i) The compound $(C_5H_5)_4Ti$ has been prepared. What does the EAN rule suggest about the attachment of the C_5H_5 rings to Ti? (ii) From the results in a and b(i), rationalize the appearance of the ^{1}H NMR spectrum as the temperature changes.

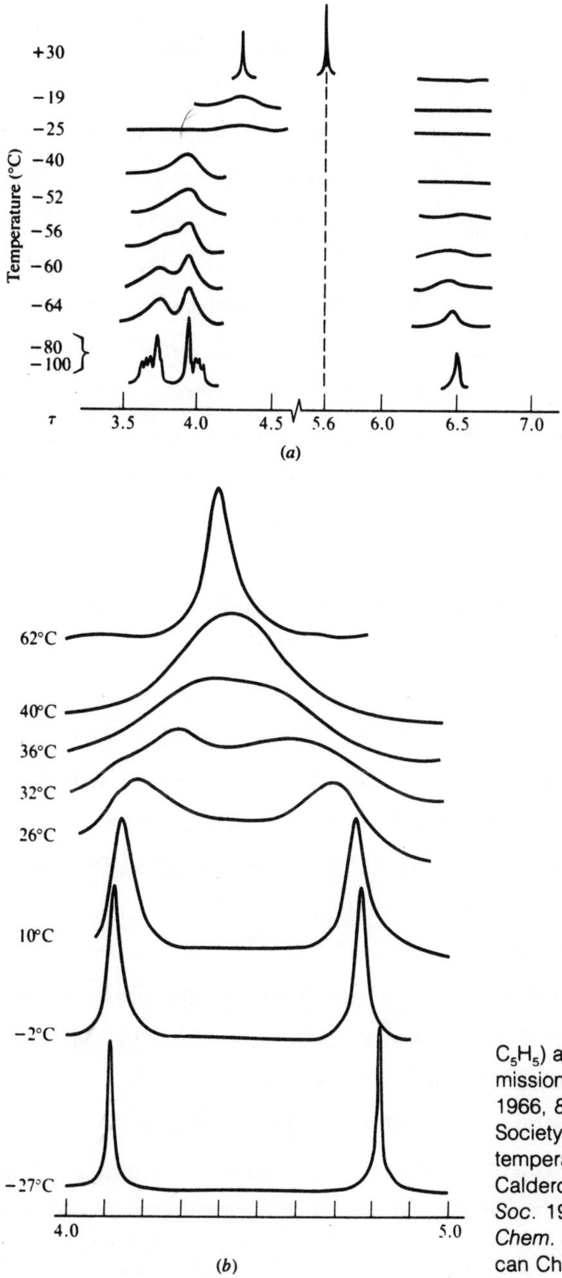

a. ^{1}H NMR spectrum of $CpFe(CO)_2(\eta^1\text{-}C_5H_5)$ at several temperatures. (Reproduced with permission from M. J. Bennett, *et. al., J. Am. Chem. Soc.* 1966, *88,* 4371. Copyright 1966, American Chemical Society.) *b.* ^{1}H NMR spectrum of $(C_5H_5)_4$ Ti at several temperatures. (Reproduced with permission from J. L. Calderon, F. A. Cotton, and J. Takats, *J. Am. Chem. Soc.* 1971, *93,* 3587; J. L. Calderon, *et. al., J. Am. Chem. Soc.* 1971, *93,* 3592. Copyright 1971, American Chemical Society.)

15.17 Solution: (a) At low T both η^1-C_5H_5(3.5-4.1τ) and η^5-C_5H_5(6.5τ) groups are present. The η^1-C_5H_5 signal broadens and collapses to a single peak because of a 1,2-migration of the $CpFe(CO)_2$ group around the ring which makes all the C's equivalent. At high enough temperatures only a single C signal appears at τ = 5.6 indicating a role exchange which equalizes both η^1- and η^5-C_5H_5's. (b) (i) If all ligands were η^5-C_5H_5, the complex would have 24e in violation of the EAN Rule. If three ligands were η^5-C_5H_5 and one were η^1-C_5H_5, the EAN Rule would be obeyed. (ii) The ^1H NMR spectrum at the low-temperature limit shows the presence of only two kinds of protons in equal numbers corresponding to $(\eta^5\text{-}C_5H_5)_2(\eta^1\text{-}C_5H_5)_2Ti$, a 16e species. This is not unreasonable since Ti occurs at the beginning of the first transition series. Moreover, only one signal is seen for η^1-C_5H_5, indicating that migration of Ti around each ring is already fast at -27°C. As the temperature increases, the two kinds of C_5H_5 ligands exchange roles and by 62°C this process is fast relative to the NMR time scale.

15.18 Problem: The compound $Co_3Rh(CO)_{12}$ is a tetrahedral cluster. (a) At -85°C, its ^{13}C NMR spectrum displays seven signals in intensity ratio 1:2:2:2:3:1:1. The second and the last two signals are coupled to $^{103}Rh(I = \frac{1}{2})$. What is the structure of the species "frozen out" at this temperature? (b) On warming to +10°C, two signals appear - a single line of relative intensity 2 showing coupling to ^{103}Rh and one of relative intensity 10 which is somewhat broadened. At +30°C, only a single broad resonance is visible. Account for these observations. (See B.F.G. Johnson, J. Lewis, and T.W. Matheson, Chem. Comm. 1974, 441.)

15.18 Solution: (a) At -85°C, a static structure is seen. Three non-equivalent CO's are coupled to ^{103}Rh. The signal of intensity two corresponds to two μ_2-CO's and the other two non-equivalent terminal CO's. There is a signal of relative intensity 3 which must correspond to CO's on the unique Co. Hence, the following static structure accounts for the spectrum. The other signals of intensity 2 would correspond to c and c' while the remaining one is due to b'.

(b) As the temperature is raised to $-30^{o}C$ a process of CO interchange among Co atoms interconverts a,b,b', c and c'. The Rh-based CO's do not participate in this interchange and appear as a single signal. However, d and d' are rendered symmetry equivalent by equivalence of the 3 Co's. At $+30^{o}$ all carbonyl ligands are involved in the interchange.

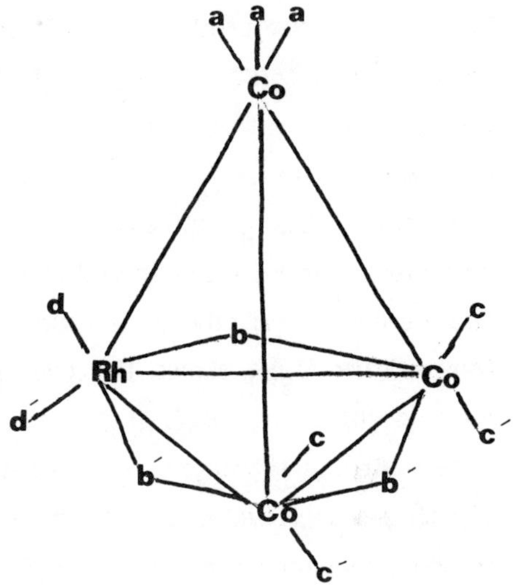

15.19 Problem: (a) Give the electron count and sketch the geometry for $1,2\text{-}B_9C_2H_{11}^{2-}$ and $1,7\text{-}B_9C_2H_{11}^{2-}$. (b) Both of these ligands form a number of compounds with transition metals. Their common names are 1,2- and 1,7-dicarbollide, respectively. Give the electron count and predict the structure for $CpCo(1,2\text{-}C_2B_9H_{11})$. (c) MO calculations indicate that six electrons are available on the open faces of the dicarbollide ligands for donation to transition metals. This makes them formally equivalent to Cp ligands. Show that the following species conform to the EAN for the metal.
$(\eta^4\text{-}Ph_4C_4)Pd(1,2\text{-}Me_2C_2B_9H_9)$, $[(1,2\text{-}B_9C_2H_{11})Re(CO)_3]^-$,
$[(1,2\text{-}B_9C_2H_{11})Mo(CO)_3W(CO)_5]^{2-}$

15.19 Solution: (a) Both have 13 pairs of framework electrons and exhibit a nido structure obtained by removing one vertex of an icosahedron.

(b) Cp^- + Co(III) + $[1,2\text{-}C_2B_9H_{11}]^{2-}$
6e + 6e + 6e = 18e

If we view the compound as a cluster, then the electron count is:

$CpCo^{2+}$ + 2(CH) + 9 (BH) + charge
0e + 2 x 3e + 9 x 2e + 2e = 26e

$CpCo^{2+}$ contributes 0 electrons since Co is Co(III) here. We have p + 1 pairs where p = 12 and a closo structure (an icosahedron) is predicted by Wade's rules (and observed).

(c) $(\eta^4\text{-}Ph_4C_4)Pd(1,2\text{-}Me_2C_2B_9H_9)$

Pd(IV)	+ $\eta^4\text{-}C_4Ph_4{}^{2-}$	+ $1,2\text{-}Me_2C_2B_9H_9{}^{2-}$		
6e	+ 6e	+ 6e		= 18e

$[(1,2\text{-}B_9C_2H_{11})Re(CO)_3]^-$

Re(I)	+ 3 CO	+ $1,2\text{-}B_9C_2H_{11}{}^{2-}$	
6e	+ 3 x 2e	6e	= 18e

$[(1,2\text{-}B_9C_2H_{11})Mo(CO)_3W(CO)_5]^{2-}$

Mo(0)	+ $1,2\text{-}B_9C_2H_{11}{}^{2-}$	+ 3 CO	
6e	+ 6e	+ 3 x 2e	= 18e

The $[(1,2\text{-}B_9C_2H_{11})Mo(CO)_3]^{2-}$ group is equivalent to $[CpMo(CO)_3]^-$ and can be considered to be a 2e-donor toward $W(CO)_5$ giving it 18e.

15.20 Problem: Using the approach developed by Mingos for cage and ring compounds (Nature Phys. Sci. 1972, 236, 99), predict plausible structures for the following:

(a) S_8 (b) $P_4(C_6H_{11})_4$ (c) P_4 (d) $[Fe(NO)_2]_2(SEt)_2$

15.20 Solution: (a) S_8: Each S atom contributes four skeletal electrons (v + x - 2, see Problem 15.6) for a total of 32 = 16 pairs. Starting with a cube of S atoms bonded via S-S bonds gives 12 edge bonds requiring 12 e pairs. The four "extra" electron pairs can occupy four S-S antibonding orbitals breaking opposite pairs of bonds on the upper and lower cube faces leading to the S_8 ring structure.

(b) $P_4(C_6H_{11})_4$: Total number of framework electron pairs is eight. Arranging the four $P(C_6H_{11})$ groups at the vertices of a tetrahedron, we have sufficient electron pairs to break two of the six P-P bonds to afford a butterfly structure.

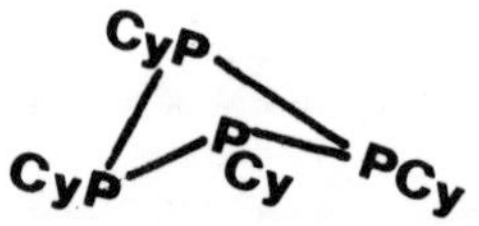

Cy: C_6H_{11}

(c) P_4: Number of framework e is 4 x 3e = 12e = 6 pairs, precisely enough to fill all six bonding orbitals for four P's arranged at the vertices of a tetrahedron.

(d) $[Fe(NO)_2]_2(SEt)_2$: Each $Fe(NO)_2$ group contributes 2e; each SEt group contributes 5e (v + x - 2) for a total of fourteen framework electrons. Arranging the groups at the corners of a tetrahedron, we have the six pairs need to form single bonds along all tetrahdral edges plus one to occupy an antibonding orbital breaking one bond giving a butterfly structure.

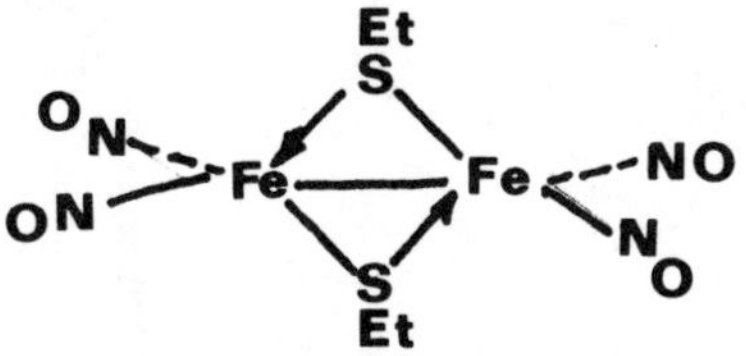

15.21 Problem: Three recently prepared clusters are depicted in the figures (a) to (d). Rationalize these structures according to Wade's rules. With what neutral boranes are $Fe_2(CO)_6B_3H_7$ and $[Fe(CO)_4B_7H_{12}]^-$ isoelectronic? (See Problem 15.13 for Wade's rules.)

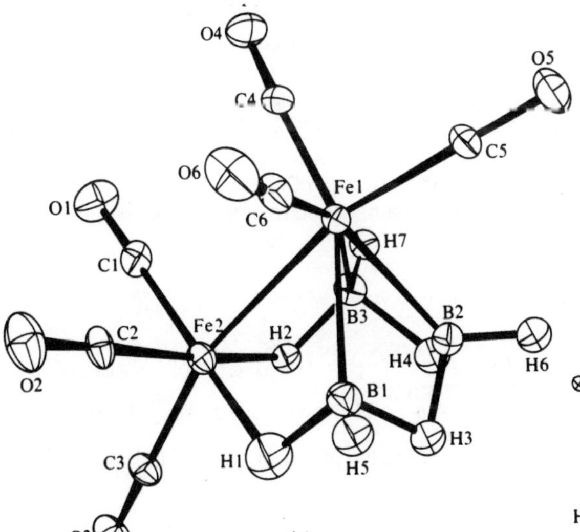

(a)

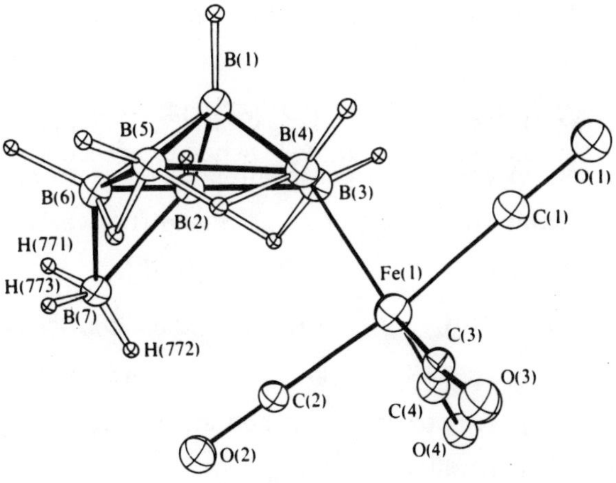

(b)

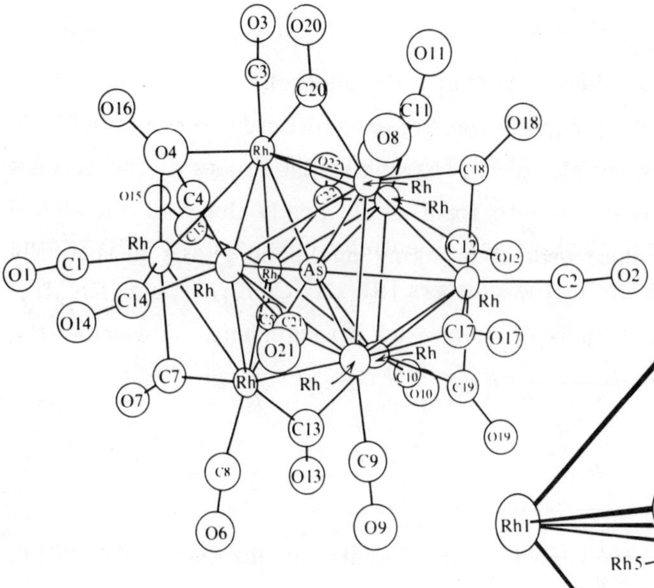

(c)

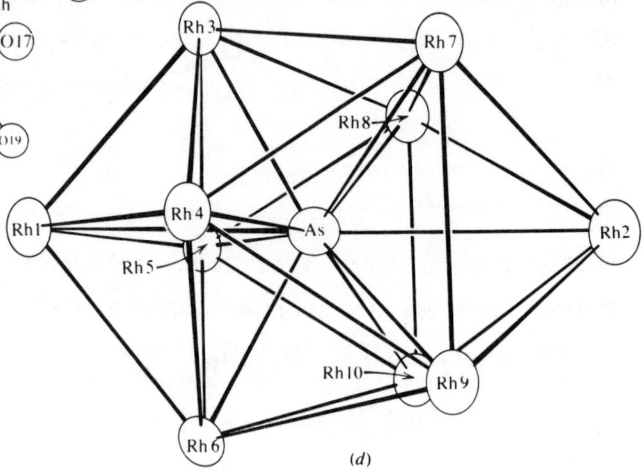

(d)

a. The structure of $Fe_2(CO)_6B_3H_7$. (Reproduced with permission from K. J. Haller, E. L. Andersen, and T. P. Fehlner, *Inorg. Chem.* 1981, *20*, 309. Copyright 1981, American Chemical Society.) *b*. The structure of $Fe(CO)_4B_7H_{12}^-$. (Reproduced with permission from M. M. Mangion, W. R. Clayton, O. Hollander, and S. G. Shore, *Inorg. Chem.* 1977, *16*, 2110. Copyright 1977, American Chemical Society.) *c* and *d*. The structure of $Rh_{10}As(CO)_{22}^{3-}$ with (*c*) and without (*d*) the carbonyl ligands. (Reproduced with permission from J. L. Vidal, *Inorg. Chem.* 1981, *20*, 243. Copyright 1981, American Chemical Society.)

15.21 Solution: (a) Each $Fe(CO)_3$ group contributes 2e for a total of 4e. Each of the three BH groups contributes 2e for a total of 6e. Four μ-H contribute 4e. Hence, the framework count of 7e pairs and five vertices leads to prediction of a nido structure. This complex is isoelectronic with B_5H_9.

(b) 6(BH)	+ μ-BH_3	+ μ-$Fe(CO)_4$	+ 3μ-H	+ charge	
6 x 2e	+ 0e	+ 0e	+ 3 x 1e	+ 1e	= 16e

Eight framework pairs and six vertices predict a nido structure. This Fe compound is isoelectronic with B_6H_{10}.

(c)	10Rh(CO)	+ 12 CO	+ As	+ charge	
	10 x (-1e)	+ 12 x 2e	+ 5e	+ 3e	= 22e

Eleven framework electron pairs and ten vertices (As is in cluster center) lead to prediction of the closo structure, a bicapped Archimedian antiprism.

15.22 Problem: Some metal carbonyl clusters are essentially close-packed units of the metal stabilized by coordination of CO. In the clusters $[Rh_{13}H_3(CO)_{24}]^{2-}$ and $[Rh_{13}H_2(CO)_{24}]^{3-}$ the Rh atoms are in a hexagonal close-packed arrangement. Describe the expected arrangement of atoms within the very symmetrical Rh_{13} cluster.

15.22 Solution: This is the simplest hcp unit giving the full C.N. 12 of one atom. The Rh at the center has a hexagonal arrangement of six Rh in one layer with triangular arrangements of three Rh above and below this layer. The triangular groups are eclipsed giving the ABA or hcp arrangement. The unit has D_{3h} symmetry.

15.23 Problem: Diborane will react with $NaBH_4$ in diglyme (a polyether) to form NaB_2H_7 in solution. No reaction occurs without the solvent present. Propose a role for the solvent.

15.23 Solution: The polyether complexes the metal ion, thus reducing the ion-ion interaction in $NaBH_4$. There is little loss in ion pair energy when the larger $B_2H_7^-$ ion forms from the BH_4^- ion. In contrast, with solid $NaBH_4$ the loss of lattice energy on forming the $B_2H_7^-$ ion is greater than the energy released in forming the (isolated) ion, so the reaction does not occur.

15.24 Problem: B_4Cl_4 is a yellow liquid. Its molecules are of T_d symmetry having a B-Cl group at each vertex of a tetrahedron. Rationalize the electronic structure of this species.

15.24 Solution: Each B and each Cl contribute one electron and one orbital to the formation of the B-Cl bond. This leaves two electrons and three orbitals contributed to the bonding of the tetrahedral framework by each B. This total of twelve orbitals and eight electrons does not conform to Wade's rules, nor do we have enough electrons to regard the molecule as electron-precise with a B-B bond along each of the six tetrahedral edges. (This would require twelve electrons.) Since we are aware that B can form three-center bonds, we could regard the molecule as having four two electron-three center bonds formed by overlap of B orbitals on each of the four tetrahedral faces. A more formalistic view could be taken and a set of styx numbers (See Problem 15.4) computed for $(BH)_4$ which would be isoelectronic as far as framework bonding is concerned.

$$p = 4 = s + t$$

$$q = 0 = s + x$$

$$p - 1/2q = 4 = t + y$$

We must have s = x = 0 since no groups are available for bridging or formation of additional terminal bonds to B. Thus, we have t = 4, y = 0.

XVI Some Aspects of Bioinorganic Chemistry

16.1 Problem: For each of the following elements, identify one significant role in biological processes: Fe, Mn, Cu, Zn, I, Mg, Co, Ca, and K.

16.1 Solution: Fe - hemoglobin - O_2 transport

Fe is present in Fe–S proteins and cytochromes, both are important in oxidation-reduction processes.

Mn - redox process producing O_2 in photosynthesis

Cu - hemocyanin - O_2 carrier in invertebrates

There are many Cu enzymes, for example cytochrome oxidase.

Zn - active center in carboxypeptidase and many other enzymes

I - thyroxin (secreted by the thyroid gland)

Mg - chlorophyll - photosynthesis

Mg^{2+} is a major cation within cells

Co - Vitamin B_{12}

Ca - bones

Ca^{2+} plays an important role in the transmission of nerve impulses

K^+ - major cation within cells.

16.2 Problem: What prevents simple iron porphyrins from functioning as O_2 carriers? How has this problem been avoided in successful models of Fe–porphyrin O_2 carriers?

16.2 Solution: Simple iron porphyrins tend to be oxidized irreversibly through dimerization. Successful models have substituents (picket fence or canopy) which prevent dimerization.

16.3 Problem: How is iron stored and transported in mammals? What is the oxidation state of iron for storage and for transfer?

16.3 Solution: Fe is stored as ferritin and transported as transferrin. Fe(III) is stored, and transported, but it seems to be transferred as Fe(II). (See DMA p. 732.)

16.4 Problem: Give an example of each of two types of reactions brought about by vitamin B_{12}.

16.4 Solution: B_{12} can undergo one-, two-, and three-electron transfers:

$$B_{12}\text{-Co-R} \begin{cases} B_{12}\text{-Co(I)} + R^+ \\ B_{12}\text{-Co(II)} + \,^{\cdot}R \\ B_{12}\text{-Co(III)} + :R^- \end{cases}$$

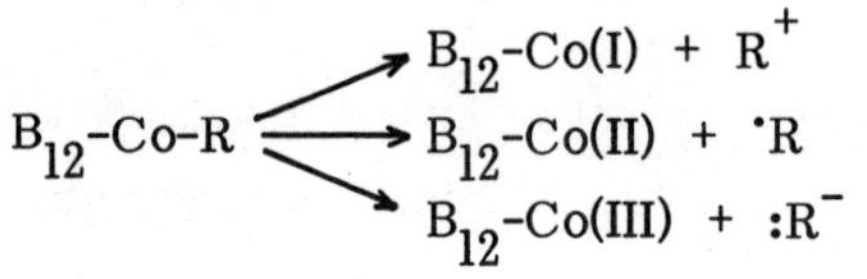

One carbon transfer – The introduction or transfer of one carbon unit.

$$\underset{\substack{| \\ NH_2 \\ \text{homocysteine}}}{HO_2CCHCH_2CH_2SH} \xrightarrow[\text{+ enzyme}]{\text{coenzyme } B_{12}} \underset{\substack{| \\ NH_2 \\ \text{methionine}}}{HO_2CCHCH_2CH_2SCH_3}$$

$$\underset{\text{glycine}}{H_2N\text{-}CH_2\text{-}CO_2H} \xrightarrow[\text{+ enzyme}]{\text{coenzyme } B_{12}} \underset{\substack{| \\ CH_2OH \\ \text{serine}}}{H_2N\text{-}CH\text{-}CO_2H}$$

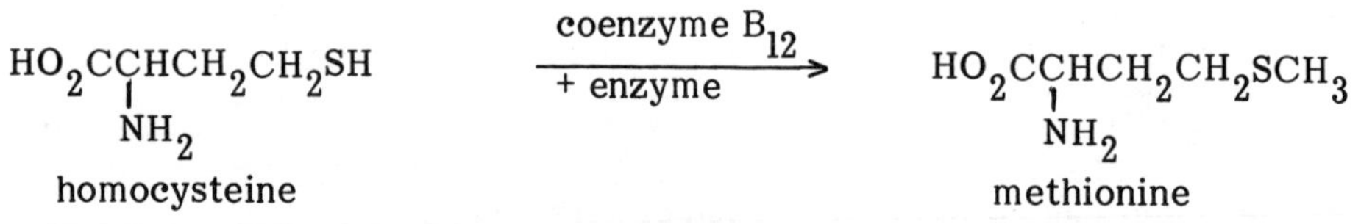

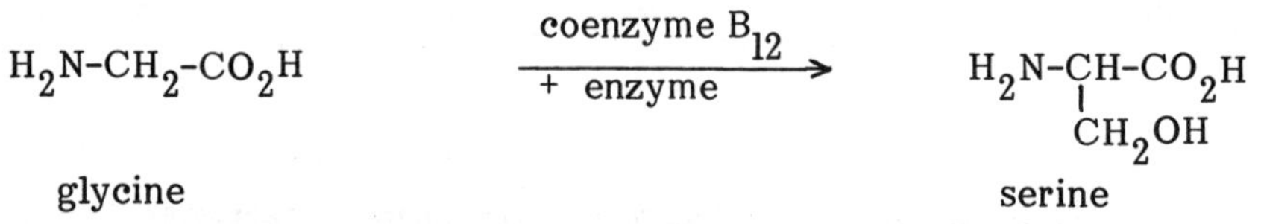

Isomerization – moving a substituent along a carbon chain.

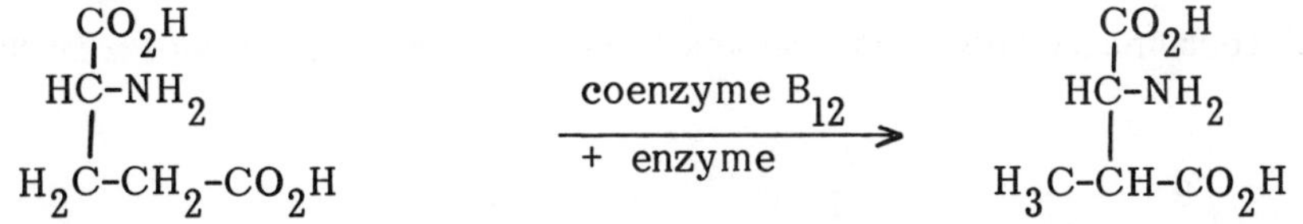

$$
\begin{array}{ccc}
\begin{array}{c}
CO_2H \\
| \\
HC\text{-}NH_2 \\
| \\
H_2C\text{-}CH_2\text{-}CO_2H \\
\text{glutamic acid}
\end{array}
&
\xrightarrow[\text{+ enzyme}]{\text{coenzyme } B_{12}}
&
\begin{array}{c}
CO_2H \\
| \\
HC\text{-}NH_2 \\
| \\
H_3C\text{-}CH\text{-}CO_2H \\
\text{methyl aspartic acid}
\end{array}
\end{array}
$$

16.5 Problem: What is the cytochrome chain? What are the advantages of such a complex system?

16.5 Solution: The cytochromes are a group of electron-transferring proteins that act in sequence to transfer electrons to O_2. The prosthetic group in cytochromes is heme, which undergoes reversible Fe(II)–Fe(III) oxidation. The cytochromes in a sequence differ from one another in electrode potentials by about 0.2 volt or less, with a total potential difference of about 1 volt. Differences in potentials result from changes in porphyrin substituents or axial ligands.

The many steps in the chain serve to break down the large amount of energy involved in the reduction of O_2 into smaller units that can be stored as ATP and keep the reactants well separated. High biological specificity is achieved by the chain. Photosynthesis uses a similar cytochrome chain.

16.6 Problem: What are the prosthetic groups of cytochromes and hemoglobin?

16.6 Solution: Heme is the prosthetic group of cytochromes and of hemoglobin.

16.7 Problem: What are the two important systems for biological electron-transfer processes?

16.7 Solution: Cytochromes and Fe–S non-heme proteins are among the most important systems for biological electron transfer.

16.8 Problem: Identify two chemical types of siderophores.

16.8 Solution: The siderophores are very stable iron complexes used by lower organisms for Fe transport. Fe is coordinated to O in hydroxamate and catechol type complexes:

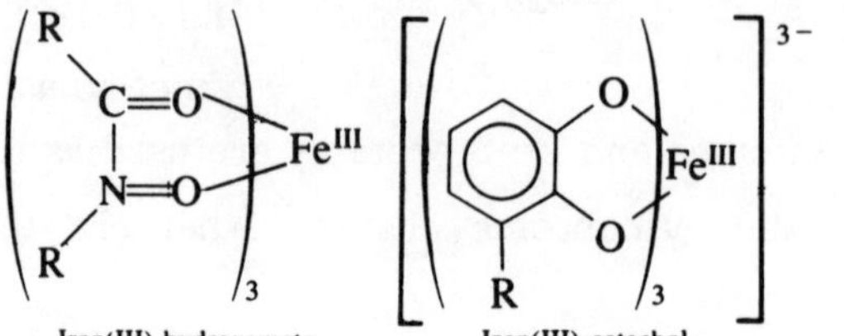

Iron(III) hydroxamate complex

Iron(III) catechol complex

16.9 Problem: The formation constant of Fe(III) enterobactin is about 10^{56}. Calculate the concentration of Fe^{3+} in equilibrium with 10^{-4} M Fe(III) enterobactin and 10^{-4} M ligand. This corresponds to how many liters per Fe^{3+}? The volume of the hydrosphere (all bodies of water, snow, and ice) is ca. 1.37 x 10^{21} L.

16.9 Solution: $$Fe^{3+} + \text{enterobactin} \rightleftharpoons Fe^{III}(\text{enterobactin})$$

$$K = 10^{56} = \frac{[Fe^{III}(\text{enterobactin})]}{[Fe^{3+}][\text{enterobactin}]} = \frac{(10^{-4})}{[Fe^{3+}]\,(10^{-4})}$$

$$[Fe^{3+}] = 10^{-56}\,\text{mole/L or } 6 \times 10.^{-33}\ \text{ions/L or ca. } 10^{32}\,\text{L per ion!}$$

16.10 Problem: Wilson's disease causes the accumulation of what element in the body? How can symptons be relieved?

16.10 Solution: Wilson's disease is hereditary, resulting in a deficiency of ceruloplasmin. Cu accumulates in the liver, brain, and kidneys. A strong chelating agent, such as edta or penicillamine, is used to form a stable complex to remove the accumulated Cu.

16.11 Problem: What chemical properties of Fe and Cu make them suitable for redox processes in biological systems?

16.11 Solution: Fe and Cu have two easily accessible oxidation states, making them suitable for redox processes in biological systems.

16.12 Problem: What electron transport systems are used in photosynthesis?

16.12 Solution: Cytochromes and Fe-S proteins are used as electron transport systems in photosynthesis.

16.13 Problem: The conversion of carbonic acid to $CO_2 + H_2O$ is a natural process; why is carbonic anhydrase needed?

16.13 Solution: Uncatalyzed dehydration of H_2CO_3 is too slow for respiration. Carbonic anhydrase (a Zn enzyme) accelerates the process greatly.

16.14 Problem: Give an example of the substitution of Co(II) for Zn(II) in an enzyme to provide a "spectral probe" for study of the enzyme.

16.14 Solution: Co(II) replaces Zn(II) in carboxypeptidase with an increase in activity. The Co(II) absorbs in the visible region so that spectral studies provide direct information about the active site.

16.15 Problem: The direct reduction products of water, H_2O_2 (or HO_2^-) and O_2^-, are toxic. How are these handled in biological systems?

16.15 Solution: H_2O_2 is decomposed by peroxidase and O_2^- is decomposed by superoxidase. Oxygen is toxic to organisms lacking these enzymes.

16.16 Problem: What metals are at the active centers of nitrogenase? Name some reduction processes other than that of N_2 that are accomplished by nitrogenase?

16.16 Solution: Nitrogenase is an Fe-Mo enzyme. It reduces N_2, C_2H_2, N_3^-, and N_2O. Nitrogenase activity of a preparation of the enzyme is usually monitored by following the reduction of acetylene, since this can be followed easily.

16.17 Problem: High-spin iron(II) is too large for the opening of the porphyrin ring, but low-spin iron(II) can be accommodated in the opening. Why does the high-spin ion have a larger radius?

16.17 Solution: Low-spin octahedral Fe(II) has the t_{2g}^6 configuration with the e_g orbitals, directed along the x, y, and z axes, empty. High-spin octahedral Fe(II) has the $t_{2g}^4 e_g^2$ configuration. The metal e_g electrons are antibonding, or, in other terms, the electrons in these orbitals along the axes provide screening of the positive metal center from the ligands, giving a larger radius.

16.18 Problem: The enzyme alkaline phosphatase contains Zn^{2+} and catalyzes the hydrolysis of orthophosphate monoesters. In order to elucidate the coordination geometry about the metal ion, Zn^{2+} can be substituted by Co^{2+}. The $d \rightarrow d$ spectrum of the resulting complex contains the following peaks:

Wavelength (nm)	ε
640	260
605 (shoulder)	220
555	708
510	335

What can you say about the coordination of Co^{2+} from these data?

16.18 Solution: Obvious candidates for consideration would be tetrahedral and octahedral geometries. Low-spin tetrahedral complexes are not known. Low-spin d^7 octahedral complexes are unlikely — especially because the O-containing groups found in biological systems are not high in the spectrochemical series. Consulting the Orgel diagram in Problem 7.16, we see that three bands are possible for either tetrahedral or octahedral high-spin d^7. Any of these bands might not be observed because they are too high or too low in energy. In particular, the first band of some tetrahedral Co^{2+} complexes is sometimes located in the infrared. The number of bands makes it likely that there are two distinct coordination environments. The rather large intensities indicate tetrahedral or coordination of lower symmetry than octahedral. (The actual situation is somewhat more complex. A more complete treatment, as well as a good example of the kinds of arguments usually made from such data can be found in M. L. Applebury and J. E. Coleman, J. Biol. Chem. 1969, 244, 709).

16.19 Problem: In birds air flows in one direction through the lungs, air sacks and hollow bones during inhalation and exhalation. Why should this offer an advantage at high altitude over our "batch" breathing process?

16.19 Solution: At high altitude the partial pressure of O_2 is low and a continuous process provides more air flow for more efficient utilization of O_2. (See K. Schmidt-Nielsen, "How Birds Breathe", Sci. Am., December, 1971, 73.)

16.20 Problem: Why might we expect some elements essential for life at low concentrations to be toxic at higher concentrations?

16.20 Solution: At high concentration, in addition to serving its essential role, an element can compete with and displace other essential elements from their essential roles. (See D. E. Carter and Q. Fernando, "Chemical Toxicology. Part II Metal Toxicity", J. Chem. Educ. 1979, 56, 490.).

16.21 Problem: Radiopharmaceuticals containing gamma-ray-emitting nuclides such as ^{99m}Tc, ^{201}Tl, ^{43}K, etc., have been used to locate areas of bone cancer, to produce heart images showing areas of myocardial infarct, and to aid in the location of specific tumor growth. To be used in the above fashion, the radiopharmaceutical must, of course, concentrate in the region of interest. Explain why ^{99m}Tc is a promising nuclide for such use for reasons other than its radioactivity. How might a bone imaging agent differ chemically from a heart imaging agent?

16.21 Solution: The metal ion should be complexed with ligands that are lipophilic (fat loving) or ossophilic (bone loving) so that the complex would seek out fatty tissue or bone tissue. Pyrophosphate or diphosphonates would form ossophilic complexes. Amino acids such as glycine, cystine, etc., would give lipophilic complexes. Tc shows a wide variety of stable oxidation states, and thus permits a wider variety of complexes to be prepared than the other metals mentioned.